ADSORPTION OF ORGANIC COMPOUNDS ON ELECTRODES

ADSORPTION OF ORGANIC COMPOUNDS ON ELECTRODES

Boris B. Damaskin, Oleg A. Petrii, and Valerii V. Batrakov

Department of Electrochemistry
Moscow State University
Moscow, USSR

With a Foreword by Academician A. N. Frumkin

Translated from Russian by
E. Boris Uvarov
Senior Scientific Translator

Translation Editor
Roger Parsons
Reader in Electrochemistry
University of Bristol
Bristol, England

ℚ PLENUM PRESS · NEW YORK–LONDON · 1971

CHEMISTRY

Boris Borisovich Damaskin was born in 1932; in 1956 he was graduated from the Faculty of Chemistry, Moscow State University. He presented his candidate's dissertation in 1959 and his doctoral thesis in 1965. Since 1967 he has been Professor of Electrochemistry, Moscow State University.

Oleg Aleksandrovich Petrii was born in 1937 and was graduated from the Faculty of Chemistry, Moscow State University in 1959. His candidate's dissertation was presented in 1962. Since 1966 he has been a senior scientist in the Department of Electrochemistry, Moscow State University.

Valerii Vladimirovich Batrakov was born in 1937 and was graduated from the Faculty of Chemistry, Moscow State University in 1959. He presented his candidate's dissertation in 1965. Since 1968 he has been a lecturer in the Department of Electrochemistry, Moscow State University.

The original Russian text, published by Nauka Press in Moscow in 1968 for the Institute of Electrochemistry of the Academy of Sciences of the USSR, has been corrected by the authors for the present edition. The English translation is published under an agreement with Mezhdunarodnaya Kniga, the Soviet book export agency.

Б. Б. Дамаскин, О. А. Петрий, В. В. Батраков

АДСОРБЦИЯ ОРГАНИЧЕСКИХ СОЕДИНЕНИЙ НА ЭЛЕКТРОДАХ

ADSORBTSIYA ORGANICHESKIKH SOEDINENII NA ELEKTRODAKH

Library of Congress Catalog Card Number 69-17533

SBN 306-30432-5

© 1971 Plenum Press, New York
A Division of Plenum Publishing Corporation
227 West 17th Street, New York, N. Y. 10011

United Kingdom edition published by Plenum Press, London
A Division of Plenum Publishing Company, Ltd.
Davis House (4th Floor), 8 Scrubs Lane, Harlesden, NW10 6SE, England

Printed in the United States of America

Foreword

The systematic study of the adsorption of organic compounds on electrodes began with the comprehensive survey of adsorption on mercury carried out by Gouy in the first decade of this century. His studies with the capillary electrometer are still useful but do not lend themselves to detailed quantitative analysis. A more detailed study of a few systems by Frumkin in his thesis (1919) led him to propose a quantitative phenomenological theory of organic adsorption (1925, 1926) at almost the same time as Stern proposed the model of the electrical double layer which remains the picture accepted in general terms today. The attempt at a molecular model made by Butler (1929) should be more satisfying but up to the present the formidable difficulties of a molecular theory of interfacial phenomena have prevented the full interpretation of experimental results along these lines.

In his work with Proskurnin (1935), Frumkin is also responsible for the major experimental advance in the demonstration that reliable measurements of the capacity of an electrode-solution interface can be obtained provided that the work is carried out under conditions of scrupulous cleanliness. Even so, precise measurements of double layer capacities were not obtained until Grahame (1941) showed how convenient and reliable the dropping mercury electrode was the for these studies. This method and the hanging drop electrode remain the preferred methods for study of adsorption on mercury. Solid electrodes present a more difficult problem. Reliable measurements on platinum were obtained by Butler et al. (1932) and by Frumkin et al. (1934) following the technique developed by Bowden (1928), but it is only in the last two decades that

v

systematic study of the adsorption of organic compounds on solid metal electrodes has been possible. A large fraction of the reliable work has been carried out by the coworkers of Frumkin in his Institute of Electrochemistry in Moscow.

Partly as a result of the recent acquisition of reliable results on solid electrodes and partly because of their greater intrinsic complication, the interpretation of these results is much less clear than that for the data on mercury electrodes. Two important guidelines were developed by Frumkin. The first is the principle of the point of zero charge as a characteristic constant of the electrode material. This is now so widely accepted that an effort of historical imagination is required to realize how brilliant was the innovation. The second remains controversial and is the application of thermodynamics to systems like that of platinum adsorbing oxygen.

It will be clear from the brief survey given above that the time is ripe for a work surveying progress and current thought on the problem of organic adsorption on electrodes. It is also evident how appropriate it is that this work should be written by the colleagues and close collaborators of Academician Frumkin. They have written an extremely useful account of this field which will be read with appreciation by many electrochemists concerned both with the theoretical problems discussed here and their manifold practical applications. That they will also read it with pleasure is due to a considerable extent to the excellent English translation by Mr. Uvarov. I am delighted to have been associated with this project.

Roger Parsons

Preface to the English Edition

The Russian edition, published in 1968, has been revised substantially for the version presented to American readers. This is mainly due to the fact that in the rapidly growing branch of electrochemistry concerned with adsorption effects on electrode surfaces a number of new investigations of considerable interest have appeared during the short time since the publication of the Russian edition. For example, we refer to a new method, developed in the Institute of Electrochemistry, Academy of Sciences of the USSR, for studying the dependence of interfacial tension between solid metals and electrolyte solutions on potential; a new theoretical treatment, with supporting experimental data, of photoelectrochemical phenomena; new results on the influence of heat treatment on electrochemical and adsorption properties of metals of the iron group; and a new branch — thermodynamics of surface phenomena on metals of the platinum group. This last, in my opinion, will become of definite significance in interpretation of adsorption of organic compounds on these metals. The account of determinations of the parameters characterizing adsorption of organic compounds at mercury—solution interfaces, given in Part I, may seem excessively detailed; however, we considered it desirable to present it in this form so that it could be of practical use in studies of new systems.

The expansion of the book made it necessary to subdivide Part II of the Russian edition into two parts, relating to metals with higher and lower hydrogen overpotentials, respectively. In the presentation we attempted to include, as far as possible, all work published up to mid-1968. However, we naturally focused the main attention on work done in the USSR, which is less accessible to the American reader.

Chapters I-IV were written by Prof. B. B. Damaskin, Chapter V jointly by Candidates of Chemical Sciences O. A. Petrii and V. V. Batrakov, Chapters VI and VII by V. V. Batrakov, and Chapters VIII-X by O. A. Petrii.

Academician A. N. Frumkin

Preface to the Russian Edition

Adsorption of organic compounds at interfaces between metals and electrolyte solutions is attracting considerable attention of everyone concerned with theoretical and applied electrochemistry. Indeed, unless adsorption effects are taken into account it is impossible to understand the mechanism of most of the processes occurring at the mercury electrode and therefore to interpret the results of polarographic determinations.

Adsorption of organic compounds is widely used for regulating processes of metal electrodeposition. It determines the behavior of organic compounds at positive electrodes of fuel cells and therefore their suitability as electrochemical fuel. The action of corrosion inhibitors is based on adsorption effects, and they must also be taken into consideration in searches for new routes of organic electrochemical synthesis. Adsorption effects are also met in the general electrochemical industry. Finally, investigations of adsorption phenomena at electrode−solution interfaces are of theoretical interest, extending our knowledge of the structure of the electric double layer.

Nevertheless, there are no monographs either in Soviet or foreign literature dealing at all fully with information relating to adsorption of organic compounds on surfaces of various metals immersed in electrolyte solutions.

The aim of the authors of this book was to fill this gap, using the large amount of experimental data now accumulated in the periodical literature and also the results of their own original researches. It should be noted that the question of the influence of adsorption on the kinetics of electrode processes is considered only to the extent necessary to obtain information on the adsorption process itself.

In the evaluation of such extensive experimental material some omissions were inevitable, but nevertheless the book offered to the reader gives a fairly complete picture of the relationships of adsorption at interfaces of the type in question, ranging from the relatively simple cases of physical adsorption on mercury and other metals to the extensive chemical changes which often accompany adsorption of metals of the platinum group.

Chapters I-IV were written by Prof. B. B. Damaskin, Chapter V by candidates of chemical sciences O. A. Petrii and V. V. Batrakov, Chapter VI by V. V. Batrakov, and Chapter VII by O. A. Petrii.

Academician A. N. Frumkin

Contents

Part I

Adsorption of Organic Compounds on Liquid Electrodes

Chapter 1

Methods for Studying Adsorption of Organic Substances on Liquid Electrodes

The principal methods for studying adsorption of organic compounds on liquid electrodes are based on measurements of interfacial tension or of the differential capacity of the double layer in relation to the electrode potential. On the basis of the quantitative theory of polarographic maxima of the second kind [1-3] it is also possible to investigate adsorption of organic compounds on liquid metals with the aid of data on inhibition, by organic molecules, of the tangential movements of a dropping electrode operating under conditions favoring appearance of maxima of the second kind. Since the charging current of the dropping electrode is directly proportional to the surface charge density, it is also possible to determine the adsorption characteristics of organic substances from the polarographic charging current in the case of liquid electrodes [4-6]. However, the last two methods have not been widely used in studies of adsorption of organic substances.

A method has recently been proposed [7] for direct electrical measurement of adsorption on electrodes, based on recording of secondary harmonics arising as the result of adsorption of organic substances. The potentialities of this method have not been fully studied as yet.

If oxidation or reduction of the adsorbed organic substance is possible as the result of changes of the electrode potential, the chronopotentiometric method [8-10] can also be used for studying adsorption. This method will be described in Chapter V, where adsorption of organic substances on solid electrodes is discussed, as in this chapter we confine ourselves to an account of methods for measurement of interfacial tension and differential capacity.

3

It may be noted that indirect data on adsorption of organic compounds at electrode surfaces may be obtained from kinetic data, since any change in the structure of the electric double layer has an appreciable influence on the rates of electrochemical processes limited by one of the heterogeneous stages, e.g., the discharge stage [11–16].

1. Methods of Measuring
Interfacial Tension

Interfacial tension can be measured by various methods. Some of these are based on the capillary depression of a mercury meniscus, which occurs because the interfacial tension of a liquid metal, such as mercury, contained in a narrow capillary tends to displace the convex meniscus into a wider part of the capillary. The pressure exerted on the mercury meniscus is

$$F = \frac{2\sigma \cos \vartheta}{R}, \tag{I.1}$$

where R is the radius of curvature of the meniscus and ϑ is the contact angle between the mercury and the capillary wall. The instrument whereby F can be connected with measurable parameters is known as the capillary electrometer.

The first two designs of the capillary electrometer were devised by Lippmann [17]. In Lippmann's first capillary electrometer the capillary was part of a U-tube, so that the pressure F was balanced by the pressure of the mercury in the other (wide) leg of the U-tube. In the narrow leg the mercury meniscus was covered with the solution under investigation connected through a siphon to the same solution in a large beaker with a layer of mercury on the bottom. When the mercury in the U-tube and the mercury on the bottom of the beaker were connected to the poles of an external current supply, only the mercury meniscus in the capillary became polarized, while the potential of the mercury at the bottom of the solution remained almost unchanged owing to its large area.

Polarization of the mercury alters the interfacial tension σ; in accordance with Eq. (I.1), this leads to a change of the pressure F. The mercury meniscus therefore moves in the capillary. By determining the position of the meniscus with the aid of a cathetometer at each potential φ, it is possible to determine the relationship between the interfacial tension and the applied potential difference, i.e., to obtain the σ vs φ curve. An improved form of Lippmann's capillary electrometer was later used by Karpachev and Stromberg [18] for measurement of electrocapillary curves of liquid molten metals in electrolyte melts.

In the second variant of Lippmann's capillary electrometer [17] the mercury was in a long vertical tube terminating at the lower end in a conical capillary. The capillary was immersed in a beaker containing the solution, with mercury which served as the second (nonpolarizable) electrode on the bottom of the beaker. In this case F was balanced by the pressure of the mercury column

$$P = hg\rho, \qquad (I.2)$$

where h is the total height of the mercury column, ρ is the density of mercury, and g is the acceleration due to gravity.

This capillary electrometer operates on the principle that the height of the mercury column required to support the mercury meniscus in the capillary in a given position is proportional to the interfacial tension of mercury. It follows from Eqs. (I.1) and (I.2) that the proportionality factor is

$$k = \frac{\sigma}{h} = \frac{\rho g R}{2 \cos \vartheta}. \qquad (I.3)$$

The sensitivity of this type of electrometer depends on the height of the mercury column and on the degree of constriction of the capillary. In Lippmann's experiments the sensitivity was very low, approximately 1 cm/V.

Gouy [19, 20] introduced considerable improvements into the second variant of Lippmann's capillary electrometer. Gouy connected the mercury column by means of a rubber side tube to a movable reservoir, making it much easier to vary the height of the mercury column. Moreover, Gouy placed the nonpolarizable mercury electrode in a separate vessel joined through a siphon to a beaker in which the electrometer capillary was immersed. With this arrangement of the auxiliary electrode it was possible to use the same nonpolarizable electrode for determination of electrocapillary curves in different solutions. Finally, by suitable selection of the geometrical capillary characteristics Gouy greatly raised the accuracy in measurements of interfacial tension with the aid of the capillary electrometer. The improved capillary electrometer of this type was subsequently named the Gouy capillary electrometer, although in some publications it is still described as the Lippmann electrometer.

Several aspects of the use of the Gouy capillary electrometer are discussed in detail by Frumkin [21-23]. For example, Frumkin examined the question of the magnitude of the ohmic potential drop ($\Delta\varphi_{ohm}$). In the case of 1 and 0.1 N solutions of inorganic salts and acids, $\Delta\varphi_{ohm}$ due to the reduction current of the dissolved oxygen may be neglected. As we pass to 0.001 N solutions $\Delta\varphi_{ohm}$ increases, so that either appropriate corrections must be applied or the dissolved oxygen removed. Frumkin also noted that in determinations of small changes of the interfacial tension better results are obtained by measurement of displacements of the meniscus in relation to the potential at constant h rather than by measurement of the change of h with φ for a fixed position of the meniscus.

In the Gouy capillary electrometer described above the height of the mercury column was simply determined visually·or with the aid of a cathetometer. Koenig [24] determined the position of the upper mercury meniscus by the closing of an electrical contact. After calibration of the apparatus it was possible to find the dependence between the positions of the upper mercury meniscus and the electrical contact. According to Koenig, the height of the mercury column can be measured to an accuracy of 0.1 mm by this method. Recently Koenig's method has been supplemented by an ac bridge circuit whereby the position of the lower meniscus can be found from the resistance of the electrolyte in the capillary [25]. However, the apparatus is very cumbersome and complicated. Subsequent improvements of the Gouy capillary electrometer involved thermostatic control of the measuring cell [26-28], both air and water thermostats being used.

It should be noted that the capillary electrometer method gives reliable values of the interfacial tension only if mercury and glass are wetted satisfactorily by the solution, with a thin layer of solution between the mercury and the capillary wall, so that the angle $\vartheta = 180°$ at all potentials. If the wetting is poor (e.g., with very dilute solutions), rupture of the film of solution between the mercury and the capillary wall occurs, the mercury meniscus ad-

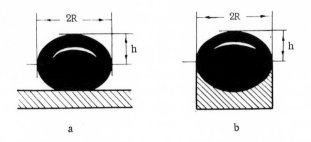

Fig. 1. Schematic diagrams of sessile mercury drops, explaining the calculation of interfacial tension by formulas (I.4), (I.5), and (I.6).

heres to the glass [29], and measurement of interfacial tension with the aid of the capillary electrometer becomes unreliable or even virtually impossible [30]. This effect is apparently the cause of the differences between the results of capacity and electrocapillary measurements in dilute solutions [31]. Another serious defect of the capillary electrometer method is slow establishment of adsorption equilibrium at low concentrations of surface-active substances.

Another independent method for measurement of interfacial tension is based on study of the geometrical form of a sessile drop of the liquid metal. This method was first described by König [32], who determined σ by measurement of the radius of curvature in the upper part of a mercury drop extruded into the solution from an acute-angled capillary. However, the results obtained by this method were less exact than those obtained with the use of Lippmann's capillary electrometers. In 1885 Worthington [33] derived a formula for the relationship between the geometry of a mercury drop lying flat on a surface and the interfacial tension. However, this formula proved to be inaccurate, and, as was noted by Nicholas et al. [34], even after correction it gives accurate results only when the drop radius $R \geq 2$ cm. The corrected Worthington formula is

$$\sigma = \frac{\rho g h^2}{2} \cdot \frac{1.641R}{1.641R + h},$$ (I.4)

where the meaning of h and R is clear from Fig. 1a.

The sessile drop method was also used incorrectly by Berget [35], who reached the erroneous conclusion that the interfacial tension is independent of the composition of the solution on the basis of this method.

Gouy [36] was the first to obtain correct results by the sessile drop method. Gouy ensured that the mercury drops assumed regular shape by placing them in a special hemisphere made from carefully polished glass (Fig. 1b). Under these conditions the radius of the mercury drop was assumed equal to the radius of the glass hemisphere. The interfacial tension was calculated from the formula

$$\sigma = \frac{a^2}{2} (\rho_{Hg} - \rho_{soln}), \qquad (I.5)$$

in which a was found from the equation

$$a^2 + \frac{2\sqrt{2}-1}{3} \left[\frac{a^3}{2\pi R} - \frac{a^4}{2\,(2\pi R)^2} \right] = h^2. \qquad (I.6)$$

The drop radius in Gouy's experiments was about 6.5 mm, and the dimensions of the drops were determined correct to $1\,\mu$. According to Gouy's data [36] the maximum value of σ at the interface between mercury and dilute sulfuric acid solution at 18°C is 426.7 dyn/cm. Gouy used this method for determining the maximum interfacial tension in solutions of various compositions and found that the ratios of the σ values obtained in this way are in very good agreement with the corresponding data determined with the aid of the capillary electrometer.

In a recent investigation Butler [37] solved the equation for the form of a sessile drop in an electrolyte solution with the aid of an electronic computer; up to 40 measurements of the sessile drop contour were fed into the computer. Naturally, this method diminishes errors involved in determination of the vertical and horizontal drop diameters. However, the probability of contamination of the electrolyte surface by traces of surface-active substances is increased, because determination of 40 measurements of the drop contour with the aid of a horizontal microscope takes considerable time.

Another version of this method for determination of interfacial tension is based on study of the form of a drop hanging from the tip of a capillary [38-40]. The practical application of this method to hanging mercury drops of small dimensions became possible only after the work of Melik-Gaikazyan [40, 136] who, on the basis of numerical solution of the Laplace equation with the aid of a computer, compiled extensive tables relating the geometrical parameters of the drop to the densities of the metal and electrolyte and to the interfacial tension.

As it is rather difficult to determine with sufficient accuracy the capillary radius at a definite distance from the tip in the Gouy capillary electrometer, the constant k in Eq. (I.3) is usually found experimentally from a known value of σ for a particular solution. At the present time the interfacial tension at the maximum of the electrocapillary curve (σ^{max}) in an acidified dilute Na_2SO_4 solution, determined by Gouy [36] by the flat drop method and given the expression

$$\sigma^{max} = 426.7 - 0.17\,(t - 18°) \tag{I.7}$$

is generally used for determination of k.

It should be pointed out that the sessile drop method requires very careful purification of the solution to ensure removal both of surface-active substances and of impurities which can be electrochemically reduced or oxidized on the drop. For example, in the presence of traces of dissolved atmospheric oxygen the passage of an electric current may lead to nonuniform polarization in different regions of the sessile drop and to occurrence of tangential movements on its surface [1, 2], and would consequently result in erroneous values of the interfacial tension owing to the inevitable deformation of the drop under these conditions.

The third independent method for measurement of interfacial tension is based on the assumption that, in the case of a dropping electrode, the drop weight Q corrected for the weight of solution displaced is equal to the force F retaining the drop, i.e., $\alpha Q = F$, where $\alpha = (\rho_{Hg} - \rho_{soln})/\rho_{Hg}$.

In the first approximation

$$Q \approx m\tau, \tag{I.8}$$

where m is the rate of flow of mercury from the capillary, τ is the drop time, and

$$F \approx 2\pi r \sigma, \tag{I.9}$$

where r is the internal capillary radius. Therefore

$$\sigma \approx \frac{\alpha m}{2\pi r}\,\tau. \tag{I.10}$$

It follows from Eq. (I.10) that in the first approximation the interfacial tension is proportional to the weight of the falling drop; at a constant rate of flow it is proportional to the drop time τ.

It was shown by Kučera [41] that Eq. (I.8) does not take into account the back pressure due to the curvature of the mercury drop surface. On the other hand, it is assumed in Eq. (I.9) that the hanging drop is held entirely by the interfacial tension at the capillary tip, and when the drop falls all the mercury below this tip becomes detached. Lohnstein [42, 43] and Harkins et al. [44, 45] estimated the degree of approximation of Eq. (I.9) (the deviations of the absolute value of σ may reach 35%) and proposed methods for applying the appropriate corrections.

Introduction of corrections for back pressure and incomplete detachment of the mercury below the capillary tip complicates the relationship between interfacial tension and the drop time considerably. It has been shown [46-48] that the relationship between σ (dyn/cm) and τ (sec) can be expressed to an accuracy of $\sim 1\%$ by the equation

$$\sigma = 6.70\cdot 10^8 \cdot \frac{r^3}{l}\left(1 + 19.3\frac{r^{2/3}}{\sigma^{1/3}}\right)\left(h - 4.65\cdot 10^{-3}\cdot\frac{\sigma^{2/3}}{r^{1/3}}\right)\tau, \qquad (I.11)$$

where the numerical coefficients correspond to 25°, and the capillary length l, the total height h of the mercury column, and the internal capillary radius r are in cm. The fairly complex functional relationship between the interfacial tension and the drop time represented by Eq. (I.11) can be used for determination of σ with the aid of computers [49].

The drop-counting method was first used for determination of electrocapillary curves by Kučera [41], but the results were not reproducible. The experimental technique was subsequently improved considerably by Craxford and McKay [50]. As an example, they used a drop weight method for determining the interfacial tension of mercury in 1 N KNO$_3$ solution; the result agreed within 0.2% with the value of σ obtained with the capillary electrometer.

However, as will be shown in Section 3, in studies of adsorption of organic substances it is necessary to know the relatively small changes of σ due to changes in the concentration of the organic substance rather than the absolute values of the interfacial tension. As follows, for example, from Meibuhr's work [51], under these conditions the approximate formula (I.10) gives results virtually coinciding with the aid of the Gouy capillary electrometer.

Two important disadvantages of the drop-counting method must be noted. First, when even a small faradaic current flows through the electrode, different portions of the mercury drops acquire different potentials and therefore different values of interfacial tension. Therefore the value of σ at the neck of the drop, which is included in Eqs. (I.10) or (I.11), does not represent the average interfacial tension. On the other hand, differences of σ between individual regions of the drop surface may lead to tangential movements of the mercury surface [2], when the drop-counting method becomes entirely inapplicable. Thus, an essential condition for the use of the drop-counting method is careful removal from the solution of all possible impurities which can be reduced at the mercury electrode, and, in particular, thorough removal of dissolved atmospheric oxygen.

The second disadvantage of the drop-counting method is that the apparatus must be assembled in such a manner that the influence of possible vibrations is excluded. Otherwise the drops will be detached somewhat earlier by the action of random shocks and erroneous values of the interfacial tension will be obtained.

On the other hand, the drop-counting method is apparently preferable to the use of the capillary electrometer in the case of nonaqueous solvents which do not wet glass well, because under such conditions the angle $\vartheta < 180°$ and the accuracy in determinations of σ with the aid of the capillary electrometer falls sharply.

2. Methods of Measuring the Capacity of the Double Layer

In contrast to the electrocapillary curves, which can be obtained only by a very limited number of methods, the capacity of the electric double layer can be measured by a great variety of methods. This is explained as follows.

If no electrochemical reactions occur at the electrode studied (an ideally polarized electrode), and the area and hence the capacity of the auxiliary electrode are considerably greater than the area and capacity of the tested electrode, the equivalent circuit of the measuring cell may be represented by the capacity of the double layer (C) and the resistance of the solution (R)* connected in series. The general equation connecting the current (I) flowing through this circuit with the variation $d\varphi/dt$ of the potential difference between its ends is

$$\frac{d\varphi}{dt} = R\,\frac{dI}{dt} + \frac{I}{C}. \qquad (I.12)$$

Thus, by supplying current programmed in a definite manner and studying the dependence of φ on the time t, or by setting up a programmed potential difference and studying the dependence of I on t, it is possible in principle to obtain any number of methods for measuring the double layer capacity. However, the methods used in practice are based only on relationships from which C can be determined mostly simply. Some of these methods are examined below.

If a current pulse which does not subsequently change with time is applied to the cell, then I = const, dI/dt = 0, and it follows from Eq. (I.12) that

$$\varphi = \varphi_0 + \frac{I}{C}\,t, \qquad (I.13)$$

i.e., the electrode potential at constant capacity varies linearly with time. The slope of the resultant charging curve gives the capacity. This method was first used by Bowden and Rideal [52] and by Erdey−Gruz and Kromrey [53]; however, owing to surface contamination by traces of organic substances, they obtained erroneous values of the capacity. The same method was used by Barclay and Butler [54] for studying adsorption of tertiary amyl alcohol on mercury. Subsequently this method was improved by Hackerman et al., who used periodic square pulses [55, 56] or single brief ($\geq$10 μsec) current pulses [57]. An even more perfect two-pulse galvanostatic method of capacity measurement, where the duration of the first pulse was varied from 2 to 10 μsec, was devised by Costa [58].

The method of capacity measurement known as the potential decay method is also based on determination of the dependence of

---

*If the test electrode is not ideally polarized, the experimentally determined values of C and R have more complex physical meaning, and may be used for calculation of the double layer parameters with the aid of the appropriate equivalent circuit.

electrode potential on time. If an irreversible electrochemical reaction occurs at the electrode, on the basis of the Tafel equation
its current can be represented by the expression

$$I = k \exp (\eta / b),$$ (I.14)

where $\eta = \varphi_e - \varphi$ is the overpotential; k and b are constants.
When the circuit is broken the charge of the double layer is lost
as the result of the electrode process, so that the current represented by Eq. (I.14) can be substituted into Eq. (I.12). Taking
into account that R = 0 and solving the resultant differential equation, we obtain the following formula [59]:

$$\Delta\eta = \eta_0 - \eta = b \ln \left(\frac{I_0 t}{bC} + 1\right),$$ (I.15)

where η_0 and I_0 are the overpotential and the current before the
circuit is broken. For sufficiently small t, when $I_0 t/bC \ll 1$, we
obtain the following expression after expansion of the logarithmic
term into series:

$$\Delta\eta \approx \frac{I_0}{C} t.$$ (I.16)

Thus, the slope of the initial region of the $\Delta\eta$ vs t curve directly
determines the capacity of the double layer. This method for capacity measurement has been used by Fedotov, Past, and others
[60-62].

In the commutator method of capacity measurement, developed
by Borisova and Proskurnin [63], the test electrode and a standard
capacity are connected in turn to a charged condenser, and the potential difference produced on them is amplified and then recorded
with the aid of a ballistic galvanometer. The ratio of the ampli-
tudes on the electrode and reference standard is inversely proportional to their capacities. The advantage of the commutator method
is that no current flows through the circuit during the measurement.
This excludes the influence of ohmic resistance, which is substantially greater than the capacitive reactance in dilute solutions, and
which therefore sharply lowers the accuracy of the capacity measurements. Measurements in dilute solutions, down to $10^{-5} N$,
could be performed by the commutator method [63]. The commutator method, which is a coulostatic method by its physical nature, was subsequently developed further by Delahay et al. [64, 65].
This method was used by Delahay for determining differential ca-

pacity curves in 10^{-4} and 10^{-5} N NaF solutions, which confirmed
the theory of the electric double layer in very dilute solutions of
surface-inactive electrolytes [64, 65].

With a linear variation of potential with time

$$\varphi = \varphi_0 + vt, \tag{I.17}$$

where v is the rate of change of the potential, we have $d\varphi/dt = v$
and solution of the differential equation (I.12) for this special case
gives

$$I = vC\left[1 - \exp\left(-\frac{t}{RC}\right)\right]. \tag{I.18}$$

At a high electrolyte concentration, when the resistance of the solution is low enough to satisfy the condition RC $\ll$ t, we have from
Eq. (I.18):

$$I \approx vC, \tag{I.19}$$

i.e., the current flowing through the cell under these conditions is
directly proportional to the capacity of the double layer. This principle was applied by Loveland and Elving [66]. In their method triangular pulses are applied to the test (ideally polarized) electrode,
and the current is indicated on the screen of an oscillograph. With
a high pulse amplitude (~2 V) two symmetrical differential capacity
curves, corresponding to the rising and falling branches of the triangular pulse, are obtained simultaneously on the screen. On the
other hand, in Barker's method [67] which is based on the same
principle, the amplitude of the triangular pulses is small (~10-30
mV) and the average electrode potential alters slowly with time.
In this case the differential capacity curve is recorded automatically on the instrument chart. Naturally, both methods are limited
to fairly concentrated solutions (≥ 0.1 N), as the resistance of the
solution is not taken into account. Deformations of the differential
capacity curves obtained by methods of this type in the case of
dilute solutions (0.001 N) were discussed by Damaskin and Petrii
[68].

Several methods of capacity measurement are based on the
use of sinusoidal alternating current. Defining the alternating
current by the expression

$$I = I_0 \exp(j\omega t), \tag{I.20}$$

where $j = (-1)^{\frac{1}{2}}$ and ω is the angular frequency, and substituting for I and dI/dt in Eq. (I.12), we easily find that the impedance of the circuit consisting of a capacity and a resistance in series is

$$Z = R - \frac{j}{C\omega}. \tag{I.21}$$

Equation (I.21) forms the basis of all ac methods for measuring the capacity of an ideally polarized electrode.

One of the first ac methods used for measurement of double layer capacity was based on comparisons with a reference capacity. It was by this method that Proskurnin and Frumkin [69] succeeded for the first time in obtaining the true values of the capacity of a mercury electrode free from errors due to contamination of the solution by traces of organic substances. A more detailed description of the method of comparison with a reference capacity may be found in papers by Proskurnin, Borisova, and Vorsina [70, 71]. The principle of this method is that the potential difference at the cell ($\Delta\varphi_x$) and at the reference capacity ($\Delta\varphi_r$) are determined at constant ac strength (I), and the required double layer capacity is found from the expression

$$C_{exptl} = C_r \frac{\Delta\varphi_r}{\Delta\varphi_x} = C_r \frac{\Delta I \cdot \frac{1}{C_r\,\omega}}{\Delta I \sqrt{R_x^2 + \frac{1}{C_x^2\omega^2}}} = \frac{C_x}{\sqrt{R_x^2 C_x^2 \omega^2 + 1}}, \tag{I.22}$$

where C_r is the reference capacity; R_x and C_x are the series-connected ohmic and capacity components of the over-all cell impedance which, in the absence of an electrochemical reaction, are respectively equal to the resistance of the cell and to the capacity of the double layer.

It is seen from Eq. (I.22) that C_{exptl} determined by the comparison method can be equated to the capacity of the double layer C_x only if the ac angular frequency ω and the solution resistance R_x are low, when the following inequality is satisfied:

$$R_x^2 C_x^2 \omega^2 \ll 1. \tag{I.23}$$

The condition (I.23) is a serious limitation of the comparison method.

Another ac method for measurement of double layer capacity, the "tensammetric" method proposed by Breyer and Hacobian [72],

is subject to the same limitation. This method consists of measurement of the dependence of the amplitude of the alternating
current passing through the cell on the potential; for an ideally
polarized electrode this amplitude is

$$I_0 = \frac{V_0}{\sqrt{R_x^2 + 1/C_x^2\omega^2}} = V_0\,\omega\frac{C_x}{\sqrt{R_x^2 C_x^2\omega^2 + 1}}, \qquad (I.24)$$

where V_0 is the amplitude of the alternating potential difference
applied to the cell.

Comparison of Eqs. (I.22) and (I.24) shows that the "tensammetric" method gives, in principle, the same results as the comparison method [73] and is therefore limited by the condition (I.23).

The resonance method proposed by Watanabe et al. [74, 75] for
measurement of differential capacity is based on the resonance
effect in a circuit where the electric double layer on an ideally
polarized electrode acts as a condenser. This method is also suitable only for determinations of capacity in fairly concentrated solutions, as at low concentrations the resonance effect does not occur
owing to the high ohmic resistance of the circuit.

Among the various methods which have been devised, whereby
the resistance of the solution and the phase shift between the capacity and the resistance can be taken into account [76-81], the
most accurate data on double layer capacity can be obtained by the
impedance bridge method, first used by Dolin and Ershler [82] and
applied to measurement of double layer capacity in presence of organic compounds by Grahame [83].

The general calculation and design principles of ac bridges are
described in a number of Soviet books [84-86] and in a paper by
Gerischer [87]. The specific design characteristics of impedance
bridges for electrochemical measurements are determined, first,
by inclusion of the electrochemical cell, which has a number of
specific properties, in one of the bridge arms; second, by the additional complication of the circuit due to connection of a dc circuit to the bridge. A general circuit diagram of this type of bridge
is given in Fig. 2.

It follows from ac bridge theory [84, 86] that the bridge is
balanced, i.e., the potentials at the points a and c are equal, when

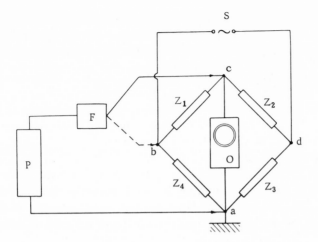

Fig. 2. Circuit diagram of an impedance bridge for electrochemical measurements. S) Sinusoidal wave source;
Z_1) impedances of the bridge arms (Z_4 represents the impedance of the electrochemical cell); O) oscillograph
balance indicator (null instrument); P) dc potentiometer;
F) filter preventing penetration of alternating current into
the potentiometer circuit.

the following condition is satisfied:

$$Z_1 Z_3 = Z_2 Z_4. \qquad (I.25)$$

The bridge arms Z_1 and Z_2 are usually the same (two ohmic re-
sistances, two condensers, or two inductances), so that $Z_1 = Z_2$.
The bridge balance condition (I.25) is then simplified:

$$Z_3 = Z_4. \qquad (I.26)$$

If both these impedances can be represented by Eq. (I.21), we conclude from the condition of equality of two complex numbers (the
real parts and the imaginary parts both equal) that when the bridge
is balanced the following equations are valid:

$$C = C_b \text{ and } R = R_b, \qquad (I.27)$$

where C_b and R_b are the capacity and resistance of boxes connected in series.

Development work on techniques of measurement of differential capacities with the aid of the impedance bridge has been

concerned mainly with extension of the ac frequency range, which is especially important in studies of the kinetics of adsorption of organic substances, and with the best design of the test electrode.

If the formation time of the double layer is less than the ac half-period, the measured capacity of an ideally polarized electrode must be independent of the frequency. Calculations by Grahame [83] and Ferry [88] showed that the formation time of a double layer in 1 N salt solution is less than 10^{-6} sec, while variations of the double layer capacity as the result of fluctuations of ion concentrations caused by the alternating current are much smaller than the experimental error in capacity measurements. Nevertheless, various investigators [89-91] observed inexplicable dependence of the capacity on frequency, described as dispersion of capacity. The frequency range in which the capacity could be measured correctly was narrowed sharply by the capacity dispersion.

Grahame [83] showed that one of the causes of capacity dispersion is nonsymmetrical positioning of the electrodes. In such cases the current lines between different parts of the electrodes had different lengths and, on increase of frequency, surface elements connected to elements of high ohmic resistance were virtually excluded from the measurements, which led to decrease of capacity. Grahame succeeded in eliminating this cause of capacity dispersion by using an electrode in the form of a small drop of mercury symmetrically surrounded by a platinum cylinder or a sphere made from platinized platinum gauze. It was then possible to measure the capacity accurately in 0.1 N solutions at frequencies in the range from 240 to 5000 Hz.

Melik—Gaikazyan and Dolin [92] showed that at higher frequencies a special matching transformer of very low output resistance must be connected between the source and the bridge. Otherwise the current consumed for charging the double layer is very small at high frequency, and the accuracy in capacity measurements is sharply lowered. On the other hand, at frequencies above 5000 Hz the small inductances in the bridge circuit (mainly in the resistance box) must be balanced by a specially chosen self-inductor ($L \approx 1$ μH). With the aid of these improvements the upper frequency limit in 1 N solutions could be raised to 0.5 MHz. At the same time, by replacing the choke coil, convenient for use as the

filter F (see Fig. 2) at frequencies above 250 Hz, by a high resistance (1-2 MΩ), Melik—Gaikazyan and Dolin [92] lowered the
bottom frequency limit to 20 Hz. It was subsequently shown by
Tedoradze [93] that an LC circuit tuned to the given frequency is
much more convenient than a high resistance as the filter at low
frequencies.

Damaskin [94] examined additional causes of capacity dispersion in dilute solutions, which are caused by connection of a
parasitic leakage to the ground in parallel to Z_4. The variations
are especially large when the dc circuit is connected to the point b
(see Fig. 2). Indeed, whereas in concentrated solutions, when Z_4
is small, ac leakage through F and P may be neglected, calculations show [94] that in dilute solutions the errors in the measured
values become very large. Therefore the dc circuit should be
connected to the point c, i.e., in parallel to the bridge diagonal.
In that case at the instant of bridge balance the potential of the
point c is equal to the potential of the point a, i.e., to the ground
potential (see Fig. 2). Therefore when the bridge is balanced
possible leakage to ground through the dc circuit, which is shunted
with the null indicator, is automatically excluded. With this bridge
design the filter F is included not to separate the dc and ac circuits
but to raise the resistance of the shunt, so that the sensitivity of the
null instrument is maintained at a sufficiently high level.

The parasitic leakage to ground is not reduced to zero by connection of the dc circuit to the point c. This is because all components of the circuit are screened in order to avoid induction from
electric fields, and the screens are grounded, so that parasitic
capacities to ground are inherent in the impedance bridge circuit.
It was shown by Damaskin [94] that this leakage can be represented
with a satisfactory degree of approximation by a small capacity C_l,
which is connected in parallel to Z_4 and which can be calculated
from the formula

$$C_l \approx \frac{C_1 - C_2}{R_1^2 C_1^2 \omega^2 - 1}, \tag{I.28}$$

where C_1 is the capacity measured at a low frequency (before dispersion begins); C_2 is the capacity measured at a fairly high frequency ω (when the dispersion is appreciable); R_1 is the measured
resistance of the solution, which does not vary appreciably with

the frequency. If the calculated $C_l > 0$, the leakage to ground is compensated by connection of a condenser of capacity C_l in parallel with Z_3. When $C_l < 0$, the leakage to ground is compensated by a similar condenser in parallel with Z_4.

The use of these improvements of the impedance bridge circuit made it possible to measure double layer capacities up to a frequency of 10,000 Hz in 0.01 N KCl solutions, and up to 1000 Hz in 0.001 N KCl solutions [94].

Differential or transformer bridges in which the center-tapped secondary winding of a matching transformer forms the arms Z_1 and Z_2 have recently come into use for measurement of differential capacity at high frequencies (above 10,000 Hz) [95-98]. At the Eighth CITCE Conference Cole and Hoar [95] described a differential bridge with two interchangeable matching transformers (for high and low frequencies), for capacity measurements in the frequency range from 20 Hz to 3 MHz. However, the upper frequency limit was evidently overestimated, because in a subsequent paper the same authors [96] describe a similar bridge but with a more modest frequency range, from 10 Hz to 1.7 MHz. According to Lorenz [99], reliable data on differential capacities can be obtained with the aid of ordinary or transformer ac bridges only at frequencies not exceeding 0.5 MHz, owing to the variable residual inductance of the resistance boxes. Much more consistent results at frequencies in the megahertz range are obtainable with the use of T-bridges [99]. A bridge of this type was used by Lorenz [99] for recording differential capacity curves in 1 N NaClO$_4$ solution at frequencies of 0.5, 0.795, and 1 MHz. A significant feature is that the capacities determined at different frequencies fit almost exactly on a common curve. A critical review of the use of various bridge methods for determination of electrode impedances over a wide range of ac frequencies was published recently by Armstrong, Race, and Thirsk [134].

Another trend in improvement of the bridge method for measurement of double layer capacity was the use of a dropping electrode. Grahame [100] was the first to propose the use of an electrode of this type for capacity measurement.

Owing to continuous renewal of its surface, the errors caused by surface contamination by traces of organic substances are largely eliminated when the dropping electrode is used for capacity measurement. On the other hand, inclusion of such an electrode in the bridge arm complicates the technique of capacity measurement. The electrode capacity increases continuously with growth of the drop, while the resistance R falls. Therefore with correctly chosen values of C_b and R_b the bridge is balanced only at a certain instant during the "life" of the drop. In order to determine the area of the drop electrode at the instant of bridge balance it is necessary to measure accurately the time interval between detachment of the "old" drop and the instant of balance (the balance time).

In this first investigation [100], Grahame determined the balance time with the aid of an electrode clock which was switched on manually when the drop became detached and switched off at the instant of bridge balance. Even after considerable practice, the balance time could not be determined to an accuracy better than 0.05 sec by this technique. Subsequently Grahame [101] used a cathode-ray oscillograph with a long-persistence screen for determination of the balance time. If the potential between points a and c is amplified and applied to the vertical plates of this oscillograph (Fig. 2), then with a fairly slow linear sweep the balance time is determined by the path traversed by the beam over the screen in that time, which is not difficult to determine, because at the instant of fall of the drop there is an abrupt rise of potential across the bridge, whereas when the bridge is balanced this potential falls to zero. The precision of this method is 0.01 sec, which corresponds to 0.2% with a balance time of 5 sec.

If the concentration of the solution is not too low ($\geq 0.1\,N$), the balance time can be conveniently determined with the aid of a special electronic relay which, using the output voltage from the null instrument, automatically connects the electric timer at the instant of drop detachment and disconnects it when the bridge is balanced. The first apparatus for determination of the balance time by this method was described by Boguslavskii and Damaskin [102]. The precision of this method is approximately 0.02 sec.

Considerably more precise and at the same time much more complicated methods for determining the balance time have recently been described in the literature [103-105]. The use of such complex apparatus is probably justified only in the case of very short balance times (of the order of 0.5-1.0 sec), when the error resulting from the use of simpler methods [101, 102] is appreci-able.

Determination of differential capacity curves with the aid of the ordinary ac bridge takes considerable time. If any processes leading to changes in the structure of the electric double layer, such as slow adsorption of an organic substance, occur on the electrode, the capacity is a function of time and determination of C vs φ curves becomes difficult. To avoid these difficulties, apparatus with automatic recording of the dependence of capacity on the time or the potential may be used [107-114]. Such apparatus generally operates on the phase detector principle. The oscillo-graphic vector polarograph recently described [111], which can be used for recording complete C vs φ and R vs φ curves in the fre-quency range from 32 to 159,000 Hz in a fraction of a second is probably one of the most effective instruments of this type. How-ever, it should be pointed out that the precision in capacity and resistance measurements with the aid of automatic devices is lower than with the use of ordinary ac bridges.

These are the main methods for determining the differential capacity of the double layer. The potentialities of some of these methods have been compared by Schubert [115].

3. Thermodynamic Methods for Calculating Adsorption of Organic Substances from the Results of Measurements of the Interfacial Tension and Double Layer Capacity

Several workers [21, 22, 116-119] have shown by independent methods that the thermodynamic laws of Gibbs, which give the re-lation between the interfacial tension σ, the electrode potential φ, and the adsorption and activity of ions and molecules (Γ_i and a_i respectively) in the solution are applicable to the electrode−solu-tion interface. At constant pressure and temperature, the funda-mental electrocapillary equation can be written in the form

$$d\sigma = -\varepsilon\,d\varphi - \sum_i \Gamma_i\,d\mu_i = -\varepsilon\,d\varphi - RT\sum_i \Gamma_i\,d\ln a_i, \qquad (I.29)$$

where μ_i is the chemical potential of component i, R is the gas constant, and T is the absolute temperature. In the general case, the quantity ε in Eq. (I.29) is the Gibbs adsorption of potential-determining ions, expressed in electrical units. In the case of a mercury electrode, which may be regarded with adequate accuracy as ideally polarized over a wide range of potentials, ε coincides with the surface charge.*

The fundamental electrocapillary equation (I.29) can be used for quantitative determination of the adsorption of solution components on the surface of mercury. However, it should be taken into account that the physical meaning of Γ_i depends on the conditions of choice of the Gibbs plane [120]. We examine this point for a two-component system consisting of a solution of alcohol (i) in water (w). At constant φ, it follows from Eq. (I.29) that

$$d\sigma = - \Gamma_i \, d\mu_i - \Gamma_w \, d\mu_w.^\dagger \qquad (I.30)$$

If the Gibbs plane is chosen so that $\Gamma_w = 0$, the experimental value $\Gamma_i^{exptl} = -d\sigma/d\mu_i$ defined by this condition represents the excess number of moles of alcohol in a portion of solution containing unit interfacial area over a portion of solution in the homogeneous volume with exactly the same molar content of water.

Another method of choosing the Gibbs plane corresponds to the condition

$$v_i \Gamma_i^{(v)} + v_w \Gamma_w^{(v)} = 0, \qquad (I.31)$$

where v_i and v_w are the partial molar volumes of alcohol and water.

The value $\Gamma_i^{(v)}$ defined by this condition represents the excess number of moles of alcohol in a portion of solution containing unit interfacial area in comparison with a portion of homogeneous solution having exactly the same volume. In this case the Gibbs plane is closest to the real electrode−solution interface, and the physical meaning of $\Gamma_i^{(v)}$ can be illustrated by Fig. 3.

---

*The physical meaning of ε will be discussed more fully in Chapter VII.

†Equation (I.30) remains valid if the solution also contains an indifferent electrolyte the activity of which is constant.

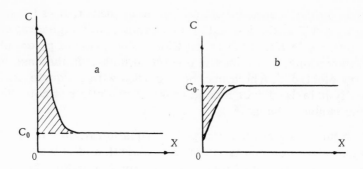

Fig. 3. Schematic interpretation of the surface excess $\Gamma_i^{(v)}$, apparently equal to the hatched areas in the plots of the dependence of the concentration of component i on the distance from the electrode. a) $\Gamma_i^{(v)} > 0$; b) $\Gamma_i^{(v)} < 0$.

Equations (I.30) and (I.31), in conjunction with the Gibbs − Duhem equation

$$N_i\, d\mu_i + N_w\, d\mu_w = 0, \tag{I.32}$$

where N_i and N_w are the mole fractions of alcohol and water in the solution volume, give the relation between Γ_i^{exptl} and $\Gamma_i^{(v)}$. Eliminating $\Gamma_w^{(v)}$ and $d\mu_w$ from these equations, we obtain

$$-\frac{d\sigma}{d\mu_i} = \Gamma_i^{exptl} = \Gamma_i^{(v)}\left(1 + \frac{v_i N_i}{v_w N_w}\right). \tag{I.33}$$

It follows from Eq. (I.33) that in sufficiently dilute solutions, when $v_i N_i/v_w N_w$ is negligible in comparison with unity $\Gamma_i^{exptl} \approx \Gamma_i^{(v)}$. However, in more concentrated solutions $\Gamma_i^{exptl} > \Gamma_i^{(v)}$.

It follows from the foregoing that the physical meaning of Γ_i^{exptl} and $\Gamma_i^{(v)}$ is not consistent with the concept of adsorption as the surface concentration of the adsorbate. In order that Γ_i and Γ_w in Eq. (I.30) should have the physical meaning of surface concentrations, it must be assumed that the heterogeneous region at the electrode surface is limited by a monolayer of adsorbate, while the Gibbs plane separates this monolayer from the homogeneous solution. We denote the surface concentrations of alcohol and water defined in this way by Γ_i' and Γ_w'. Then from Eqs. (I.30) and (I.32) we have

$$-\frac{d\sigma}{d\mu_i} = \Gamma_i^{exptl} = \Gamma_i' - \Gamma_w'\frac{N_i}{N_w}. \tag{I.34}$$

Equation (I.34) contains two unknowns, Γ_i' and Γ_w', which therefore cannot be determined by a purely thermodynamic method.

For determination of Γ_i' (or Γ_w') it is necessary to make certain model assumptions with regard to the structure of the interface, from which it is possible to calculate the areas corresponding to 1 mole of the organic substance and 1 mole of water in the surface layer (S_i and S_w respectively). Then Γ_i' and Γ_w' are found by solving the system of equations

$$\left. \begin{aligned} \Gamma_i^{exptl} &= \Gamma_i' - \Gamma_w' \frac{N_i}{N_w}, \\ \Gamma_i' S_i + \Gamma_w' S_w &= 1. \end{aligned} \right\} \qquad (I.35)$$

Suppose, for example, that the structure of the surface layer is as shown schematically in Fig. 4, and that the areas occupied by a molecule of the organic substance and a molecule of water are 22 and 10 Å² respectively. Then

$$S_i = 22 \cdot 10^{-16} \cdot 6.02 \cdot 10^{23} = 1.32 \cdot 10^9 \ (cm^2/M)$$

and

$$S_w = (1/3) \cdot 10 \cdot 10^{-16} \cdot 6.02 \cdot 10^{23} = 2.01 \cdot 10^8 \ (cm^2/M).$$

Suppose, further, that $\Gamma_i^{exptl} = 5 \cdot 10^{-10}$ mole/cm² and $N_i/N_w = 0.01$ (as the concentration of pure water is 55.5 M, this corresponds to approximately 0.56 M concentration of the organic substance).

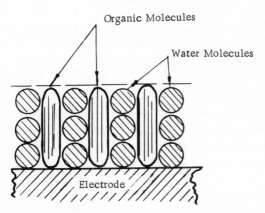

Organic Molecules

Water Molecules

Electrode

Fig. 4. Simplified model of a surface layer in presence of an organic substance, required for calculation of the surface concentrations Γ_i' and Γ_w'.

Solving the system of equations (I.35) under these conditions, we find $\Gamma_i' = 5.15 \cdot 10^{-10}$ mole/cm^2, which differs relatively little from Γ_i^{exptl}. The concentrations of organic compounds in aqueous solutions used in electrocapillary measurements are generally lower than our chosen value of 0.56 M, and therefore the differences between Γ_i^{exptl} and Γ_i' must be even smaller. It is therefore permissible to take experimental adsorption values as the surface concentrations of the organic substances. However, in the case of organic substances having low surface activity, such as C_2H_5OH solutions in water, adsorption saturation is reached only at very high concentrations, and the difference between the surface concentrations and Γ_i^{exptl} becomes very substantial [120-123].

For simplicity we will henceforth denote the experimentally determined Γ_i^{exptl} by Γ_i, although it must be remembered that these adsorption values relate to the plane where $\Gamma_w = 0$.

Adsorption at a given potential can easily be determined from the results of electrocapillary measurements with the aid of the fundamental electrocapillary equation (I.29), from which it follows that

$$\Gamma_i = -\frac{1}{RT}\left(\frac{\partial\sigma}{\partial\ln a_i}\right)_\varphi. \tag{I.36}$$

If it can be assumed that the activity coefficient varies little with the concentration of component i in the concentration range studied, the approximate relation

$$\Gamma_i = -\frac{1}{RT}\left(\frac{\partial\sigma}{\partial\ln c_i}\right)_\varphi, \tag{I.37}$$

where c_i is the concentration of component i, may be used instead of Eq. (I.36).

In order to find the value of adsorption at a given electrode charge, it is necessary to introduce, following Parsons [124], the new function $\xi = \sigma + \varepsilon\varphi$ the total differential of which is

$$d\xi = d\sigma + \varepsilon\,d\varphi + \varphi\,d\varepsilon. \tag{I.38}$$

From Eqs. (I.29) and (I.38) it follows that

$$d\xi = \varphi\,d\varepsilon - RT\sum_i \Gamma_i\,d\ln a_i, \tag{I.39}$$

hence

$$\Gamma_i = -\frac{1}{RT}\left(\frac{\partial \xi}{\partial \ln a_i}\right)_\varepsilon, \tag{I.40}$$

or, approximately, if the activity coefficient varies little with the concentration

$$\Gamma_i = -\frac{1}{RT}\left(\frac{\partial \xi}{\partial \ln c_i}\right)_\varepsilon. \tag{I.41}$$

The adsorption isotherms obtained from experimental data with the aid of Eqs. (I.36) and (I.40) are self-consistent in the sense that, having calculated the Γ vs c curve at φ = const from Eq. (I.36) and knowing the dependence of ε on φ at various concentrations of the adsorbate, we can plot the Γ vs c curve at ε = const, which exactly coincides with the adsorption isotherm calculated from Eq. (I.40). Thus, with a purely thermodynamic approach to investigation of adsorption isotherms the choice of the electrical variable is of no significance in principle, and is determined by considerations of convenience.

In addition to electrocapillary measurements, measurements of the differential capacity may be used for obtaining data on adsorption of organic substances at electrode–solution interfaces. The thermodynamic method of using capacity data is based on double integration of the differential capacity curves; this in accordance with the Lippmann equation gives the dependence of the interfacial tension on the electrode potential:

$$\varepsilon = \int_{\varphi_{\varepsilon=0}}^{\varphi} C\, d\varphi \tag{I.42}$$

and

$$\sigma = \sigma^{max} - \int_{\varphi_{\varepsilon=0}}^{\varphi} \varepsilon\, d\varphi. \tag{I.43}$$

The σ vs φ curves (or ξ vs ε curves) obtained in this way are subsequently used for calculating adsorption of the components as described above.

Two integration constants must be known in this method. In Eq. (I.42) this constant is the potential of zero charge, $\varphi_{\varepsilon=0}$; in Eq. (I.43) it is the interfacial tension at the maximum of the electrocapillary curve, i.e., σ^{max}. Various methods for determining potentials of zero charge of liquid electrodes have been described in detail by Grahame et al. [125, 126] and by other authors [127, 128]. Various methods for determinations of potentials of zero charge have also been reviewed by Frumkin [129]. Determination of the second integration constant requires electrocapillary measurements in the region of the potential of zero charge.

The capacity method is simplified considerably when it is used for studying adsorption of an organic substance which is added in small concentrations to a solution of an indifferent (supporting) electrolyte and is desorbed at fairly negative electrode potentials, so that the differential capacity curves recorded in the presence of the added organic substance merge with the C vs φ curve of the supporting electrolyte. Since under these conditions the structure of the electric double layer is determined only by the composition of the indifferent electrolyte, it is quite reasonable to assume that at this negative potential (e.g., φ_1) not only the differential capacities but also the magnitudes of the charges and interfacial tension in the pure solution of the supporting electrolyte and in solutions containing the organic substance should coincide. Thus, the integration constant in Eq. (I.42) in this case is the electrode charge in a solution of the supporting electrolyte at $\varphi = \varphi_1$, while the integration constant in Eq. (I.43) is the value of σ in the supporting electrolyte at $\varphi = \varphi_1$. Under these conditions the integration is performed from negative potentials (starting with φ_1) to positive (the "back integration" method [126]).

As only the change of interfacial tension with increasing concentration and not the absolute value of σ need be known for subsequent calculation of the adsorption of the organic substance with the aid of Eq. (I.37), the constant σ^{max} may be arbitrarily taken as zero in calculation of the electrocapillary curve of the supporting electrolyte. Therefore the only constant needed for calculation of adsorption from the results of measurements of the differential capacity in this case is the potential of zero charge in a pure solution of the supporting electrolyte.

It should be noted that the use of this method is restricted to not very high concentrations of the organic substance, when negative adsorption, leading to increase of σ with increasing concentration of the adsorbate, may be neglected (see, e.g., [135]).

On the other hand, at low concentrations of the organic substance equilibrium in the double layer is not reached during the ac half-period and it becomes necessary to extrapolate the experimentally determined nonequilibrium capacities to zero frequency. The errors arising from such extrapolation substantially lower the accuracy of the results obtained by this method. This is probably why the thermodynamic method for studying adsorption of organic substances with the aid of capacity data has not been widely used. It was used for studying adsorption of thiourea from solutions in $0.1\,N$ NaF on mercury [130] and, with other methods, for studying adsorption of aniline, phenol, and pentafluorophenol [131-133].

References

1. A. N. Frumkin and V. G. Levich, Zh. Fiz. Khim., 21:1135, 1183 (1947).
2. V. G. Levich, Physicochemical Hydrodynamics [in Russian], 2nd ed., Fizmatgiz, Moscow (1959), p. 581.
3. T. A. Kryukova and A. N. Frumkin, Zh. Fiz. Khim., 23:819 (1949).
4. S. G. Mairanovskii, Élektrokhimiya, 1:164 (1965).
5. J. Koryta and S. Vavřička, J. Electroanal. Chem., 10:451 (1965).
6. S. Vavřička, L. Němec, and J. Koryta, Collection Czech. Chem. Commun., 31:947 (1966).
7. H. H. Bauer and A. K. Shallal, Nature, 214:381 (1967).
8. W. Lorenz and E. O. Schmalz, Z. Elektrochem., 62:301 (1958).
9. H. B. Herman and S. V. Tatwawadi, Anal. Chem., 35:2210 (1963).
10. Ya. M. Zolotovitskii, G. A. Tedoradze, and A. B. Érshler, Élektrokhimiya, 1:828, 1280 (1965).
11. A. N. Frumkin, Dokl. Akad. Nauk SSSR, 85:373 (1952).
12. P. Zuman, Chem. Zvesti, 8:789 (1954).
13. C. N. Reilley and W. Stumm, Progress in Polarography, Interscience (1962), p. 81.
14. J. Kůta, Rev. Polarogr. (Japan), 11:62 (1963).
15. E. B. Veronskii, Zh. Fiz. Khim., 36:816 (1962); 40:1695,2425 (1966).
16. H. Matsuda, Rev. Polarogr. (Japan), 14:87 (1967).
17. G. Lippmann, Ann. Chim. Phys., 5:494 (1875).
18. S. V. Karpachev and A. G. Stromberg, Zh. Fiz. Khim., 10:739 (1937); 13:1831 (1939); 18:47 (1944).
19. G. Gouy, Compt. Rend., 114:22, 211, 657 (1892).
20. G. Gouy, Ann. Chim. Phys., (7), 29:145 (1903).
21. A. N. Frumkin, Electrocapillary Phenomena and Electrode Potentials [in Russian], Odessa (1919).
22. A. N. Frumkin, Z. Phys. Chem., 103:55 (1923).
23. A. N. Frumkin, Ergebn. Exakt. Naturwiss., 7:235 (1928).

24. F. O. Koenig, Z. Phys. Chem., 154:454 (1931).
25. B. E. Conway and L. C. M. Gordon, J. Electroanal. Chem., 15:7 (1967).
26. L. A. Hansen and J. W. Williams, J. Phys. Chem., 39:439 (1934).
27. A. Murtazaev, Acta Physicochim. URSS, 12:225 (1940).
28. A. N. Frumkin, V. A. Kuznetsov, and R. I. Kaganovich, Dokl. Akad. Nauk SSSR, 155:175 (1964).
29. A. V. Gorodetskaya, A. N. Frumkin, and A. S. Titievskaya, Zh. Fiz. Khim., 21:675 (1947).
30. J. Lawrence, R. Parsons, and R. Payne, J. Electroanal. Chem., 16:193 (1968).
31. J. O'M. Bockris, K. Müller, H. Wroblowa, and Z. Kovač, J. Electroanal. Chem., 10:416 (1965).
32. A. König, Wied. Ann., 16:1 (1882).
33. A. M. Worthington, Phil. Mag., 20:51 (1885).
34. M. E. Nicholas, P. A. Joyner, B. M. Tessem, and M. D. Olson, J. Phys. Chem., 65:1373 (1961).
35. A. Berget, Compt. Rend., 114:531 (1892).
36. G. Gouy, Ann. Phys., (9), 6:5 (1916).
37. J. Butler, J. Phys. Chem., 69:3817 (1965).
38. J. Andreas, E. Hauser, and W. Tucker, J. Phys. Chem., 42:1001 (1938).
39. S. Fordham, Proc. Roy. Soc., A194:1 (1948).
40. V. I. Melik-Gaikazyan, Izv. Tomsk. Politekhn. Inst., 136:110, 133 (1965).
41. G. Kučera, Ann. Phys., 11:529, 698 (1903).
42. T. Lohnstein, Ann. Phys., 20:237, 606 (1906).
43. T. Lohnstein, Z. Phys. Chem., 64:686 (1908); 84:410 (1913).
44. W. D. Harkins and E. C. Hamphery, J. Am. Chem. Soc., 38:228 (1916).
45. W. D. Harkins and F. E. Brown, J. Am. Chem. Soc., 41:499 (1919).
46. P. Corbusier and L. Gierst, Anal. Chim. Acta, 15:254 (1956).
47. S. Sathyanarayana, Indian J. Chem., 2:474 (1964).
48. B. Kastening, Z. Elektrochem., 68:979 (1964).
49. R. G. Barradas, F. M. Kimmerle, and E. M. L. Valeriote, J. Polarog. Soc., 13:30 (1967); Can. J. Chem., 45:109 (1967).
50. S. R. Craxford and H. A. C. McKay, J. Phys. Chem., 39:545 (1935).
51. S. G. Meibuhr, Electrochim. Acta, 10:215 (1965).
52. F. P. Bowden and E. K. Rideal, Proc. Roy. Soc., A120:59, 80 (1928).
53. T. Erdey-Gruz and G. Kromrey, Z. Phys. Chem., A157:213 (1931).
54. I. M. Barclay and J. A. V. Butler, Trans. Faraday Soc., 36:128 (1940).
55. R. J. Brodd and N. Hackerman, J. Electrochem. Soc., 104:704 (1957).
56. J. J. McMullen and N. Hackerman, J. Electrochem. Soc., 106:341 (1959).
57. J. S. Riney, G. M. Schmid, and N. Hackerman, Rev. Sci. Instr., 32:588 (1961).
58. M. Costa, Compt. Rend., 254:2973 (1962).
59. G. Armstrong and J. A. V. Butler, Trans. Faraday Soc., 29:1261 (1933).
60. N. A. Fedotov, Zh. Fiz. Khim., 25:3 (1951).
61. V. É. Past and Z. A. Iofa, Dokl. Akad. Nauk SSSR, 106:1050 (1956); Zh. Fiz. Khim., 33:913, 1230 (1959).
62. U. V. Pal'm and V. É. Past, Dokl. Akad. Nauk SSSR, 146:1374 (1962).
63. T. I. Borisova and M. A. Proskurnin, Zh. Fiz. Khim., 21:463 (1947).

64. P. Delahay, R. de Levie, and A. M. Giuliani, Electrochim. Acta, 11:1141
     (1966).
65. P. Delahay and D. J. Kelsh, J. Electroanal. Chem., 16:116 (1968).
66. J. W. Loveland and P. J. Elving, J. Phys. Chem., 56:250, 255 (1952).
67. G. C. Barker and R. L. Faircloth, Advances in Polarography, Vol. 1, Pergamon
     Press (1960), p. 313.
68. B. B. Damaskin and O. A. Petrii, Zh. Fiz. Khim., 35:1862 (1961).
69. M. A. Proskurnin and A. N. Frumkin, Trans. Faraday Soc., 31:110 (1935).
70. T. I. Borisova, M. A. Proskurnin, Acta Physicochim. URSS, 4:819 (1936).
71. M. A. Proskurnin and M. A. Vorsina, Dokl. Akad. Nauk SSSR, 24:915 (1939).
72. B. Breyer and S. Hacobian, Australian J. Sci. Res., A5:500 (1952); Australian
     J. Chem., 9:7 (1956).
73. A. N. Frumkin and B. B. Damaskin, J. Electroanal. Chem., 3:36 (1962).
74. A. Watanabe, F. Tsuji, and S. Ueda, Bull. Inst. Chem. Res. Kyoto Univ., 33:91
     (1955).
75. A. Watanabe, J. Electrochem. Soc., 110:72 (1963).
76. H. H. Bauer and P. J. Elving, Australian J. Chem., 12:343 (1959); J. Am.
     Chem. Soc., 82:2091 (1960).
77. E. Niki, J. Electrochem. Soc. Japan, 23:526 (1955).
78. J. H. Sluyters and J. J. C. Oomen, Rec. Trav. Chim., 79:1101 (1960).
79. M. Breiter, H. Kammemaier, and C. A. Knorr, Z. Elektrochem., 60:37 (1956).
80. N. P. Gnusin, Zh. Fiz. Khim., 32:689 (1958); Izv. Vysshikh Uchebn. Zavedenii,
     Khim. i Khim. Tekhnol., 4:760 (1961).
81. N. P. Gnusin and D. V. Savchik, Zh. Fiz. Khim., 35:2151 (1961).
82. P. I. Dolin and B. V. Ershler, Acta Physicochim. URSS, 13:747 (1940).
83. D. C. Grahame, J. Am. Chem. Soc., 68:301 (1946).
84. K. B. Karandeev, Bridge Measurement Methods [in Russian], Gostekhizdat,
     Kiev (1953).
85. K. M. Sobolevskii and Yu. A. Shakola, Shielding of Alternating Current Bridges
     [in Russian], Izd. Akad. Nauk Ukr. SSR, Kiev (1957).
86. A. D. Nesterenko, Principles of Bridge Circuit Calculations [in Russian], Izd.
     Akad. Nauk Ukr. SSR, Kiev (1960).
87. H. Gerischer, Z. Electrochem., 58:9 (1954).
88. J. D. Ferry, J. Chem. Phys., 16:737 (1948).
89. F. Krüger, Z. Phys. Chem., 45:1 (1903).
90. N. Thon, Compt. Rend., 198:1219 (1934); 200:54 (1935).
91. T. I. Borisova and M. A. Proskurnin, Acta Physicochim. URSS, 12:371 (1940).
92. V. I. Melik-Gaikazyan and P. I. Dolin, Tr. Inst. Fiz. Khim. Akad. Nauk SSSR,
     No. 1, 115 (1950).
93. G. A. Tedoradze, Zh. Fiz. Khim., 38:334 (1964).
94. B. B. Damaskin, Zh. Fiz. Khim., 32:2199 (1958).
95. M. Cole and T. P. Hoar, CITCE Proc., 8:158 (1958).
96. G. C. Wood, M. Cole, and T. P. Hoar, Electrochim. Acta, 3:179 (1960).
97. W. Lorenz, F. Möckel, and W. Müller, Z. Phys. Chem. (N.F.), 25:145 (1960).
98. G. H. Nancollas and C. A. Vincent, Chemistry and Industry, No. 16, 506 (1961).
99. W. Lorenz, Z. Phys. Chem. (N.F.), 26:424 (1960).

100. D. C. Grahame, J. Am. Chem. Soc., 63:1207 (1941).

101. D. C. Grahame, J. Am. Chem. Soc., 71:2975 (1949); J. Phys. Chem., 61:701 (1957).

102. L. I. Boguslavskii and B. B. Damaskin, Zh. Fiz. Khim., 34:2099 (1960).

103. P. Kivalo and E. Haikola, Suomen Kem., 34:B131 (1961).

104. G. A. Tedoradze, L. L. Knots, and G. G. Dubovik, in: Advanced Technological and Production Experience [in Russian], No. 37-63-466/2, GOSINTI, Moscow (1963), p. 31.

105. G. H. Nancollas and C. A. Vincent, J. Sci. Instr., 40:306 (1963).

106. G. J. Hills and R. Payne, Trans. Faraday Soc., 61:316 (1965).

107. C. O. Anderson, F. Möhl, and E. Stenhagen, Acta Chem. Scand., 12:415 (1958).

108. J. Schön, W. Mehl, and H. Gerischer, Z. Electrochem., 59:144 (1955).

109. S. B. Tsfasman, Zavod. Lab., 25:888 (1960).

110. Z. Kowalski and J. Srzednicki, Roczniki Chem., 36:565 (1962); J. Electroanal. Chem., 8:399 (1964).

111. Ya. M. Zolotovitskii and G. A. Tedoradze, Izv. Akad. Nauk SSSR, Ser. Khim., 2133 (1964).

112. L. Němec, Collection Czech. Chem. Commun., 31:1162 (1966).

113. E. M. L. Valeriote and R. G. Barradas, J. Electroanal. Chem., 12:67 (1966).

114. J. Clavillier, Compt. Rend., C263:191 (1966).

115. P. Schubert, Z. Chem., 4:11 (1964).

116. F. O. Koenig, J. Phys. Chem., 38:111, 339 (1934).

117. S. R. Craxford, O. Gatty, and J. Philpot, Phil. Mag., (7), 19:965 (1935).

118. D. C. Grahame and R. B. Whitney, J. Am. Chem. Soc., 64:1548 (1942).

119. R. Parsons and M. A. V. Devanathan, Trans. Faraday Soc., 49:404 (1953).

120. E. A. Guggenheim and N. K. Adam, Proc. Roy. Soc., A139:218 (1933).

121. J. A. V. Butler and A. Wightman, J. Chem. Soc., 2089 (1932).

122. J. J. Kipling, J. Colloid Sci., 18:502 (1963).

123. S. I. Zhdanov and A. M. Khopin, in: Electrochemical Processes Involving Organic Substances [in Russian], Izd. "Nauka," Moscow (1969).

124. R. Parsons, Trans. Faraday Soc., 51:1518 (1955).

125. D. C. Grahame, R. P. Larsen, and M. A. Poth, J. Am. Chem. Soc., 71:2978 (1949).

126. D. C. Grahame, E. M. Coffin, J. P. Cummings, and M. A. Poth, J. Am. Chem. Soc., 74:1207 (1952).

127. D. A. Jenkins and R. J. Newcombe, Electrochim. Acta, 7:685 (1962).

128. Z. Koczorowski, J. Dabkowski, and S. Minc, J. Electroanal. Chem., 13:189 (1967).

129. A. N. Frumkin, Svensk. Kem. Tidskr., 77:300 (1965).

130. F. W. Schapink, M. Oudeman, R. W. Leu, and J. N. Helle, Trans. Faraday Soc., 56:415 (1960).

131. B. B. Damaskin, I. P. Mishutushkina, V. M. Gerovich, and R. I. Kaganovich, Zh. Fiz. Khim., 38:1797 (1964).

132. B. B. Damaskin, V. M. Gerovich, I. P. Gladkikh, and R. I. Kaganovich, Zh. Fiz. Khim., 38:2495 (1964).

133. B. B. Damaskin, M. M. Andrusev, V. M. Gerovich, and R. I. Kaganovich, Élektrokhimiya, 3:667 (1967).

134. R. D. Armstrong, W. P. Race, and H. R. Thirsk, Electrokhim. Acta, 13:215 (1968).
135. R. S. Maizlish, I. P. Tverdovskii, and A. N. Frumkin, Zh. Fiz. Khim., 28:87 (1954).
136. V. I. Melik-Gaikazyan, V. V. Voronchikhina, and É. A. Zakharova, Élektrokhimiya, 4:479 (1968).

Chapter 2

Experimental Data on Adsorption of Organic Substances on Liquid Electrodes

1. Electrocapillary Measurements on Mercury in Presence of Organic Compounds

Many organic substances are adsorbed on mercury electrodes; in accordance with the fundamental electrocapillary equation (I.29), this adsorption leads to decrease of interfacial tension (σ). Some determinations of interfacial tension in presence of organic substances were carried out even at the end of the 19th century [1, 2], but the first systematic data on the influence of organic compounds on electrocapillary curves were obtained by Gouy [3].

Gouy showed that in presence of an organic substance the electrocapillary curve always lies below the curve determined for a pure solution of the indifferent electrolyte; the decrease of σ is greatest in the region of the potential of zero charge in the solution of the inorganic electrolyte, and diminishes if the negative or positive charge of the surface is increased substantially. In the majority of cases at sufficiently high negative or positive charges no decrease of interfacial tension occurs in presence of organic substances, and the electrocapillary curve obtained in presence of an organic additive coincides with the curve for the solution of the supporting electrolyte. Another fairly common feature of electrocapillary curves in presence of organic compounds is a shift of the electrocapillary maximum, i.e., of the potential of zero charge.

Electrocapillary curves determined in presence of organic substances exhibit a great diversity of form. In presence of not unduly small amounts of aliphatic compounds characteristic curves with a sharp change of slope are obtained in most cases; sometimes this change of slope assumes the nature of a break.

Comparison of σ vs φ curves for solutions containing organic substances which form a homologous series shows that the decrease of interfacial tension and consequently the adsorption increase with increasing length of the organic chain in solutions of equal concentrations. According to Kaganovich and Gerovich [4, 5], adsorption of normal aliphatic acids, alcohols, and amines from solutions in $1N$ Na_2SO_4 on mercury conforms to the Traube rule. On the other hand, in accordance with the Gibbs equation, the decrease of σ becomes greater with increasing concentration of the organic substance [6].

Gouy [7] showed that increase of the concentration of the supporting electrolyte in presence of an organic substance leads to lowering of surface tension, but in the case of a surface-inactive electrolyte (Na_2SO_4) the decrease of σ is due not to adsorption of the electrolyte ions but to increase of the chemical potential (μ_{org}) of the organic substance. Indeed, if μ_{org} = const, as in the case of solutions saturated with the organic substance and containing different amounts of Na_2SO_4, the interfacial tension also remains approximately constant [6].

On the other hand, surface-active ions of the supporting electrolyte not only change the chemical potential of the organic substance in the solution but also influence the structure of the electric double layer, increasing or decreasing adsorption of organic molecules. For example, in presence of I^- ions, which diminish mutual repulsion of adsorbed $[(C_3H_7)_4N]^+$ cations, adsorption of these organic cations increases [6]. At the same time, specific adsorption of Br^- and I^- anions diminishes adsorption of ethyl alcohol, phenol, and salicylate anions on mercury [8].

Frumkin [9] and Butler [10, 11] attributed the shift of the point of zero charge (p.z.c.) in adsorption of organic molecules on mercury partly to the fact that these molecules have a permanent dipole moment. Thus, with oriented adsorption of these dipoles, which had been noted by Gouy [3], an adsorption potential corresponding to the shift of p.z.c. arises at the mercury−solution interface. Frumkin [12, 13] correlated the shifts of p.z.c. under the influence of adsorption of various organic compounds on mercury with adsorption potentials at solution−air interfaces.

The results of this correlation for aliphatic oxygen compounds showed that adsorption of these molecules on both interfaces leads to potential differences of the same sign and similar magnitude. These values can be interpreted qualitatively on the assumption that adsorption of organic substances involves replacement of water dipoles oriented with their negative oxygen ends outward (toward air or mercury) by organic molecules. The latter are oriented with their hydrocarbon chains toward the interface, and orientation of the C − O bond creates a positive potential difference between the external phase and the solution volume [14].

A different theory was developed in publications by Devanathan [15] and by Bockris, Devanathan, and Müller [16]. In their view, the shift of p.z.c. in presence of aliphatic compounds is produced only by displacement of water dipoles from the double layer, while the polar groups of the organic molecules, being outside the limits of the double layer, should not influence the distribution of potential. It is difficult to accept these views [17, 18]. According to Bockris, Devanathan, and Müller [16], the potential difference due to orientation of water dipoles does not exceed 0.1 V. Therefore it is impossible to explain in the light of their theory how adsorption of organic substances can produce shifts of p.z.c. reaching 0.5 V. It is interesting to note that a similar theory was put forward by Kamienski [19] for interpreting the shifts of potential resulting from adsorption of organic substances at the water − air interface. However, Frumkin [20] showed that here again this theory leads to impossible values for the chemical energy of cation and anion hydration.

Comparison of potential differences ($\Delta\varphi$) at solution − mercury and solution − air interfaces in presence of aromatic compounds reveals considerable differences in magnitude and in some cases of the sign of $\Delta\varphi$. For example, the values of $\Delta\varphi$ for o- and p- cresols are −0.20 and −0.29 V at the mercury − solution interface, and +0.01 and +0.26 at the air − solution interface.

It was initially believed that the shift of p.z.c. in the negative direction produced by adsorption of aromatic compounds on mercury is due to the more flat orientation of the molecules on the mercury surface, which facilitates interaction of negative atoms in the polar groups with the metal. However, Gerovich [21, 22] showed that compounds such as benzene, naphthalene, anthracene,

phenanthrene, and chrysene, although nonpolar, also shift the p.z.c.
in the negative direction, and that the adsorbability of these compounds at $\varepsilon > 0$ increases with increasing number of benzene rings
in the organic molecule. These results suggested that the anomalous behavior of aromatic compounds at the mercury−solution
interface is due not only to their more flat orientation but also to
the structural characteristics of the benzene ring.

In fact, comparison of the electrocapillary behavior of aromatic and the corresponding hydroaromatic compounds, and of hydrocarbons with different numbers of double bonds in the molecule,
carried out by Gerovich et al. [21-23], showed that the shift of the
p.z.c. and adsorption of aromatic compounds at high positive surface charges are due to interaction between π-electrons of the aromatic nucleus and positive charges on the mercury surface.

Gerovich and Polyanovskaya [24] showed that on replacement
of neutral by acid aniline solutions, where the aniline is present in
the form of $[C_6H_5NH_3]^+$ cations, the adsorption of the organic compound on positively charged surfaces diminishes but does not fall
to zero. This result shows that forces of interaction between
positive surface charges and π-electrons of the aromatic nucleus
predominate over electrostatic repulsions between the organic
cations and the surface.

A similar conclusion was reached by Blomgren and Bockris
[25], who used the electrocapillary curve method for studying adsorption of certain aromatic amines (aniline, o-toluidine, 2,3- and
2,6-dimethylanilines, pyridine, and quinoline) from 0.1 N HCl solutions on mercury. According to these authors, these substances
are adsorbed predominantly in the form of $[RNH_3]^+$ ions which lie
flat on the electrode surface. The amount adsorbed varies little
with increase of potential over a range of about 1 V; the adsorption
is determined predominantly by π-electron interaction along the
positive branch of the curve, and by Coulomb forces along the negative branch.

These conclusions were subsequently confirmed by Conway
and Barradas [26]. They used the results of electrocapillary
measurements for calculating isotherms of adsorption of pyridine,
its derivatives, and aniline from 1 N KCl and HCl solutions. The
standard free energies of adsorption of these compounds on the
mercury electrode were calculated from the isotherms. In neu-

tral solutions, maximum adsorption corresponds to perpendicular orientation of adsorbed dipoles relative to the surface; the greater the π-deficiency of the organic molecule the lower are the values of Γ at which such orientation occurs. At the same time, in acid solutions (1 N HCl) where aniline and pyridine and its derivatives are present in cationic form, the maximum adsorption corresponds to flat orientation of the adsorbed particles, which favors π-electron interaction with the metal surface.

The electrocapillary behavior on mercury of aromatic and heterocyclic compounds containing π-electron bonds (pyridine and its derivatives) has also been investigated by Antropov and Benergee [27], Gierst et al. [28, 29], Conway et al. [30, 31], Zwierzy-kowska [32], and Nürnberg and Wolff [33]. Their experimental data and theoretical conclusions were compared in a recent paper by Damaskin et al. [34].

In all these cases π-electron interaction between adsorbed molecules and the electrode surface has a characteristic influence on the adsorption behavior of organic substances. It should be noted that replacement of hydrogen atoms in aromatic compounds by fluorine atoms leads to π-electron exhaustion of the aromatic nucleus on account of the higher electron affinity of fluorine atoms. For this reason π-electron interaction has virtually no influence on adsorption of pentafluoroaniline, pentafluorobenzoic acid, and penta-fluorophenol molecules on mercury [35, 36].

It should be pointed out that in presence of specifically adsorbed anions, leading to surface charge reversal, adsorption of organic cations on a positively charged mercury surface may occur even in absence of π-electron interaction. Such an effect was observed by Frumkin et al. [37] in the case of cyclohexylammonium ions. Studies of adsorption of organic cations not containing π-electron bonds by the electrocapillary curve method have also been reported in other publications [38-41]. It is interesting to note that the surface activity of tetrabutyl- and tetrapropylammonium ions is considerably higher at the solution−mercury than at the solution−air interface. Damaskin et al. [40] showed that this effect can be attributed to image forces, absent at the air interface. In the case of propylammonium, butylammonium, and amylammonium ions the corresponding effect is substantially weakened as the result of strong hydration of the functional group in compounds of the RNH_3^+ type.

Blomgren, Bockris, and Jesch [42] studied the influence of functional groups in adsorption of organic molecules or ions at the surface of a mercury electrode. They used the results of electrocapillary measurements in a detailed investigation of the adsorption of butyl, phenyl, and naphthyl derivatives (containing $-OH$, $-CHO$, $-COOH$, $-CN$, $-SH$, $\rangle CO$, $-NH_3^+$, and $-SO_3^-$ substitutent groups) on mercury. The adsorption isotherms of all the compounds were calculated, and used for finding the standard free energy of adsorption, $\Delta \overline{G}_S^0$.

It was found by Blomgren et al. [42] that at constant coverage of the surface with the organic substance ($\theta = 0.25$) the $\Delta \overline{G}_A^0$ values for the derivatives of a given radical are, in the first approximation, linear functions of their standard free energies of solution ($\Delta \overline{G}_S^0$). By extrapolating the $\Delta \overline{G}_S^0$ values to $\Delta \overline{G}_S^0 = 0$ and subtracting from the results the difference, calculated with certain assumptions, between the chemical potentials of water at the surface and in the solution volume, Blomgren et al. [42] obtained what they described as the "net" standard energies of adsorption ($\Delta \overline{G}_A^0$), which characterize the interaction of mercury with the corresponding organic radicals regardless of the functional substituents. Thus, according to these workers [42], the role of functional groups of adsorption of organic molecules is manifested mainly in the energy of solution of these compounds. However, in the light of the work of Kaganovich and Gerovich [4, 5] and of Damaskin et al. [43, 44] one cannot agree completely with this simplified treatment of the problem.

In fact, the data of Kaganovich and Gerovich [4, 5] indicate that the adsorption behavior of aliphatic alcohols, acids, and amines at the solution—mercury interface does not differ very much from their behavior at the solution—air interface. However, in the cause of acids and especially of alcohols the surface activity is somewhat lower at the mercury than at the air interface, whereas in the case of amylamine and partially of butylamine the reverse is true. These differences cannot be explained other than by the assumption that, despite the fact that the polar group is oriented toward the solution and is therefore removed from the mercury surface, its nature nevertheless has some influence on the change of the work of adsorption in passing from the free surface of the solution to the interface with mercury [45].

The small difference between the adsorbabilities of normal aliphatic compounds at air and mercury interfaces indicates that the gain of free energy resulting from contact of mercury with the hydrocarbon tails of the organic molecules is approximately balanced by the expenditure of free energy of the adsorbed water molecules. However, such compensation cannot be general in character; this is confirmed by consideration of the adsorption of organic compounds the molecules of which lie flat on the interface.

Flat orientation is to be expected in the case of hydrocarbons, which contain no polar groups, at least when the degree of surface coverage is low. Kaganovich et al. [46] showed that under such conditions the adsorbability of n-hexane is much lower at the interface with mercury than with air. The decrease of σ at the solution−air interface due to saturation of $0.1\,N$ CsCl solution with hexane is $\Delta\sigma = 5.0$ dyn/cm, whereas at the interface with an uncharged mercury surface $\Delta\sigma_{Hg}$ is about one-seventh of this value. The decrease in the gain of free energy by transfer of a hydrocarbon molecule from the solution volume to the surface in this case is due to the fact that with flat orientation of the hydrocarbon molecules water is displaced from the mercury surface mainly by $\rangle CH_2$ rather than by $-CH_3$ groups. This effect is even more pronounced in adsorption of perfluoro compounds [47]. Therefore the gain of free energy when mercury is wetted by $-CF_3$ and $\rangle CF_2$ groups is much less than when it is wetted by water.

Flat orientation is apparently characteristic of saturated aliphatic compounds with two or more functional groups, as well as of hydrocarbons. Under these conditions the consequence of interaction between the polar group and the mercury surface is that the adsorbability of the organic compound is appreciably higher at the solution−mercury than at the solution−air interface [13, 48]. For example, in $1\,M$ glycerol solution $\Delta\sigma = 0.4$ dyn/cm at the solution−air interface and $\Delta\sigma_{Hg} = 9.3$ dyn/cm at the solution−mercury interface; and similar results have been obtained for $1\,M$ sucrose solution: $\Delta\sigma = -2.0$ and $\Delta\sigma_{Hg} = 23.8$ dyn/cm [13]. Flat orientation of these compounds is confirmed by the fact that, in distinction from substances with one polar group in the molecule, in this case the shift of p.z.c. in the positive direction caused by orientation of the $C-O$ linkage with carbon toward the mercury decreases almost to zero.

As already noted, the main difference in the adsorption behavior of aromatic and heterocyclic compounds at the mercury interface as compared with the air-solution interface is attributable
to π-electron interaction. However, interaction between mercury
and the polar groups also plays a substantial part in the increased
adsorbability of these compounds at the mercury interface. π-
Electron interaction tends to make orientation of aromatic molecules at the solution-mercury interface more flat; this, in its turn,
facilitates interaction of polar groups with the metal. Thus, in
addition to decreasing surface activity at the solution-air interface (by increasing the solubility), introduction of a polar group
into the molecule of an organic compound may lead to increase of
surface activity at the mercury interface. Solutions of benzene
and aniline are an example illustrating this conclusion. In 0.01 M
benzene solution $\Delta\sigma$ = 4.2 and $\Delta\sigma_{Hg}$ = 13.9 dyn/cm, while in 0.01 M
aniline solution $\Delta\sigma$ = 0.3 and $\Delta\sigma_{Hg}$ = 17.4 dyn/cm [45].

In a recent study of the dependence of adsorption of organic
compounds on mercury on the polar properties of substituents in
their molecules, Grigor'ev and Ekilik [49] found the existence of
a linear relation between the decrease $\Delta\sigma_{Hg}$ of interfacial tension under conditions when $\theta \approx 1$, and the Hammett constants
(σ_H) for the given reaction series. These results were obtained
for derivatives of aniline (para-substituents: $-OH$, $-OCH_3$, $-CH_3$,
$-F$, $-Cl$, $-Br$, $-COOCH_3$, $-COOC_2H_5$, and $-COOH$, and m-CH_3)
of 1-hydroxy-2-naphthalaniline (para-substituents: $-CH_3$, $-F$,
$-Cl$, $-Br$, and $-I$), and of benzaldehyde (para-substituents:
$-N(CH_3)_2$, $-OH$, $-CH(CH_3)_2$, $-Cl$, and $-Br$, and also o-OCH_3 and
m-OH); in the case of amines the dependence of $\Delta\sigma_{Hg}$ on σ_H is
V-shaped, with a break at σ_H = 0.

Introduction of nucleophilic (electron-donor) substituents, for
which σ_H < 0 [50], leads to increase of adsorbability both owing to
increase of the π-electron density in the aromatic nucleus and in
consequence of increase of unshared electron density at the hetero
atom (O or N). Introduction of electrophilic substituents (σ_H > 0)
has the opposite effect. Apart from these factors, introduction of
substituents also influences the magnitude of the dipole moment μ:
in the case of benzaldehyde derivatives μ increases with increasing
negative value of σ_H, whereas in the case of amines it increases
with increase of the positive value of σ_H [50, 51]. According to
Grigor'ev and Ékilik [49], this is the cause of the different charac-

ter of the dependence of $\Delta\sigma_{Hg}$ on σ_H in the case of benzaldehyde derivatives and amines respectively: in the former case $\Delta\sigma_{Hg}$ decreases linearly with increasing σ_H, whereas in the latter, as already noted, the $\Delta\sigma_{Hg}$ vs σ_H curve is V-shaped. The linear dependence of $\Delta\sigma_{Hg}$ on σ_H with the condition $\theta \approx 1$ is easily interpreted with the aid of the Shishkovskii equation [49].

The investigations discussed here indicate that the electrocapillary curve method has been used widely for studying adsorption of various organic compounds on mercury. In addition to the workers already cited, Frumkin, Gorodetskaya, and Chugunov [52] used this method in a study of multilayer formation on mercury during adsorption of caproic acid and phenol from saturated solutions. Multilayers are not formed at the solution−air interface under similar conditions. The electrocapillary curve method has also been used for quantitative determination of the adsorption of butyl alcohol on mercury [53]; for determination of the effective dipole moments [54] and dissociation constants [55] of certain organic acids; for studying adsorption on mercury of certain alkaloids [56] and wetting agents [57], camphor [58], dibenzyl sulfoxide [59], aliphatic amino ethers [60], polyethylene glycols [61], certain organic detergents [62], and coumarin [63, 64].

Investigations of the electrocapillary behavior of mixtures of two organic substances, initiated by Butler and Ockrent [65], have been continued by Pamfilov et al. [66], Kastening [67], and recently by Arakelyan and Tedoradze [68]. It is interesting to note that the laws governing joint adsorption of n-butyl alcohol and aniline on positively and negatively charged mercury electrodes differ substantially [68]. At $\varepsilon < 0$ mutual intensification of adsorption occurs, and the electrocapillary curve for a mixed solution of aniline and butanol lies below the σ vs φ curves for solutions of the individual substances.

However, when the electrode is positively charged adsorption of aniline leads to decreased adsorbability of n-C_4H_9OH. Under these conditions the σ vs φ curve in presence of the binary additive approximates the electrocapillary curve for pure aniline solution with increase of $\varepsilon > 0$. In other words, desorption of butyl alcohol from a mercury surface covered by aniline molecules in flat orientation proceeds in the same way as desorption from a clean surface carrying the same positive charge. The cause of this difference in the behavior of aniline + n-butyl alcohol mixtures along the cathodic and anodic branches of the electro-capillary curve evidently lies in the different orientation of ad-sorbed aniline molecules at $\varepsilon > 0$ and at $\varepsilon < 0$ [43].

2. Influence of Adsorption of Organic Substances on the Capacity of the Mercury Electrode

The differential capacity of the electric double layer is a much more sensitive function of the adsorption of organic mole-cules at the electrode surface than is the interfacial tension. On the other hand, this function is also much more complex so that, as already noted earlier (see Chapter I, Section 3), the purely thermodynamic method of calculating adsorption on the basis of C vs φ curves has not been widely adopted. Schapink et al. [69] used this method in a quantitative study of the adsorption of thiourea molecules from solutions in presence of 0.1 N NaF. They con-cluded from the results that the thiourea adsorption isotherms conform to the Langmuir equation, and the energy of adsorption is, in the first approximation, a linear function of the electrode potential.

In all other investigations (e.g., see Damaskin et al. [44]) the purely thermodynamic method of using capacity data was supple-mented by calculations based on certain model concepts of the structure of the electric double layer. These results will be ex-amined in Chapters III and IV; in this section we discuss the quali-tative aspects of differential capacity curves in presence of vari-ous organic compounds.

In view of the shape of the electrocapillary curves determined in presence of aliphatic organic compounds, and with Lippmann's equation

$$C = -\frac{d^2\sigma}{d\varphi^2} \qquad\qquad (II.1)$$

taken into account, one would expect that under these conditions the differential capacity curves would have a region of low ca-pacity values with adsorption−desorption peaks on each side, corresponding to sharp changes of surface charge within a narrow range of potentials. At high positive and negative surface charges the C vs φ curves determined in presence and in absence of organic substances should come closer together. Differential capacity curves of this form were first obtained by Proskurnin and Frumkin [70] for Na_2SO_4 solution saturated with n-octyl alcohol. Capacity curves of similar shape were obtained by Ksenofontov et al. [71] for solutions of ethyl and n-butyl alcohols, whereas in the case of phenol solution there was no anodic peak on the C vs φ curve. It was later shown by Damaskin et al. [44] that this effect is due to strong repulsive interaction between the adsorbed phenol molecules at $\varepsilon > 0$, as the result of which the degree of surface coverage and the surface charge alter very slowly with the electrode potential.

It should be noted that the presence of adsorption peaks on the C vs φ curves in presence of organic substances follows on thermodynamic grounds from experimental data on interfacial tension; therefore they cannot be attributed, as has been suggested by Doss et al. [72, 73], to tangential movements of mercury during adsorption and desorption processes.

Melik−Gaikazyan [74], in a study of normal alcohols of the aliphatic series, established certain general relationships for differential capacity curves in presence of aliphatic compounds. The region of the minimum in the C vs φ curves become broader with increasing concentration of the organic substance, and the minimum capacity tends to a certain limiting value of the order of 4-5 $\mu F/cm^2$ (Fig. 5). The height of the adsorption peaks also increases with increasing concentration of the organic substance, while the peaks themselves are shifted: the cathodic peak toward higher negative potentials, and the anodic peak toward more positive potentials.

In the same investigation of Melik−Gaikazyan [74] showed that in presence of n-octyl and n-hexyl alcohols multimolecular layers are formed on the mercury surface, and this is accompanied by decrease of capacity down to 0.9 $\mu F/cm^2$, whereas multilayer formation does not occur in solutions of n-butyl and n-amyl alcohols.

Gupta and Sharma [75] used the tensammetric method for studying the influence of structure of aliphatic alcohols on their surface activity. They determined the C vs φ curves for 0.1 N KCl

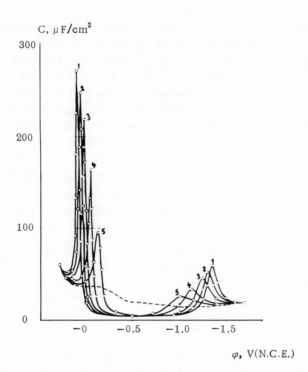

C, $\mu F/cm^2$

φ, V(N.C.E.)

Fig. 5. Differential capacity curves of a mercury electrode in pure 0.1 N Na$_2$SO$_4$ solution (dash line) and in
presence of n-butyl alcohol at concentrations (M): 1)
0.8; 2) 0.6; 3) 0.4; 4) 0.2; 5) 0.1 (frequency 400 Hz).

solution with 5% additions of the following alcohols: methyl, ethyl,
n- and isopropyl, n-, sec-, iso-, and tert-butyl, n-amyl, allyl,
propylene glycol, and glycerol. The results led to the qualitative
conclusion that adsorption of alcohols increases with increasing
length of the carbon chain, with decreased branching of the carbon
chain with the same number of carbon atoms (tert-< iso-< sec-< n-),
in passing from unsaturated to saturated alcohols, and with decrease of the number of hydroxyl groups in the molecule.

In their next series of investigations, Gupta and Sharma [76]
studied the influence of pH, the nature of the buffer, the buffer
capacity, the nature of the supporting electrolyte, and the nature
of the solvent on the form of the C vs φ curves. It was shown that
the nature of the buffer and the pH have virtually no effect on the
shape and position of the adsorption—desorption peaks. If the peaks

on the C vs φ curves correspond to sufficiently high negative charges, the nature of the supporting electrolyte has no effect on their height either; otherwise the height of the adsorption peak decreases in the series KI > KCNS > KCl > KClO$_4$. Finally, the adsorption–desorption peaks on the C vs φ curves are very much lower or vanish entirely in the case of solutions in organic solvents (methyl, ethyl, and isopropyl alcohols, pyridine, acetone, or methyl ethyl ketone) in comparison with aqueous solutions. Rise of temperature also diminishes the adsorption–desorption peaks. However, the influence of the concentration of the indifferent electrolyte on the height and position of the peaks on the C vs φ curves in presence of organic compounds was not considered in this series of investigations. According to Grigor'ev and Damaskin [148], increase of the activity of the supporting electrolyte from a_1 to a_2 should have a dual influence: 1) owing to the change of the electrode charge, it should lead to a shift of the cathode adsorption– desorption peak in the positive direction by $(RT/n_+F) \ln (a_2/a_1)$ without change of peak height; 2) owing to salting-out of the organic substance, it should lead to a shift of the cathodic peak in the negative direction and to a considerable increase of peak height.

Experimental data indicates that the first effect predominates in adsorption of thiourea on mercury, while in adsorption of tert-amyl alcohol salting-out of the organic substance has a considerable influence.

Narayan [77] showed that in the case of resorcinol and phloroglucinol the cathodic peak on the C vs φ curves is higher and narrower for alkaline than for acid solutions, but in this case the dependence on pH is due to a change in the nature of the adsorbed particles; according to Narayan [77], in alkaline solutions organic ions are adsorbed rather than neutral molecules.

In a number of investigations measurements of differential capacity were used for studying adsorption on mercury of surface-active tetraalkyl-ammonium cations [41, 78–82]. It was found that large organic cations behave similarly to neutral surface-active molecules on the surface of the mercury cathode. At sufficiently high negative potentials, despite electrostatic attraction, these cations are desorbed from the mercury surface, and this is accompanied by the appearance of a characteristic peak on the differential-capacity curve.

Several publications [57, 83-85] give C vs φ curves recorded for mercury electrodes in camphor solutions. A characteristic feature of these curves is that desorption of camphor from the mercury surface is accompanied by an abrupt change of differential capacity without distinct adsorption—desorption peaks. Similar C vs φ curves were also observed in the case of $[(C_4H_9)_4N]^+$ cations in 1 N KI [80] and in solutions of nonylic acid [86]. The theoretical aspects of this effect will be considered in Chapter III, Section 4,c.

In presence of macromolecular organic compounds the C vs φ curves also have characteristic peaks; however, according to Miller and Grahame [87-90], they are caused by desorption only of some segments of the adsorbed macromolecules and not by complete desorption of the polymer molecules. Measurements of differential capacity have been used for studying adsorption of polymethacrylic acid, polylysine, polyvinyl-pyridine, copolymers of ethylene oxide with lauryl alcohol, and polyvinyl alcohol [87-93]. If the C vs φ curve is recorded for a solution of a monomer, and a polymeric product is formed as the result of electrochemical initiation of

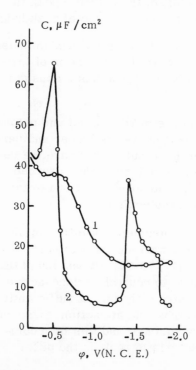

Fig. 6. Differential capacity curves for a mercury electrode in 0.1 N H$_2$SO$_4$ solution (1) and in 0.1 N H$_2$SO$_4$ solution saturated with methyl methacrylate (2). (From data in [94]).

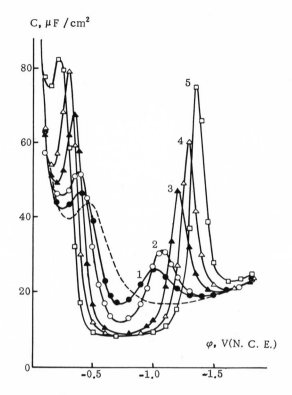

Fig. 7. Differential capacity curves for a mercury electrode in pure 1 N KCl solution (dash line) and with additions of aniline in concentrations (M): 1) 0.01; 2) 0.02; 3) 0.05; 4) 0 .1; 5) 0.2 (frequency 400 Hz). (From data in [43]).

the polymerization reaction, adsorption of the polymer under such conditions may be accompanied by a sharp fall of capacity at high negative potentials (on the negative side of the cathodic desorption peak), as is shown in Fig. 6 [94].

Differential capacity curves of mercury electrodes in aniline and pyridine solutions have an interesting peculiarity [43, 95]. In the case of aniline the curves are strongly reminiscent of the C vs φ curves in presence of aliphatic compounds (Fig. 7). However, this is merely an outward resemblance. Whereas in the presence of aliphatic compounds both maxima on the C vs φ curves are caused by adsorption–desorption processes, in the case of

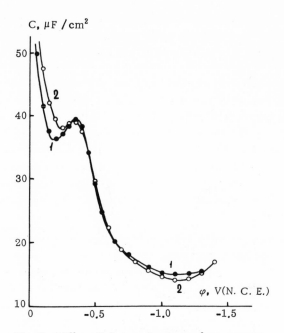

Fig. 8. Differential capacity curves for a mercury
electrode in 0.1 N HCl (1) and 0.1 N HCl + 0.1 M
aniline (2) solutions (frequency 400 Hz). (From data
in [96]).

aniline only the cathodic maxima correspond to these processes.
The anodic maxima, which are very pronounced in the case of aniline, are associated with reorientation of the adsorbed organic
molecules on the mercury surface: vertical orientation of the
molecules, characteristic of a negatively charged surface, is replaced by flat orientation, when the π-electrons of the benzene ring
interact with positive charges of the mercury surface. As the
area occupied by an aniline molecule in flat orientation is approximately double the area corresponding to vertical orientation [42],
the reorientation process is accompanied by partial desorption of
aniline molecules.

At high positive charges the C vs φ curves determined in pure
KCl solution and with additions of aniline approximately coincide
(Fig. 7), although electrocapillary measurements show that in this
potential region aniline molecules are adsorbed on the mercury
surface. Dyatkina and Damaskin [96] showed that in acid solu-

tions, where aniline is present mainly in the form of anilinium cations, this effect is observed over the entire range of potentials
studied (Fig. 8). At $\varepsilon > 0$ the aniline molecules or anilinium
cations, lying flat on the electrode surface, constitute a kind of extension of the metallic surface toward the solution; as a result, the
layers of the double layer are not pushed apart and its capacity
therefore alters very slightly if at all. However, at $\varepsilon < 0$, in addition to this cause of the small influence of adsorption of anilinium
cations on the capacity of the double layer, it is necessary also to
take into account the strong hydration of the $-NH_3^+$ group. This
follows from the fact that adsorption of butyl- and amylammonium
cations, which do not contain π-electron bonds, has little effect on
the differential capacity of the mercury electrode [40].

Differential capacity curves for aqueous solutions of benzene
and toluene also have two peaks [97, 98]; however, comparison of
these data with the results of electrocapillary measurements [46]
shows that, in distinction from aniline, the anodic peaks in the C
vs φ curves for these systems correspond mainly to adsorption –
desorption processes and not to reorientation of adsorbed molecules.

In a number of cases the form of the differential capacity
curves becomes considerably more complex in presence of surface-active compounds. For example, in the case of saturated
solutions of n-heptyl and n-octyl alcohols Loveland and Elving [99]
observed four capacity peaks in the differential capacity curves
determined by the oscillographic method. The two outer peaks
were attributed to complete desorption of alcohol molecules from
the electrode surface, while the two inner peaks were thought to be
caused by formation (or destruction) of a second layer of adsorbed
molecules, which form a micellar film at low electrode charges.
It must be pointed out that if an inert gas is blown through the solution even for a short time the alcohol concentration in the solution
may fall; in that case the differential capacity curves assume the
usual form, with two adsorption –desorption peaks.

In a tensammetric study of the adsorption of methyl orange on
mercury, Gupta [100] also concluded that condensed micellar films
are formed. Similar capacity peaks, caused by formation and
breakdown of micellar films, were observed in studies of adsorption of mercury of alkylsulfonate anions with long carbon chains

(12 or more carbon atoms) [101-103], and of sodium laurate and caprylate [104].

Ostrowski and Fischer [105, 106] obtained C vs φ curves of unusual form for certain weak bases which are also effective corrosion inhibitors. For example, ethyl dodecyl sulfoxide in $6N$ HCl gives a C vs φ curve with two deep minima where the capacity falls to approximately 3 $\mu F/cm^2$. These investigators ascribed the lefthand minimum to adsorption of the free base and secondary adsorption of the sulfoxonium cation, and the right-hand minimum to primary adsorption of that cation.

Differential capacity curves of very complex form, with several peaks of different types, are also obtained in presence of potassium xanthates of the general formula $C_nH_{2n+1}CSSK$, where n = 1-12 [107].

Thus, differential capacity curves have been widely used in qualitative studies of the adsorption of various organic substances on the surface of the mercury electrode. In addition to the investigations cited above, this method has been used for studying the structure of thin films formed by palmitic, oleic, and myrisitic acids and by cetyl alcohol at the mercury−solution interface [108, 109]; for investigation of adsorption of thiourea and its derivatives [110], pyridine and its derivatives [111, 112], cyclohexanone and its derivatives [113], vitamin B_2 [114] and DNA and RNA [115], and certain other organic compounds [116-119].

Differential capacity curves of mercury electrodes in presence of mixtures of two organic substances have been determined in a number of investigations [41, 68, 120-126]. Molecules of the two organic substances generally interact within the adsorption layer, and this is reflected in the shape of the C vs φ curves.

Gupta and Sharma [125] consider that if in the joint presence of two organic substances one common adsorption−desorption peak is obtained at higher surface charges than in the case of the individual substances (e.g., aliphatic alcohol + aliphatic acid [120, 126]; tetraalkylammonium cations + β-naphthylsulfonic acid [123]; $C_{12}H_{25}OSO_3Na$ + $C_8H_{17}OH$ [124]; n-$C_5H_{11}OH$ + o-cresol [125]), then the molecules of these substances form complexes, which are adsorbed at the electrode. With certain assumptions, the interaction between molecules of the two substances in such cases may be assessed quantitatively [126]. On the other hand, if a change in the concentration of one of the organic substances influences only the height and not the position of the adsorption–desorption peak (e.g., iso-$C_5H_{11}OH$ + cyclo-hexanol [120]; $C_{12}H_{25}OSO_3Na$ + polyethylene glycol [124]; bromthymol blue + pyridine [125]), then the adsorption of each of the two substances can probably be considered independently.

3. Relation between the Frequency Dependence of the Differential Capacity and the Kinetics of Adsorption of Organic Compounds

The form of the C vs φ curves determined by the ac method, and especially the peak height, depends on the frequency, because at sufficiently high frequencies establishment of adsorption equilibrium cannot keep pace with the change of potential. This problem was first considered by Frumkin and Melik-Gaikazyan [74, 127].

If adsorption of an organic compound at the electrode surface is not complicated by formation of multimolecular layers or micellar films, then the surface charge is a function of the potential and degree of surface coverage only, $\varepsilon = \varepsilon(\varphi, \theta)$, and consequently at a given concentration c = const of the organic substance we can write for the differential capacity:

$$C = \left(\frac{d\varepsilon}{d\varphi}\right)_c = \left(\frac{\partial\varepsilon}{\partial\varphi}\right)_\theta + \left(\frac{\partial\varepsilon}{\partial\theta}\right)_\varphi \cdot \left(\frac{\partial\theta}{\partial\varphi}\right)_c . \qquad \text{(II.2)}$$

In other words, the differential capacity determined under these conditions consists of two terms, one of which

$$C_{\text{true}} = \left(\frac{\partial\varepsilon}{\partial\varphi}\right)_\theta \qquad \text{(II.3)}$$

is called the true capacity, and the other,

$$C_{\text{add}} = \left(\frac{\partial\iota}{\partial\theta}\right)_\varphi \cdot \left(\frac{\partial\varepsilon}{\partial\varphi}\right)_c \qquad \text{(II.4)}$$

is the additional capacity (also known as the adsorption pseudo-capacity).

It was shown by Frumkin and Melik-Gaikazyan [74, 127] that C_{add} is a function of the ac frequency ω, decreasing from a certain equilibrium value corresponding to $\omega = 0$ [$C_{add(\omega=0)}$] to zero when $\omega \to \infty$; the character of the dependence of C_{add} on ω is determined by the nature of the rate-determining step in the adsorption of the organic compound. Thus, if diffusion is rate-determining, the dependence of C_{add} on ω is given by the equation

$$C_{add(\omega)} = C_{add(\omega=0)} \cdot \frac{2\,(\sqrt{\omega}/M + 1)}{(2\sqrt{\omega}/M + 1)^2 + 1}, \qquad (II.5)$$

where

$$M = \sqrt{2D} \, \Big/ \, \left(\frac{\partial \Gamma}{\partial c}\right)_\varphi \qquad (II.6)$$

(D is the diffusion coefficient of the organic compound and c is its concentration in the solution volume).

At the same time, the occurrence of the adsorption–desorption process leads to a resistive component of the electrode impedance. If the adsorption process is characterized at a given frequency by the capacity $C_{add(\omega)}$ and the resistance $R_{p(\omega)}$, which are assumed to be in parallel in the equivalent electric circuit, then for a slow diffusion step we have [127]

$$R_{p(\omega)} = \frac{[(2\sqrt{\omega}/M) + 1]^2 + 1}{2\omega C_{add(\omega=0)}\,\sqrt{\omega}/M}. \qquad (II.7)$$

From Eqs. (II.6) and (II.7) it follows that

$$\cot \delta = C_{add(\omega)}\; R_{p(\omega)}\;\; \omega = 1 + \frac{M}{\sqrt{\omega}}. \qquad (II.8)$$

The consequence of the decrease of C_{add} with frequency is that the adsorption–desorption peaks disappear from C vs φ curves determined at high frequencies. At a given frequency the decrease of peak height is more pronounced at lower concentrations of the adsorbed substance.

Melik-Gaikazyan [74], in a study of the dependence of differential capacity in the region of the adsorption–desorption peaks on frequency in solutions containing normal alcohols of the sat-

urated series, showed that this dependence conforms satisfactorily to Eq. (II.5); consequently, the slow step in adsorption of
these alcohols is the diffusion of alcohol molecules to the electrode surface.

However, this theory was concerned only with two limiting
cases: when the rate of the process is determined either by diffusion of the organic molecules to the electrode surface, or by the
rate of adsorption itself. Berzins and Delahay [128] gave a more
general solution of the problem, with both diffusion and adsorption
taken into account. In deriving the fundamental equation they assumed that adsorption of the organic compounds conforms to the
Langmuir isotherm and that the rate constants of the adsorption
and desorption processes are exponential functions of the electrode
potential. A more rigorous derivation, free from these limitations, was given later by Lorenz [86]. According to Lorenz, when
there are two rate-determining steps (diffusion and adsorption),
instead of Eq. (II.8) we have

$$\cot \delta = \frac{1 + (M / \sqrt{\omega})}{1 + ML \sqrt{\omega}} , \tag{II.9}$$

where

$$L = -1 \left/ \left(\frac{\partial \dot{v}}{\partial \Gamma} \right)_{\varphi, c_S} \right. \tag{II.10}$$

(v is the rate of the adsorption step and c_s is the concentration of
the organic substance at the electrode surface).

The exchange rate v_0 of the adsorption−desorption process
can in principle be found with the aid of Eq. (II.9) if the diffusion
coefficient and the maximum adsorption are known, with a definite
equation of the adsorption isotherm. Lorenz et al. [86, 129-131]
used this method for studying the kinetics of adsorption of a number of organic compounds on mercury. According to Lorenz, the
time constant of the adsorption stage for the substances studied
is of the order of 10^{-5} sec. However, this was subsequently re-
futed by Armstrong, Race, and Thirsk [147], according to whom
the time constant of the adsorption stage for cyclohexanol and n-
butyric acid is between 10^{-6} and 10^{-10} sec, and therefore cannot
be determined experimentally by the available techniques for
measurement of electrode impedances (see Chapter I).

According to Lorenz [86], in the case of carbon compounds with carbon chains of sufficient length, two-dimensional association occurs in the surface layer; this complicates the dependence of cot δ on frequency considerably. In such cases the dependence of cot δ on frequency can be satisfactorily represented by the following equation:

$$\cot \delta = \frac{k_1 \left(k_2 + k_1 \dfrac{\sqrt{\omega}}{M}\right) + \left(\dfrac{\omega}{w_0}\right)^2 \left(k_3 + \dfrac{\sqrt{\omega}}{M}\right)}{\left[k_1^2 + \left(\dfrac{\omega}{w_0}\right)^2\right]\left(\dfrac{\omega}{v_0} + \dfrac{\sqrt{\omega}}{M}\right) + \dfrac{\omega}{w_0}(k_1 k_3 - k_2)}, \qquad (II.11)$$

if it is assumed that the exchange rate w_0 of two-dimensional association is in the range of 10-1000 sec^{-1}. In this equation k_1, k_2, and k_3 are certain constants characterizing the process of two-dimensional association [86].

In accordance with the theory developed by Lorenz [132], specific adsorption of ions or neutral molecules, which is accompanied by covalent bonding between the electrode and the adsorbed particles, leads to partial charge transfer λ from the adsorbate to the electrode ($0 \le \lambda < 1$). It is possible to determine λ experimentally from the frequency dependence of the differential capacity in concentrated solutions of indifferent electrolytes with small additions of specifically adsorbed substances. Data obtained by Lorenz et al. [133, 134] show that partial charge transfer, and therefore covalent bonding, occurs in adsorption of thiourea and p-benzoquinone molecules on mercury, whereas no covalent bonding with partial charge transfer occurs in adsorption of various aliphatic compounds or of tetraalkylammonium cations.

The kinetic relationships involved in simultaneous adsorption of two organic substances on the electrode were considered theoretically in a recent paper by Belokolos [135].

The dependence of the differential capacity on the ac frequency leads to deformation of the experimental C vs φ curves. It follows from Eqs. (II.5) and (II.6) that the variation of capacity with frequency depends on the value of the derivative $(\partial \Gamma / \partial c)_\varphi$, and therefore on the shape of the adsorption isotherm and on the degree of coverage of the surface by the organic substance at the given potential. Some of the consequent properties of nonequilibrium differential capacity curves will be examined in detail in Chapter III, Section 4b.

The dependence of the differential capacity on the ac frequency may be due not only to adsorption–desorption processes but also to other kinetic effects associated with changes in the structure of the adsorption layer with variations of the electrode potential. Such processes occur, for example, during formation of micellar films on the electrode surface [101, 102] and during sharp changes in the orientation of aromatic or heterocyclic compounds adsorbed on mercury [43, 95]. Generally these processes are not accompanied by diffusion of organic molecules to the electrode surface and it is therefore reasonable to assume that they are much more rapid than adsorption–desorption processes. In fact, the decrease of capacity with increase of ac frequency in the potential region where a micellar layer of $C_{12}H_{25}OSO_3^-$ anions is formed [102] and at potentials where adsorbed aniline molecules change their orientation [68] is much slower than the decrease of capacity in the region of the adsorption–desorption peaks.

Equations (II.5), (II.7), and (II.8) can be used for determining equilibrium values of differential capacity in presence of organic substances from experimental data. This is usually done by extrapolation of the experimental capacities to $\omega = 0$ in a plot of C vs $\sqrt{\omega}$. Figure 9, which is based on Eq. (II.5), shows that such extrapolation can give satisfactory results (correct to 2%) only under the condition that $\sqrt{\omega}/M \leq 0.4$, i.e., at sufficiently low frequencies. Equation (II.8) can be used for approximate estimation of M in this case.

Another method for determination of $C_{add(\omega=0)}$, proposed by Tedoradze and Arekelyan [136], involves accurate determination of $C_{add(\omega)}$ and $R_{p(\omega)}$. If these quantities are known, the formula

$$C_{add(\omega=0)} = \frac{C_{add(\omega)}^2 R_{p}^2 \omega^2 + 1}{[C_{add(\omega)} R_{p(\omega)} \omega - 1] R_{p(\omega)} \omega}, \qquad (II.12)$$

which follows from Eqs. (II.5) and (II.7), can be used for calculating the equilibrium differential capacity. Formula (II.12) is strictly valid only if diffusion of the organic substance to the electrode is the rate-determining step in the adsorption process. This condition is shown to be satisfied if the calculated value of $C_{add(\omega=0)}$ is independent of the ac frequency. Certain other methods for determination of $C_{add(\omega=0)}$ have recently been discussed by Armstrong et al. [147].

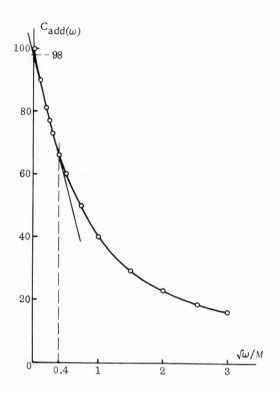

Fig. 9. Dependence of the additional capacity on
ac frequency, calculated from Eq. (II.5) with
$C_{add(\omega=0)} = 100 \ \mu F/cm^2$.

Another convenient criterion of the equilibrium character of
C vs φ curves found by extrapolation or calculation is agreement
of the magnitude of the charge ε determined by integration of the
C vs φ curve for a solution of an organic compound between potentials corresponding to total desorption of the organic compound at
the anodic and cathodic branches with the value of ε found by the
same method in absence of the adsorbate.

4. Adsorption of Organic Substances

on Other Liquid Electrodes

Apart from mercury, the liquid metallic adsorbents of greatest interest at normal temperatures, where decomposition of or-

ganic substances does not occur, are gallium and amalgams of thallium and indium (because high concentrations are accessible). Experimental data on adsorption of organic compounds on such electrodes are meager. Electrocapillary curves for thallium amalgam in presence of isoamyl alcohol, pyrogallol, and thiourea have been determined by Frumkin and Gorodetskaya [137], and for liquid gallium in presence of isoamyl alcohol, phenol, and pyrogallol by Murtazaev and Gorodetskaya [138]. Figure 10 shows electrocapillary curves for pure solutions of the supporting electrolyte and in presence of isoamyl alcohol on mercury, 57.5% indium amalgam, and gallium, determined recently by Frumkin et al. [139-141] with the use of highly pure metals.

It follows from Fig. 10 that the electrocapillary curves for indium and gallium amalgams are qualitatively similar to the corresponding curves determined with a mercury electrode: in every case the point of zero charge is shifted in the positive direction in presence of isoamyl alcohol, while at high positive and negative surface charges desorption of the organic molecules from the electrode surface occurs.

On the other hand, Fig. 10 shows that the adsorbability of iso-amyl alcohol on indium amalgam and especially on gallium is appreciably lower than on mercury. In the latter case this is due

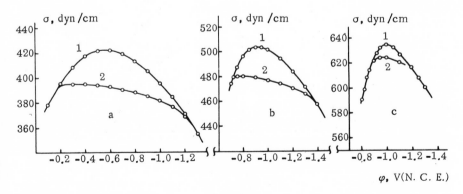

Fig. 10. Electrocapillary curves of various metals: a) Mercury: 1) 1 N KCl + 0.1 N HCl; 2) 1 N KCl + 0.1 N HCl + 0.1 M iso-$C_5H_{11}OH$; b) 57.5% indium amalgam: 1) 1 N Na_2SO_4 + 0.01 N H_2SO_4; 2) 1 N Na_2SO_4 + 0.01 N H_2SO_4 + 0.14 M iso-$C_5H_{11}OH$; c) gallium: 1) 1 N KCl + 0.1 N HCl; 2) 1 N KCl + 0.1 N HCl + 0.1 M iso-$C_5H_{11}OH$ (from data in [139-141]).

to strong chemisorption of water molecules on the surface of the gallium electrode at $\varepsilon \geq 0$ [141].

Differential capacity curves for 40% gallium amalgam in presence of n-hexyl alcohol [142] and for a liquid gallium electrode in presence of various amounts of n-propyl and tert-amyl alcohols [143] confirm the conclusions which follow from interfacial tension data. For example, the differential capacities in presence of n-C_3H_7OH and tert-$C_5H_{11}OH$ on mercury and liquid gallium at equal surface charges are virtually the same [143]. However, as the p.z.c. is approached chemisorption of water molecules occurs at the gallium electrode; this leads to appreciable decrease of the adsorbability of the organic substance and to a consequent increase of capacity in the region of the minimum of the C vs φ curve.

Comparison of the differential capacity curves for mercury and 40% thallium amalgam in presence of n-hexyl alcohol [142]

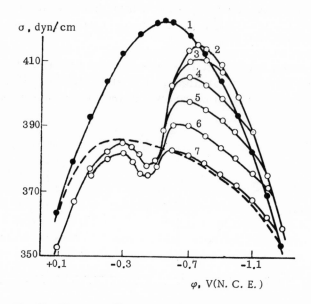

Fig. 11. Electrocapillary curves determined in 0.2 N $TlNO_3$ + 0.8 N KNO_3 solutions with various additions of n-butyl alcohol: 2) 0; 3) 0.037; 4) 0.075; 5) 0.15; 6) 0.3; 7) 0.6 M. Curve 1 corresponds to 1 N KNO_3. The dash line corresponds to 1 N KNO_3 + 0.6 M n-C_4H_9OH (from data in [146]).

shows that equal amounts of n-hexyl alcohol added to 0.1 N KCl
lead to a considerably greater decrease of capacity at the p.z.c.
on mercury than on 40% thallium amalgam. This effect is especially pronounced at low n-C_6H_{13}OH concentrations. This result shows
that the energy of hexyl alcohol adsorption is greater on mercury
than on 40% thallium amalgam. The consequence of this is that the
surface charges at which the peak on the C vs φ curves is observed are considerably higher on mercury than on thallium amalgam at the same hexyl alcohol concentrations.

On the other hand, Petrii [144] showed that the surface charges
on mercury and 40% thallium amalgam, corresponding to the desorption peaks of large organic cations, $[(C_4H_9)_4N]^+$, $[(C_5H_{11})_4N]^+$,
and $[(C_6H_{13})_4N]^+$, are approximately equal (although lower in the
case of 40% thallium amalgam). This is apparently due to negative adsorption of thallium atoms in the amalgam [137], which
lowers the surface concentration of the amalgam when the cathodic
polarization is considerable.

In a study of the system mercury$-$TlNO$_3$ + KNO$_3$ solution,
Frumkin and Polyanovskaya [145] obtained an electrocapillary
curve of unusual form, with two maxima; the left-hand maximum
corresponds to the point of zero charge of a mercury electrode
in a solution of the thallium salt, while the other corresponds to
the p.z.c. of thallium amalgam in a solution virtually free from
Tl$^+$ ions (Fig. 11, curve 2). The shift of the cathodic branch of
this curve in the negative direction in comparison with the σ vs φ
curve for a mercury electrode is due both to the different nature
of the electrode and to ohmic potential drop [145].

Adsorption of n-butyl alcohol, phenol, and aniline at the elec-
trode$-$solution interface in a system of this type was studied recently by Andrusev, Ayupova, and Damaskin [146]. It follows
from this study that, as was to be expected, in presence of aniline
and phenol the greatest lowering of interfacial tension occurs
near both p.z.c., whereas adsorption of n-butyl alcohol in the region of the left-hand maximum on the electrocapillary curve becomes appreciable only at the highest n-C_4H_9OH concentration (Fig.
11). Analysis of electrocapillary curves for mercury$-$(solution
of TlNO$_3$ + KNO$_3$ + organic substance) systems and comparison
with the corresponding σ vs φ curves for mercury$-$(KNO$_3$ solution + organic substance) systems shows that competitive adsorp-

tion of organic molecules on the one hand, and of thallium cations
or atoms on the other, occurs. The decreased adsorbability of the
organic compound as the result of adsorption of thallium is most
pronounced in the left-hand side of the σ vs φ curve in presence of
n-butanol (Fig. 11). In all the other cases studied adsorption of
the organic compounds appears to predominate, and suppresses
adsorption of thallium cations and atoms.

References

1. H. Luggin, Z. Phys. Chem., 16:677 (1895).
2. R. Luther, Z. Phys. Chem., 19:529 (1896).
3. G. Gouy, Ann. Chim. Phys. (8), 8:291 (1906); 9:75 (1906).
4. R. I. Kaganovich, V. M. Gerovich, and T. G. Osotova, Dokl. Akad. Nauk SSSR, 155:893 (1964).
5. R. I. Kaganovich and V. M. Gerovich, Élektrokhimiya, 2:977 (1966).
6. A. N. Frumkin, Electrocapillary Phenomena and Electrode Potentials [in Russian], Odessa (1919).
7. G. Gouy, Ann. Phys. (9), 6:5 (1916).
8. J. A. V. Butler and A. Wightman, J. Phys. Chem., 35:3293 (1931).
9. A. N. Frumkin and J. Williams, Proc. Natl. Acad. Sci. U.S., 15:400 (1929).
10. J. A. V. Butler, Proc. Roy. Soc., A122:399 (1929).
11. J. A. V. Butler and C. Ockrent, J. Phys. Chem., 34:2286 (1930).
12. A. N. Frumkin, A. A. Donde, and R. M. Kulvarskaya, Z. Phys. Chem., 123:321 (1926).
13. A. N. Frumkin, Ergbn. Exakt. Naturwiss., 7:235 (1928); Colloid Symp. Ann., 7:89 (1930).
14. A. N. Frumkin, Z. Phys. Chem., 111:190 (1924).
15. M. A. V. Devanathan, Proc. Roy. Soc., A267:256 (1962).
16. J. O'M. Bockris, M. A. V. Devanathan, and K. Müller, Proc. Roy. Soc., A274:55 (1963).
17. A. N. Frumkin and B. B. Damaskin, Modern Aspects of Electrochemistry, Vol. 3, Plenum Press, New York (1964), p. 149.
18. A. N. Frumkin, B. B. Damaskin, and Yu. A. Chizmadzhev, Élektrokhimiya, 2:875 (1966).
19. B. Kamienski, Electrochim. Acta, 1:272 (1959).
20. A. N. Frumkin, Electrochim. Acta, 2:351 (1960).
21. M. A. Gerovich and O. G. Ol'man, Zh. Fiz. Khim., 28:19 (1954).
22. M. A. Gerovich, Dokl. Akad. Nauk SSSR, 96:543 (1954); 105:1278 (1955).
23. M. A. Gerovich and G. F. Rybal'chenko, Zh. Fiz. Khim., 32:109 (1958).
24. M. A. Gerovich and N. S. Polyanovskaya, Nauchn. Dokl. Vysshei Shkoly, Khim. i Khim. Tekhnol., No. 4, 651 (1958).
25. E. Blomgren and J. O'M. Bockris, J. Phys. Chem., 63:1475 (1959).
26. B. E. Conway and R. G. Barradas, Electrochim. Acta, 5:319, 349 (1961).
27. L. I. Antropov and S. N. Benergee, J. Indian Chem. Soc., 35:531 (1958); 36:451 (1959).

28. L. Gierst, Transactions, Symposium on Electrode Processes, J. Wiley and Sons, Philadelphia, New York (1959), p. 294.

29. L. Gierst and P. Herman, Z. Anal. Chem., 216:238 (1966).

30. R. G. Barradas, P. G. Hamilton, and B. E. Conway, J. Phys. Chem., 69:3411 (1965); Collection Czech. Chem. Commun., 32:1790 (1967).

31. B. E. Conway, R. G. Barradas, P. G. Hamilton, and J. M. Parry, J. Electroanal. Chem., 10:485 (1965).

32. I. Zwierzykowska, Roczniki Chem., 38:663, 1169, 1195, 1367 (1964).

33. H. W. Nürnberg and G. Wolff, Collection Czech. Chem. Commun., 30:3997 (1965).

34. B. B. Damaskin, A. A. Survila, S. Ya. Vasina, and A. I. Fedorova, Élektrokhimiya, 3:825 (1967).

35. V. A. Kuznetsov and B. B. Damaskin, Élektrokhimiya, 1:1153 (1965).

36. B. B. Damaskin, M. M. Andrusev, V. M. Gerovich, and R. I. Kaganovich, Élektrokhimiya, 3:667 (1967).

37. A. N. Frumkin, R. I. Kaganovich, and É. S. Bit-Popova, Dokl. Akad. Nauk SSSR, 141:670 (1961).

38. M. A. V. Devanathan and M. J. Fernando, Trans. Faraday Soc., 58:368 (1961).

39. I. Zwierzykowska, Roczniki, Chem., 39:101 (1965).

40. B. B. Damaskin, R. I. Kaganovich, V. M. Gerovich, and S. L. Dyatkina, Élektrokhimiya (in press).

41. M. M. Andrusev, B. B. Damaskin, and R. I. Kaganovich, in: Electrochemical Processes Involving Organic Substances [in Russian], Izd. Nauka, Moscow (1969).

42. E. Blomgren, J. O'M. Bockris, and C. Jesch, J. Phys. Chem., 65:2000 (1961).

43. B. B. Damaskin, I. P. Mishutushkina, V. M. Gerovich, and R. I. Kaganovich, Zh. Fiz. Khim., 38:1797 (1964).

44. B. B. Damaskin, V. M. Gerovich, I. P. Gladkikh, and R. I. Kaganovich, Zh. Fiz. Khim., 38:2495 (1964).

45. A. N. Frumkin and B. B. Damaskin, Pure and Appl. Chem., 15:263 (1967).

46. R. I. Kaganovich, V. M. Gerovich, and O. Yu. Gusakova, Élektrokhimiya, 3:946 (1967).

47. A. N. Frumkin, V. A. Kuznetsov, and R. I. Kaganovich, Dokl. Akad. Nauk SSSR, 155:175 (1964).

48. R. I. Kaganovich, B. B. Damaskin, and I. M. Ganzhina, Élektrokhimiya, 4:(7) (1968).

49. V. P. Grigor'ev and V. V. Ékilik, in: Electrochemical Processes Involving Organic Substances [in Russian], Izd. Nauka, Moscow (1969).

50. H. H. Jaffe, Chem. Revs., 53:191 (1953).

51. O. A. Osipov and V. I. Minkin, Handbook on Dipole Moments [in Russian], Izd. "Vysshaya Shkola," Moscow (1965).

52. A. N. Frumkin, A. V. Gorodetskaya (Gorodezkaja), and P. S. Chugunov, Acta Physicochim. URSS, 1:12 (1934).

53. T. A. Kryukova and A. N. Frumkin, Zh. Fiz. Khim., 23:819 (1949).

54. R. Grand, Ann. Phys., 10:738 (1955).

55. R. Grand, J. Chim. Phys. et Phys. Chim. Biol., 60:1315 (1963).

56. B. E. Conway, J. O'M. Bockris, and B. Lovreček, CITCE Proc., 6:207 (1955).

57.   M. A. Loshkarev and M. A. Sevryugina, Dokl. Akad. Nauk SSSR, 108:111
      (1956); Nauchn. Tr. Dnepropetr. Khim.-Tekhnol. Inst., No. 12, 97 (1961).
58.   A. G. Stromberg and L. S. Zagainova, Zh. Fiz. Khim., 31:1042 (1957).
59.   W. Palczewska and H. Wroblowa, Roczniki Chem., 32:1333 (1958).
60.   É. A. Aikazyan, Izv. Akad. Nauk Arm. SSR, Khim. Nauki, 12:9 (1959).
61.   P. J. Hillson, J. Phot. Sci., 11:225 (1963).
62.   J. Pasciak, Roczniki Chem., 37:1233 (1963).
63.   V. S. Griffiths and J. B. Westmore, J. Chem. Soc., 1704 (1962).
64.   L. K. Partridge, A. C. Tansley, and A. S. Porter, Electrochim. Acta, 11:517
      (1966).
65.   J. A. V. Butler and C. Ockrent, J. Phys. Chem., 34:2297, 2841 (1930).
66.   A. V. Pamfilov, V. S. Kuzub, and N. V. Palamarchuk, Dokl. Akad. Nauk Ukr.
      SSR, No. 6, 813 (1960); Ukr. Khim. Zh., 28:528 (1962).
67.   B. Kastening, Ber. Bunsenges, 68:979 (1964).
68.   R. A. Arakelyan and G. A. Tedoradze, Élektrokhimiya, 4:144 (1968).
69.   F. W. Schapink, M. Oudeman, K. W. Leu, and J. N. Helle, Trans. Faraday
      Soc., 56:415 (1960).
70.   M. A. Proskurnin and A. N. Frumkin, Trans. Faraday Soc., 31:110 (1935).
71.   A. Ksenofontov, M. A. Proskurnin, and A. V. Gorodetskaya, Zh. Fiz. Khim.,
      12:408 (1938).
72.   K. S. G. Doss and V. K. Venkatesan, Proc. Indian Acad. Sci., A49:129 (1959).
73.   K. S. G. Doss, Bull. (India Sect.) Electrochem. Soc., 8:84 (1959).
74.   V. I. Melik-Gaikazyan, Zh. Fiz. Khim., 26:560, 1184 (1952).
75.   S. L. Gupta and S. K. Sharma, Kolloid-Z. Z. Polymere, 190:40 (1963).
76.   S. L. Gupta and S. K. Sharma, J. Indian Chem. Soc., 41:384, 663, 668 (1964);
      43:53 (1966).
77.   R. Narayan, Electrochim. Acta, 7:111 (1962).
78.   A. N. Frumkin and B. B. Damaskin, Dokl. Akad. Nauk SSSR, 129:862 (1959).
79.   N. V. Nikolaeva-Fedorovich, B. B. Damaskin, and O. A. Petry, Collection
      Czech. Chem. Commun., 25:2981 (1960).
80.   B. B. Damaskin and N. V. Nikolaeva-Fedorovich, Zh. Fiz. Khim., 35:1279
      (1961).
81.   Tza Chuan-sin and Huang Tě-tung, Hua Hsüeh Hsüeh Pao, 28:5 (1962).
82.   R. J. Meakins, J. Appl. Chem., 17:157 (1967).
83.   A. A. Moussa, H. M. Sammour, and H. A. Ghaly, Egypt. J. Chem., 2:169
      (1959).
84.   H. Jehring, Z. Phys. Chem., 226:59 (1964).
85.   S. Sathyanarayana, J. Electroanal. Chem., 10:56 (1965).
86.   W. Lorenz, Z. Electrochem., 62:192 (1958).
87.   I. R. Miller and D. C. Grahame, J. Am. Chem. Soc., 78:3577 (1956); 79:3006
      (1957).
88.   I. R. Miller and D. C. Grahame, J. Colloid Sci., 16:23 (1961).
89.   I. R. Miller, J. Phys. Chem., 64:1790 (1960).
90.   I. R. Miller, Trans. Faraday Soc., 57:301 (1961).
91.   S. Ueda, F. Tsuji, and A. Watanabe, Bull. Inst. Chem. Res. Kyoto Univ.,
      38:59 (1960).

92. A. Watanabe, F. Tsuji, and S. Ueda, Kolloid-Z. Z. Polymere, 193:39 (1963); 198:87 (1964).
93. W. Wojciak and E. Dutkievicz, Roczniki Chem., 38:271 (1964).
94. I. V. Shelepin, A. N. Frumkin, A. I. Fedorova, and S. Ya. Vasina, Dokl. Akad. Nauk SSSR, 154:203 (1964).
95. L. D. Klyukina and B. B. Damaskin, Izv. Akad. Nauk SSSR, Otd. Khim. Nauk, No. 6, 1022 (1963).
96. S. L. Dyatkina and B. B. Damaskin, Élektrokhimiya, 2:1340 (1966).
97. J. Dojlido and B. Behr, Roczniki Chem., 37:1043 (1963).
98. B. A. Shenoi and K. R. Narasimhan, J. Sci. Ind. Res., B21:262 (1962).
99. J. W. Loveland and P. J. Elving, J. Phys. Chem., 56:935, 941, 945 (1952).
100. S. L. Gupta, Proc. Indian Acad. Sci., A47:254 (1958).
101. K. Eda, J. Chem. Soc. Japan, 80:349, 708 (1959); 81:689 (1960).
102. B. B. Damaskin, N. V. Nikolaeva-Fedorovich, and R. V. Ivanova, Zh. Fiz. Khim., 34:894 (1960).
103. H. Jehring, Chem. Zvesti, 18:313 (1964).
104. K. Eda, J. Chem. Soc. Japan, 80:461, 465 (1959).
105. Z. Ostrowski and H. Fischer, Electrochim. Acta, 8(1):37 (1963).
106. Z. Ostrowski, H. A. Brune, and H. Fischer, Electrochim. Acta, 9:175 (1964).
107. W. Pötsch and K. Schwabe, J. Prakt. Chem., 18:1 (1962).
108. A. V. Gorodetskaya and A. N. Frumkin, Dokl. Akad. Nauk SSSR, 18:649 (1938).
109. A. V. Gorodetskaya, Zh. Fiz. Khim., 14:371 (1940).
110. R. Narayan and V. K. Venkatesan, Proc. Indian Acad. Sci., A54:109 (1961).
111. Ya. M. Zolotovitskii and G. A. Tedoradze, Izv. Akad. Nauk SSSR, Ser. Khim., No. 12, 2133 (1964); Élektrokhimiya, 1:1339 (1965).
112. G. A. Tedoradze, Ya. M. Zolotovitskii, and A. B. Érshler, Élektrokhimiya, 1:1280 (1965).
113. K. Tsuji, Rev. Polarogr., 11:233 (1964).
114. J. Sancho, J. C. Hurtado, and P. Salmeron, Ann. Real Sco. Espan. Fis. y Quim., B58:511 (1962).
115. J. Miller, J. Mol. Biol., 3:229, 357 (1961).
116. H. Jehring, Z. Phys. Chem., 225:116 (1964); 229:39 (1965).
117. G. Palyi and H. Jehring, Collection Czech Chem. Commun., 30:4339 (1965).
118. H. Jehring and H. Mehner, Z. Anal. Chem., 224:136 (1967).
119. A. Pomianowski, Roczniki Chem., 41:775 (1967).
120. G. A. Dobren'kov and R. K. Bankovskii, Proceedings of Electrochemical Conference [in Russian], Kazan' (1959), p. 74.
121. P. A. Kirkov, Dokl. Akad. Nauk SSSR, 135:651 (1960).
122. T. Kambara, A. Hayashi, and J. Joshimi, Rev. Polarogr., 10:131 (1962).
123. Tza Chuan-sin, Chou Yuen-hong, Chou Ging-yuen et al., Scienta Sinica, 14:63 (1965).
124. K. Eda and K. Takahasi, J. Chem. Soc. Japan, 85:828 (1964).
125. S. L. Gupta and S. K. Sharma, Electrochim. Acta, 10:151 (1965).
126. G. A. Tedoradze, R. A. Arakelyan, and E. D. Belokolos, Élektrokhimiya, 2:563 (1966).
127. A. N. Frumkin and V. I. Melik-Gaikazyan, Dokl. Akad. Nauk SSSR, 77:855 (1951).

128. T. Berzins and P. Delahay, J. Phys. Chem., 59:906 (1955).
129. W. Lorenz and F. Möckel, Z. Electrochem., 60:507, 939 (1956).
130. W. Lorenz, Z. Phys. Chem. (N.F.), 18:1 (1958); 26:424 (1960).
131. W. Lorenz and E. O. Schmalz, Z. Electrochem., 62:301 (1958).
132. W. Lorenz, Z. Phys. Chem., 218:272 (1961); 219:421 (1962).
133. W. Lorenz and G. Krüger, Z. Phys. Chem., 221:231 (1962).
134. W. Lorenz and U. Gaunitz, Collection Czech. Chem. Commun., 31:1389 (1966).
135. E. D. Belokolos, Élektrokhimiya, 1:498 (1965).
136. G. A. Tedoradze and R. A. Arakelyan, Dokl. Akad. Nauk SSSR, 156:1170 (1964).
137. A. N. Frumkin and A. V. Gorodetskaya (Gorodetskaja), Z. Phys. Chem., 136:451 (1928).
138. A. Murtazaev (Murtazajev) and A. Gorodetskaya (Gorodetskaja), Acta Physico-chim. URSS, 4:75 (1936).
139. A. N. Frumkin, N. S. Polyanovskaya, and N. B. Grigor'ev, Dokl. Akad. Nauk SSSR, 157:1455 (1964).
140. N. S. Polyanovskaya and A. N. Frumkin, Élektrokhimiya, 1:538 (1965).
141. A. Frumkin, N. Polyanovskaya (Poljanovskaja), N. Grigor'ev (Grigorjev), and I Bagotskaya (Bagotskaja), Electrochim. Acta, 10:793 (1965).
142. A. N. Frumkin, O. A. Petrii, and N. V. Nikolaeva-Fedorovich, Dokl. Akad. Nauk SSSR, 147:878 (1962).
143. N. B. Grigor'ev and I. A. Bagotskaya, Élektrokhimiya, 2:1449 (1966).
144. O. A. Petrii, Candidate's Dissertation [in Russian], Mosk. Gos. Univ. (1962).
145. A. N. Frumkin and N. S. Polyanovskaya, Zh. Fiz. Khim., 32:157 (1958).
146. M. M. Andrusev, N. Kh. Ayupova, and B. B. Damaskin, Élektrokhimiya, 2:1480 (1966).
147. R. D. Amstrong, W. P. Race, and H. R. Thirsk, J. Electroanal. Chem., 16:517 (1968).
148. N. B. Grigor'ev and B. B. Damaskin, Advances in the Electrochemistry of Organic Compounds [in Russian], Izd. Nauka, Moscow (1968), p. 66.

Chapter 3

Quantitative Theory of the Influence of an Electric Field on Adsorption of Organic Substances at a Mercury Electrode

The first quantitative theory of the influence of an electric field on adsorption of neutral molecules at mercury surfaces was advanced by Frumkin in 1925-1926 [1, 2] (see also [3]). In this theory the attractive forces between the adsorbed molecules were taken into account, and the interface between the electrode and the solution was regarded as consisting of two condensers in parallel, with only molecules of the organic substance between the plates of one, and only solvent (water) molecules between the plates of the other. Electrocapillary curves for a mercury electrode in 1 *N* NaCl solution with various additions of tert-amyl alcohol served as the experimental basis of the theory.

The theory of the influence of an electric field on adsorption of organic compounds could be developed further on the basis of the results of experimental determinations of the differential capacity of the mercury electrode in presence of organic additives, since the differential capacity is much more sensitive than the interfacial tension to changes in the structure of the double layer. Investigations in this direction were carried out by Damaskin, Tedoradze, and their co-workers [4-34]. The principal results of these investigations, which essentially constitute generalization and further development of Frumkin's theory, are presented below

1. Selection of the Electrical Variable

If the adsorption of an organic compound at the electrode surface is reversible, and not complicated by formation of multimolecular layers or micellar films, the electrode potential φ, the electrode charge ε, and the degree of coverage θ of the surface by the organic substance are interconnected in a perfectly definite manner. At present there is no agreed opinion in the literature whether it is preferable to use φ or ε as the independent electrical variable. For example, Frumkin, Damaskin, and Tedoradze [1-41], Hansen et al. [42, 43], and Lorenz et al. [44-47] chose the electrode potential as the independent electrical variable, whereas the electrode charge was chosen by Parsons [48-53], Devanathan and Tilak [54, 55], and Bockris et al. [56, 57]. The controversy which has arisen on this subject [11, 15, 40, 49, 53, 57] prompts a more detailed examination of the problem.

If the potential is taken as the independent electrical variable, then for a dilute solution of an organic substance the fundamental electrocapillary equation may be written in the form

$$d\sigma = -\varepsilon \, d\varphi - RT\Gamma \, d\ln c = -\varepsilon \, d\varphi - A\theta \, d\ln c, \qquad \text{(III.1)}$$

where $A = RT\Gamma_m$ (Γ_m is the maximum adsorption). Since $d\sigma$ in this equation is an exact differential, the reciprocal relation gives

$$\frac{1}{A}\left(\frac{\partial \varepsilon}{\partial \ln c}\right)_\varphi = \left(\frac{\partial \theta}{\partial \varphi}\right)_c. \qquad \text{(III.2)}$$

From this equation, which was orginally derived by Gouy [58], it follows that

$$\left(\frac{\partial \ln c}{\partial \varphi}\right)_\theta = -\left(\frac{\partial \ln c}{\partial \theta}\right)_\varphi \left(\frac{\partial \theta}{\partial \varphi}\right)_c = -\frac{1}{A} \cdot \left(\frac{\partial \varepsilon}{\partial \theta}\right)_\varphi. \qquad \text{(III.3)}$$

If the electrode charge is the independent electrical variable, the fundamental electrocapillary equation for a dilute solution of an organic substance becomes [see Eq. (I.39)]

$$d\xi = \varphi \, d\varepsilon - RT\Gamma \, d\ln c = \varphi \, d\varepsilon - A\theta \, d\ln c. \qquad \text{(III.4)}$$

By the property of the exact differential

$$-\frac{1}{A}\left(\frac{\partial\varphi}{\partial\ln c}\right)_{\varepsilon} = -\left(\frac{\partial\theta}{\partial\varepsilon}\right)_{c}. \tag{III.5}$$

From this equation it follows that

$$\left(\frac{\partial\ln c}{\partial\varepsilon}\right)_{\theta} = -\left(\frac{\partial\ln c}{\partial\theta}\right)_{\varepsilon}\left(\frac{\partial\theta}{\partial\varepsilon}\right)_{c} = \frac{1}{A}\left(\frac{\partial\varphi}{\partial\theta}\right)_{\varepsilon}. \tag{III.6}$$

Equation (III.6) was originally derived by Parsons [48].

As already noted in Chapter I, Section 3, with a purely thermodynamic approach to investigation of adsorption isotherms the choice of the electrical variable is not significant in principle and is determined by considerations of convenience. For example, in studies of adsorption of inorganic ions it is more convenient to consider isotherms at ε = const, since by the Gouy–Chapman theory the structure of the diffuse layer is determined by the magnitude of its charge.

However, the situation is different when the thermodynamic approach to adsorption at electrodes is supplemented by certain model assumptions; in particular, the assumption of congruence of the adsorption isotherm with respect to one of the electrical variables. The condition for congruence of the isotherm with respect to the potential may be written in the form of the equation

$$Bc = f(\theta), \tag{III.7}$$

and the analogous condition for congruence of the isotherm with respect to the electrode charge as

$$Gc = f(\theta), \tag{III.8}$$

where $B = B(\varphi)$ and $G = G(\varepsilon)$ are adsorption equilibrium constants which are functions of φ and ε respectively, while $f(\theta)$ is a certain function of θ. Similarity of isotherms determined at different electrode potentials is assumed in Eq. (III.7), and similarity of isotherms determined at different charges is assumed in Eq. (III.8). Thus, the controversy regarding the "choice of the electrical variable" is essentially concerned with the question which of the two equations, (III.7) or (III.8) is in better agreement with experimental data on adsorption of organic compounds rather than with the choice of the variable. Under such conditions arguments in favor of the electrode charge as the more convenient or appropriate choice of the independent electrical variable cannot be taken into consideration.

Let us first assume that experimental data on adsorption of an organic substance on an electrode surface conform to Eq. (III.7). The choice of the electrical variable is the consequence of this assumption: since the form of the isotherm (III.7) in reduced coordinates (θ vs $c/c_{\theta=\text{const}}$) is independent of φ, it is evident that under these conditions the potential should be chosen as the independent variable.

From Eq. (III.7) it follows that

$$\frac{d \ln B}{d\varphi} = - \left(\frac{\partial \ln c}{\partial \varphi}\right)_{\theta}. \tag{III.9}$$

From Eqs. (III.3) and (III.9) we have

$$\left(\frac{\partial \varepsilon}{\partial \theta}\right)_{\varphi} = A \frac{d \ln B}{\partial \varphi}, \tag{III.10}$$

hence

$$\varepsilon = A \left(\frac{d \ln B}{d\varphi}\right) \theta + \varepsilon_0, \tag{III.11}$$

where ε_0 is the integration constant, equal to the electrode charge when $\theta = 0$.

We denote the charge at $\theta = 1$ and a given potential by ε'; then, by Eq. (III.11)

$$\varepsilon' = A \left(\frac{d \ln B}{d\varphi}\right) + \varepsilon_0. \tag{III.12}$$

From Eqs. (III.11) and (III.12) it follows that

$$\varepsilon = \varepsilon_0 (1 - \theta) + \varepsilon' \theta. \tag{III.13}$$

Thus, the assumption that the adsorption isotherm is congruent with respect to the electrode potential is strictly equivalent to the model of two parallel condensers proposed earlier by Frumkin [2, 3]. The respective capacities of these condensers are

$$C' = \frac{d\varepsilon'}{d\varphi} \text{ and } C_0 = \frac{d\varepsilon_0}{d\varphi}. \tag{III.14}$$

If we assume in the first approximation that the capacities C_0 and C' are independent of the potential, the dependences of ε_0 and ε' on φ are two straight lines; Fig. 12 shows that the equations of these lines may be written in the form

$$\varepsilon_0 = C_0\varphi; \quad \varepsilon' = C'(\varphi - \varphi_N), \tag{III.15}$$

or in the form

$$\varepsilon_0 = \varepsilon_m + C_0(\varphi - \varphi_m);$$
$$\varepsilon' = \varepsilon_m + C'(\varphi - \varphi_m). \tag{III.16}$$

Here and subsequently, the potential φ is taken from the point of zero charge (p.z.c.) at $\theta = 0$. The quantity φ_N represents the shift of p.z.c. in the transition from $\theta = 0$ to $\theta = 1$; φ_m and ε_m are the abscissa and ordinate of the intersection of the ε_0 vs φ and ε' vs φ lines.

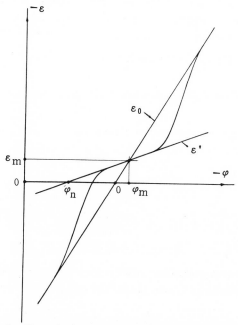

Fig. 12. Schematic representation of the dependence of the electrode charge on the potential under the condition C_0 = const and C' = const (explanation in text).

Inserting Eqs. (III.15) or (III.16) into (III.13), we obtain

$$\varepsilon = C_0 \varphi \, (1 - \theta) + C' \, (\varphi - \varphi_N) \, \theta \qquad \text{(III.17)}$$

or

$$\varepsilon = \varepsilon_m + [C_0 \, (1 - \theta) + C'\theta] \, (\varphi - \varphi_m). \qquad \text{(III.18)}$$

It follows from Eq. (III.17) that the dependence of the p.z.c. ($\varphi_{\varepsilon=0}$) on the degree of surface coverage is expressed by the equation

$$\varphi_{\varepsilon=0} = \varphi_N \theta \left/ \left[\frac{C_0}{C'} (1 - \theta) + \theta \right] \right. \qquad \text{(III.19)}$$

It follows from Eq. (III.19) that if the adsorption isotherm is congruent with respect to the potential the dependence of $\varphi_{\varepsilon=0}$ on θ (or on Γ) is not linear, and the deviation from linearity increases with increase of the C_0/C' ratio.

Now suppose that the experimental data may be represented by Eq. (III.8). In this case the electrode charge should be chosen as the independent electrical variable, as the form of the isotherm (III.8) in reduced coordinates will be independent of ε.

From Eq. (III.8) it follows that

$$\frac{d \ln G}{d\varepsilon} = - \left(\frac{\partial \ln c}{\partial \varepsilon} \right)_\theta. \qquad \text{(III.20)}$$

From Eqs. (III.6) and (III.20) we find

$$\left(\frac{\partial \varphi}{\partial \theta} \right)_\varepsilon = - A \, \frac{d \ln G}{d\varepsilon}, \qquad \text{(III.21)}$$

hence

$$\varphi = - A \left(\frac{d \ln G}{d\varepsilon} \right) \theta + \varphi_0, \qquad \text{(III.22)}$$

where φ_0 is the integration constant, equal to the value of φ at $\theta = 0$, corresponding to the given charge ε.

If the potential at a given ε at $\theta = 1$ is designated by φ' we have

$$\varphi' = - A \left(\frac{d \ln G}{d\varepsilon} \right) + \varphi_0. \qquad \text{(III.23)}$$

From Eqs. (III.22) and (III.23) it follows that

$$\varphi = \varphi_0 (1 - \theta) + \varphi'\theta. \tag{III.24}$$

Differentiating (III.24) with respect to ε at θ = const, we obtain

$$\frac{1}{C} = \frac{1-\theta}{C_0} + \frac{\theta}{C'}. \tag{III.25}$$

Therefore the assumption that the adsorption isotherm is congruent with respect to the electrode charge is equivalent to the model of two condensers with capacities $C_0/(1 - \theta)$ and C'/θ, connected in series. Parsons showed that physical interpretation of this model is possible if it is assumed that the decrease of capacity on adsorption of organic molecules is due only to increased separation of the charged layers of the double layer while the dielectric constant between them remains unchanged.

If the capacities C_0 and C' are independent of ε, it follows from Fig. 12 that $\varphi_0 = \varepsilon/C_0$ and $\varphi' = \varepsilon/C' + \varphi_N$. In this case Eq. (III.24) may be written in the form

$$\varepsilon = (\varphi - \theta\varphi_N) \Big/ \left(\frac{1-\theta}{C_0} + \frac{\theta}{C'}\right). \tag{III.26}$$

Putting $\varepsilon = 0$ in Eq. (III.26), we obtain a linear dependence of the p.z.c. on θ (or on Γ):

$$\varphi_{\varepsilon=0} = \theta\varphi_N. \tag{III.27}$$

Thus, the dependence of the shift of potential due to adsorption (p.z.c. at a given θ less p.z.c. at θ = 0) makes it possible to decide whether condition (III.7) or (III.8) is satisfied, and therefore which of the electrical variables (φ or ε) is the more suitable choice for study of the adsorption of an organic substance at the interface.

Experimental dependences, found by two methods, of $\varphi_{\varepsilon=0}$ on Γ in adsorption of n-propyl alcohol and n-caproic acid on mercury are shown in Fig. 13. It is seen that they deviate appreciably from linearity and are in good agreement with Eq. (III.19). The same result was obtained by Frumkin et al. [15] for adsorption of n-valeric acid and n-amylamine on mercury. It should be noted that $\varphi_{\varepsilon=0}$ vs Γ curves of similar form were obtained for camphor

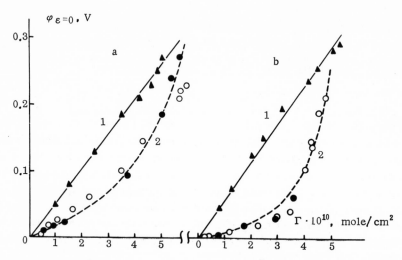

Fig. 13. Dependence of the change of adsorption potential at solution−air (1) and solution−mercury (2) interfaces on the amount adsorbed for: (a) n-propyl alcohol; (b) caproic acid. White circles represent data obtained from maxima on the electrocapillary curves; black circles represent data from minima on the differential capacity curves for dilute solutions; the dash lines are calculated for the following conditions: a) $\varphi_N = 0.31$ V; $C_0/C' = 3.5$; $\Gamma_m = 6 \cdot 10^{-10}$ mole/cm^2; b) $\varphi_N = 0.29$ V; $C_0/C' = 7$; $\Gamma_m = 5 \cdot 10^{-6}$ mole/cm^2; supporting electrolyte 0.01 N H$_2$SO$_4$ (from data in [26], Chap. III).

[59], n-butylamine [60], and certain other aliphatic compounds [61], although they were explained differently, in terms of a change in the orientation of adsorbed dipoles. Since the dependence of the change of potential due to adsorption on Γ at the solution−air interface in the same system is linear (see Fig. 13), this explanation cannot be accepted as correct.

Recently a nonlinear dependence of $\varphi_{\varepsilon=0}$ on Γ at the solution−mercury interface was also obtained in presence of tetrapropyl- and tetrabutylammonium cations [62].

The good agreement of the experimental $\varphi_{\varepsilon=0}$ vs Γ curves with Eq. (III.19) shows that the interface between an electrode and an aqueous solution containing a small amount of an organic substance may be represented in the first approximation by two condensers connected in parallel. Additional experimental confirmation of Eq. (III.13) can be found in a recent paper by Frumkin et al. [40] and in publications by Breiter and Delahay [63] and by Hansen et al. [43]. At the same time, Hansen et al. [43] noted the following point, with a reference to Overbeek. When $\varepsilon = 0$, the terms $\varepsilon_0(1 - \theta)$ and $\varepsilon'\theta$ in Eq. (III.13) are not individually equal to zero, i.e., parts of the surface are charged. Therefore for quantitative justification of Eqs. (III.7) and (III.13) it should be additionally assumed that the adsorption layer is confined on both sides by strictly equipotential surfaces and that the lines of force to them are perpendicular throughout [26]. These conditions may be approximately satisfied if dipoles of water and the organic substance, perpendicularly oriented to the surface, are situated between two phases with sufficiently high dielectric constants. For exact fulfillment of these conditions the dielectric constants of both phases must be infinitely large. Naturally, these conditions are not satisfied in the case of the solution−air interface.

It should be noted that variation of the concentration of the supporting electrolyte has virtually no influence on the dependence of $\varphi_{\varepsilon=0}$ on Γ [15]. It follows that equalization of the potential in the aqueous phase is determined by the high dielectric constant of water itself rather than by the ion concentration in the diffuse layer (cf. [64]). This conclusion is confirmed by the good agreement [26] between the experimental differential capacity curve for a solution of $0.002\ N$ NaF $+ 0.2\ M$ n-C_3H_7OH and the C vs φ curve calculated by Grahame's method [65] from experimental data for a solution of $0.1\ N$ NaF $+ 0.2\ M$ n-C_3H_7OH. The observed agreement between calculation and experiment is possible only under the condition that the presence of adsorbed dipoles of the organic substance does not interfere appreciably with the ion distribution in the diffuse part of the double layer.

It follows from these results that experimental data on adsorption of organic substances on mercury are represented better by isotherms congruent with respect to the potential than by isotherms congruent with respect to the charge; therefore under these conditions the electrode potential rather than its charge should be chosen as the electrical variable.

2. Certain General Relationships, Independent of the Particular Form of the Adsorption Isotherm

a. Dependence of the Energy of Adsorption on the Electrode Potential. Suppose that conditions corresponding to Eq. (III.7) exist on the electrode surface during adsorption of an organic substance, and that adsorption equilibrium is established at each value of the potential. In this case it follows from Eqs. (III.10) and (III.18) that

$$\frac{d \ln B}{d\varphi} = -\frac{C_0 - C'}{A}(\varphi - \varphi_m) = -2\alpha(\varphi - \varphi_m), \qquad \text{(III.28)}$$

where

$$\alpha = \frac{C_0 - C'}{2A}. \qquad \text{(III.29)}$$

Integration of Eq. (III.28) followed by conversion to exponential form gives

$$B = B_m \exp[-\alpha(\varphi - \varphi_m)^2], \qquad \text{(III.30)}$$

where B_m is the value of B when $\varphi = \varphi_m$.

On the other hand, the relation between the adsorption equilibrium constant B and the free energy of adsorption of the organic substance ($\Delta \overline{G}_A$) is given by the expression

$$B = \frac{1}{55.5} \exp\left(-\frac{\Delta \overline{G}_A}{RT}\right). \qquad \text{(III.31)}$$

Thus, if the form of the adsorption isotherm is independent of the electrode potential, and the capacity of the double layer decreases to a certain limiting value C' with increasing concentration of the organic substance, then the energy of adsorption of that substance is a quadratic function of the electrode potential.

Frumkin [2, 3] showed that the same result can be obtained from electrostatics if adsorption of neutral molecules is regarded as equivalent to separation of the plates of a condenser and replacement of one dielectric in it by another. When water is replaced by an organic substance at $\varphi = $ const, the quantity of electricity to be released from the condenser plates is $(C_0 - C')\varphi$. The work required for this is $(C_0 - C')\varphi^2$. On the other hand, when the dielectric is changed the energy of the double-layer condenser is decreased by $\frac{1}{2}(C_0 - C')\varphi^2$. Thus, as the result of these two opposite effects the work of replacement of the dielectric is

$$w_1 = (C_0 - C')\varphi^2 - \frac{1}{2}(C_0 - C')\varphi^2 = \frac{1}{2}(C_0 - C')\varphi^2. \qquad \text{(III.32)}$$

As the dielectric constant of water is higher than that of the organic substance, $C_0 - C' > 0$ and $w_1 > 0$. In other words, the work required for replacement of water in the double layer by the organic substance increases with increasing $|\varphi|$. The reverse replacement occurs spontaneously. This, strictly speaking, is the explanation of the desorption of organic molecules at sufficiently high positive or negative surface charges.

If the organic molecules have a dipole moment, oriented adsorption of these molecules produces an additional change of potential, φ_N. Thus, when an already polarized dielectric of this kind is introduced into the double layer, an additional quantity of electricity, $C'\varphi_N$ Coulombs, must be removed. The work required for this at $\varphi = \text{const}$ is

$$w_2 = C'\varphi_N\varphi. \qquad \text{(III.33)}$$

The quantity w_2 may be positive or negative, dependent on the signs of φ and φ_N; the sign of φ_N is determined by the molecular structure and by the properties of the contiguous phases.

If we refer the over-all work of adsorption $(w_1 + w_2)$ to 1 mole of the adsorbed substance (i.e., to Γ_m), then in accordance with the Boltzmann equation we obtain the following expression for the adsorption equilibrium constant:

$$B = B_0 \exp\left[-\frac{\frac{1}{2}(C_0 - C')\varphi^2 + C'\varphi_N\varphi}{RT\Gamma_m}\right], \qquad \text{(III.34)}$$

where B_0 is the value of B when $\varphi = 0$.[*]

---

[*] A more detailed derivation of Eq. (III.34), based on the principles of electrostatics, is given elsewhere [3].

Putting the symbols

$$\alpha = \frac{C_0 - C'}{2RTT_m} \; ; \; \varphi_m = -\frac{C'\varphi_N}{C_0 - C'} \text{ and } B_m = B_0 \exp\left[\frac{(C'\varphi_N)^2}{2RTT_m(C_0 - C')}\right],$$

we at once obtain Eq. (III.30) from Eq. (III.34).

Butler [66] derived Eq. (III.30) on the basis of a molecular picture of the adsorption process, in which the molecules of water and of the organic compound were regarded as volume elements having polarizabilities p_W and p_A. The permanent dipole moments of the water and organic molecules, μ_W and μ_A were also taken into account. Thus, according to Butler [66], the electrical component of the work of adsorption is given by the expression

$$W = \left[\frac{1}{2}(p_W - p_A) X^2 + (\mu_W - \mu_A) X\right]\delta V, \qquad (\text{III.35})$$

where X is the electrical field strength and δV is an element of volume. With Butler's assumption that X is proportional to φ, Eq. (III.30) is easily obtained from Eq. (III.35). Indeed, Eq. (III.35) is similar to the corresponding equation in Frumkin's theory for the sum of the work terms w_1 and w_2; the only difference is that Eq. (III.35) involves the volume properties of the dielectrics, which is hardly permissible in relation to the behavior of the substances in the double layer, whereas Frumkin's theory utilizes the experimentally determined properties of the double layer.

On the other hand, certain investigators [56, 57, 67, 68] obtained a different relationship between the energy of adsorption and the electrode potential. Breyer and Hacobian [67] and Miller [68] took into account only the interaction of the permanent water and organic dipoles with the charged surface in calculation of the electrical energy of adsorption. Under these conditions the first term in Eq. (III.35) vanishes and consequently, if X is proportional to φ, the energy of adsorption becomes a linear function of the electrode potential [67, 68].

According to the theory of Bockris et al. [56, 57] the principal factor determining the course of adsorption of organic substances at the mercury–electrode interface in relation to the electrode charge is competition between the organic molecules and water molecules. The water molecules, located at random in the vicinity of an uncharged mercury surface, become oriented under the influence of the field; this hinders adsorption of organic substances. Thus, the potential of maximum adsorption of the organic substance corresponds to the potential at which the number of water dipoles with their positive ends toward the electrode is equal to the number of water dipoles having the opposite orientation. Owing to some degree of specific interaction between the negative ends of the water dipoles and the mercury surface, this potential is shifted slightly in the negative direction from p.z.c. ($\varepsilon_m \approx -2 \ \mu C/cm^2$). Quantitative calculations based on this assumption, but with the energy of formation of the ionic double layer ignored, show that in the first approximation the energy of adsorption is a linear function of

$$\frac{n\mu_W X}{kT} \ \tanh\left(\frac{\mu_W X}{kT}\right),$$

where n is the number of water molecules displaced by one organic molecule from the surface, and k is the Boltzmann constant.

It was shown above that in the case of adsorption of aliphatic compounds on mercury the form of the adsorption isotherm is, in the first approximation, independent of the electrode potential, while the capacity of the double layer decreases to a certain limiting value C' with increasing concentration of the organic substance. Since a quadratic dependence of the adsorption energy on the potential follows from these conditions on thermodynamic grounds, the model concepts leading to a different dependence of $\Delta \overline{G}_A$ on φ are contrary to the principles of thermodynamics.

b. Characteristics of Equilibrium C vs φ Curves with Considerable Adsorption of Organic Substances. Differentiating Eq. (III.18) with respect to potential and taking Eq. (III.29) into account, we obtain, for the conditions $C_0 = $ const and $C' = $ const, the following expression for the differential capacity:

$$C = C_0 - 2A\alpha\theta - 2A\alpha\left(\varphi - \varphi_m\right)\left(\frac{\partial\theta}{\partial\varphi}\right)_c . \qquad (III.36)$$

From Eqs. (III.7) and (III.30) it follows that

$$B_m c \exp\left[-\alpha\left(\varphi - \varphi_m\right)^2\right] = f\left(\theta\right). \qquad (III.37)$$

Taking logarithms and differentiating with respect to φ at c = const, we obtain, after transformations

$$\left(\frac{\partial \theta}{\partial \varphi}\right)_c = -2\alpha\,(\varphi - \varphi_m)\,h, \qquad (III.38)$$

where

$$h = \left[\frac{\partial \ln f\,(\theta)}{\partial \theta}\right]^{-1}. \qquad (III.39)$$

The derivative $(\partial\theta/\partial\varphi)_c = 0$ when $\varphi = \varphi_m$, and therefore φ_m is the potential of maximum adsorption, because the plot of θ vs φ passes through a maximum at $\varphi = \varphi_m$.

Putting Eq. (III.38) into Eq. (III.36), we obtain

$$C = C_0 - 2A\alpha\theta + 4A\alpha^2\,(\varphi - \varphi_m)^2\,h. \qquad (III.40)$$

To find the position of the extremal points on the C vs φ curve we determine the value of the derivative $(dC/d\varphi)_c$ from Eq. (III.40). Taking Eq. (III.38) into account, we obtain the expression

$$\left(\frac{\partial C}{\partial \varphi}\right) = 12A\alpha^2\,(\varphi - \varphi_m)\,h\left[1 - \frac{2}{3}\alpha\,(\varphi - \varphi_m)^2\,\frac{\partial h}{\partial \theta}\right]. \qquad (III.41)$$

It follows from Eq. (III.41) that the condition $(dC/d\varphi)_c = 0$, corresponding to extremal points on the C vs φ curve, is satisfied in three cases: 1) when h = 0; 2) when $\varphi = \varphi_m$; 3) when the expression in square brackets is zero.

According to Eq. (III.7), the curve for the dependence of $f(\theta)$ on θ at a given potential is the adsorption isotherm, the only difference being that in this case the quantity proportional to the concentration is taken along the ordinate axis and not along the abscissa. Therefore, without taking a definite equation for the function $f(\theta)$ as yet, but using the form of the experimental isotherms, we can represent the dependence of $f(\theta)$ on θ by the curve shown in Fig. 14a, where $\theta \to 1$ when $f(\theta) \to \infty$ (i.e., when $c \to \infty$). The presence of an inflection on curves of this type reflects attractive interaction between the adsorbed particles. It follows from the shape of the $f(\theta)$ vs θ curve that the dependence of the derivatives of the functions $\ln f(\theta)$, $\partial \ln f(\theta)/\partial\theta$, $h = \partial\theta/\partial \ln f(\theta)$, and $\partial h/\partial\theta$ on the surface coverage θ must be of the form shown in Fig. 14(b-e).

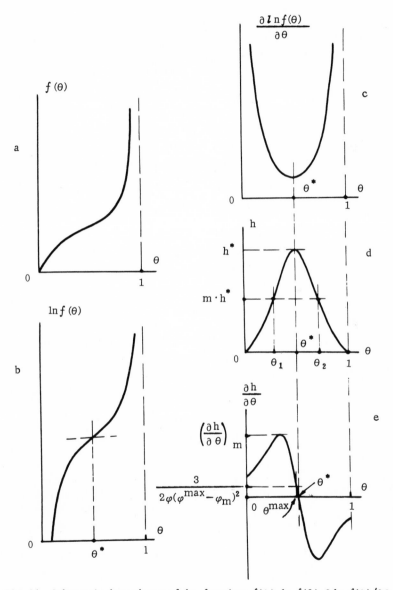

Fig. 14. Schematic dependence of the functions $f(\theta)$, $\ln f(\theta)$, $\partial \ln f(\theta)/\partial\theta$, h, and $\partial h/\partial\theta$ on the degree of coverage of the surface by the organic substance, θ (explanations in text).

It is seen from Fig. 14d that $h = 0$ either when $\theta = 0$ which, in accordance with Eq. (III.37), corresponds to $c = 0$ or $|\varphi - \varphi_m| \to \infty$, or when $\theta = 1$; this, by the conditions of selection of the function $f(\theta)$, corresponds to $c \to \infty$. Thus, the first condition for equality of the derivative $(\partial C / \partial \varphi)_c$ to zero is not achieved in practice.

The second condition, $\varphi = \varphi_m$, determines the position of the minimum on the differential capacity curve in presence of an organic substance, which therefore corresponds to the potential of maximum adsorption. The third condition defines the positions of the maxima φ^{max} on the C vs φ curves in presence of an organic substance. In this case it follows from Eq. (III.41) that

$$\frac{\partial h}{\partial \theta} = \frac{3}{2\alpha (\varphi^{max} - \varphi_m)^2}. \qquad \text{(III.42)}$$

We first suppose that the maximum value of the derivative $\partial h / \partial \theta$ is considerably greater than $^3/_2\alpha(\varphi^{max} - \varphi_m)^2$, i.e.,

$$\left(\frac{\partial h}{\partial \theta}\right)_m \gg \frac{3}{2\alpha (\varphi^{max} - \varphi_m)^2} \qquad \text{(III.43)}$$

(see Fig. 14e). Increase of the concentration of the organic substance favors conformity to the condition (III.43), because it leads to increase of $(\varphi^{max} - \varphi_m)^2$ and of the attractive interaction between the adsorbed organic particles. In this case the adsorption isotherm assumes a characteristic sigmoid shape, which leads to a sharp increase of the maximum value of h and hence of the maximum value of the derivative $\partial h / \partial \theta$ (see Fig. 14).

It is seen in Fig. 14e that when the condition (III.43) is satisfied the degree of coverage θ^{max} of the electrode at the peak potentials on the C vs φ curves is approximately equal to the degree of coverage $\theta *$ corresponding to the maximum of the function h: $\theta^{max} \approx \theta *$. Thus, when $\varphi = \varphi^{max}$ we have from Eq. (III.37)

$$Bc = B_m c \, \exp\left[-\alpha (\varphi^{max} - \varphi_m)^2\right] \approx f(\theta^*) = \text{const.} \qquad \text{(III.44)}$$

Converting Eq. (III.44) to logarithmic form and combining the constants, we obtain

$$K + \log c \approx \frac{\alpha}{2.3} (\varphi^{max} - \varphi_m)^2, \qquad \text{(III.45)}$$

where

$$K = \log [B_m / f(\theta^*)].$$

Thus, when the condition (III.43) is satisfied, i.e., with fairly strong attractive interaction between the adsorbed particles and at not too low concentrations of the organic substance, the peak potentials on the C vs φ curves should conform to a square-law variation with the logarithm of the concentration of the organic substance. This result is illustrated in Fig. 15 by the data from the paper by Lorenz and Möckel [44] for various organic compounds.

When the condition (III.43) is satisfied for $\varphi = \varphi^{\max}$, Eq. (III.40) becomes

$$C^{\max} \approx C_0 - 2A\alpha\theta^* + 4A\alpha^2 (\varphi^{\max} - \varphi_m)^2 h^*. \qquad (III.46)$$

where $h^* = f(\theta^*)/f'(\theta^*)$ represents values of the function h when $\theta = \theta^*$. Inserting into Eq. (III.46) the expression for $(\varphi^{\max} - \varphi_m)^2$ from Eq. (III.45) and combining all the constants, we obtain

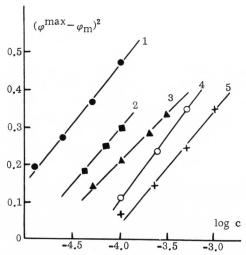

Fig. 15. Dependence of the peak potentials ($\varphi^{\max}$) on the differential capacity curves on the concentration (c) of the organic substance. 1) Cyclohexanol; 2) isoamyl alcohol; 3) n-butyl alcohol; 4) methyl ethyl ketone; 5) n-butyric acid (from data in [44], Chap. III).

$$C^{\max} \approx K_1 \log c + K_2, \tag{III.47}$$

where

$$K_1 = 4.6(C_0 - C')h^* \text{ and } K_2 = K_1 \log [B_m/f(\theta^*)] + C_0(1-\theta^*) + C'\theta^*.$$

Thus, with sufficiently strong attractive interaction between the adsorbed particles and at not too low concentrations of the organic substance a linear relation should exist between the height of the peak on the differential capacity curve and the logarithm of the concentration of the organic substance. This conclusion is in good agreement with experimental data [5, 34, 69] (Fig. 16).

If the function h, which determines the peak height on the C vs φ curve, is a certain fraction m ($0 < m < 1$) of h^*, solution of the equation

$$h = mh^* \tag{III.48}$$

gives two values, θ_1 and θ_2, for the surface coverage (see Fig. 14d). If θ_1 and θ_2 are known, it is possible in the first approximation to determine the breadth of the peak ($\Delta\varphi$) on the C vs φ curve for a definite relative peak height. From Eqs. (III.38) and (III.39) it follows that at constant concentration

$$-2\alpha (\varphi - \varphi_m) \, d\varphi = d \ln f(\theta). \tag{III.49}$$

We now integrate the right-hand side of this equation between θ_1 and θ_2, and the left-hand side between φ_1 and φ_2, where the potentials φ_1 and φ_2 correspond to θ_1 and θ_2. Taking into account that $\varphi_1 - \varphi_2 = \Delta\varphi$, while $(\varphi_1 + \varphi_2)/2 \approx \varphi^{\max}$, we obtain

$$2\alpha (\varphi^{\max} - \varphi_m) \Delta\varphi \approx \ln \left[\frac{f(\theta_2)}{f(\theta_1)} \right]. \tag{III.50}$$

Equations (III.39) and (III.50) enable us to connect the shape of the peak on the C vs φ curve with the shape of the adsorption isotherm and therefore with the attractive interaction between the adsorbed particles. The steeper the isotherm in its middle region, where the $\ln f(\theta)$ vs θ curves have an inflection, the greater is the value of h^* and the less is the value of $\Delta\varphi$, as the difference between $f(\theta_1)$ and $f(\theta_2)$ diminishes (see Fig. 14). Thus, the stronger the attraction between the adsorbed particles, the higher and narrower is the peak on the differential capacity curve. In the limiting case, when at $\theta = \theta^*$ $\partial \ln f(\theta)/\partial\theta = 0$, $h^* \to \infty$, $\theta_2 = \theta_1$ and $\Delta\varphi \to 0$, i.e., the peak on the C vs φ curve degenerates into a vertical line.

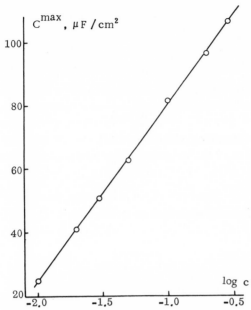

Fig. 16. Dependence of the cathodic peak height on the differential capacity curve on the logarithm of the concentration (c) of tert-amyl alcohol (from data in [5], Chap. III).

If the inequality (III.43) is not satisfied, i.e., if the forces of attractive interaction between the adsorbed particles are not large and the concentration of the organic substance is so low that the peak potentials on the C vs φ curves are not too far from φ_m, then θ^{max} depends on φ^{max} and is not a constant equal to θ^*: $\theta^{max} \neq \theta^* =$ const. Under such conditions, in order to find the dependence of the height and position of the peaks on the C vs φ curves on the concentration of the organic substance, we must choose a definite equation for the adsorption isotherm, i.e., the function $f(\theta)$.

3. Selection and Verification of the Adsorption Isotherm

We will now compare the results which follow from the equations for various adsorption isotherms with experimental data, in order to find the isotherm in best agreement with the experimental results (see [70, 71]). In this context we will examine certain simple adsorption isotherms which have been used by various investigators for representing adsorption on electrodes; for convenient comparison, we present them in analogous form as far as possible.

1. The Henry isotherm

$$Bc = \theta. \tag{III.51}$$

2. The Freundlich isotherm [72]

$$Bc^n = \theta, \tag{III.52}$$

where $0 < n < 1$.

3. The Langmuir isotherm [73]

$$Bc = \frac{\theta}{1-\theta}. \tag{III.53}$$

4. The Volmer isotherm [74]

$$Bc = \frac{\theta}{1-\theta} \exp\left(\frac{\theta}{1-\theta}\right). \tag{III.54}$$

5. The Amagat isotherm [see 48, 71]

$$Bc^n = \frac{\theta}{1-\theta} \exp\left(\frac{\theta}{1-\theta}\right). \tag{III.55}$$

6. The Helfand−Frisch−Lebowitz isotherm [75]

$$Bc = \frac{\theta}{1-\theta} \exp\left[\frac{2-\theta}{(1-\theta)^2}\right]. \tag{III.56}$$

7. The Frumkin isotherm [1]

$$Bc = \frac{\theta}{1-\theta} \exp\left(-2a\theta\right), \tag{III.57}$$

where a is a certain quantity characterizing interaction between the adsorbed particles.

8. The Hill−de Boer isotherm [76, 77]

$$Bc = \frac{\theta}{1-\theta} \exp\left(\frac{\theta}{1-\theta}\right) \exp\left(-2a\theta\right). \tag{III.58}$$

9. The Parsons isotherm [49]

$$Bc = \frac{\theta}{1-\theta} \exp\left[\frac{2-\theta}{(1-\theta)^2}\right] \exp(-2a\theta). \qquad \text{(III.59)}$$

10. The isotherm with virial coefficients [see 48, 71]:

$$Bc = \theta \exp(-2a\theta). \qquad \text{(III.60)}$$

In contrast to Eqs. (III.57)-(III.59), in this case a is always less than zero, i.e., it represents repulsive interaction between the adsorbed molecules. If repulsion can be reduced to incompatibility of rigid molecules having circular symmetry, then a in Eq. (III.60) is determined by double the area per mole of adsorbed molecules.

11. The "square root" isotherm (see [48, 71]).

In this case the relation between the concentration and the surface coverage can be expressed parametrically with the aid of two equations

$$\frac{1}{\theta} = \frac{A}{\Delta\sigma} + \sqrt{-\frac{aA}{\Delta\sigma}}; \; \ln(Bc) = \ln(\Delta\sigma) + 2\sqrt{-\frac{a\Delta\sigma}{A}}, \qquad \text{(III.61)}$$

with the surface pressure $\Delta\sigma$ as the parameter. It is seen from Eqs. (III.61) that the molecular interaction factor is again negative throughout $(a < 0)$, and the equation therefore represents only repulsive interaction between the adsorbed molecules.

12. The Temkin isotherm [78]:

$$Bc = \frac{\exp(a\theta) - 1}{1 - \exp[-a(1-\theta)]}. \qquad \text{(III.62)}$$

13. The Lorenz isotherm [46]

$$\theta = \frac{(1 + 1/Bc)^{\nu-1} + K_{as}}{(1 + 1/Bc)^{\nu} + K_{as}}, \qquad \text{(III.63)}$$

where $K_{as} = v_{as}/v_{dis}$ is the equilibrium constant for association of the adsorbed molecules:

$$\nu M \underset{v_{dis}}{\overset{v_{as}}{\rightleftarrows}} M_{\nu};$$

ν is the "degree of association," i.e., the average number of organic molecules in an associated group.

14. The Blomgren−Bockris isotherms [79]:

$$Bc = \frac{\theta}{1-\theta} \exp{(p\theta^{3/2} - q\theta^3)} \tag{III.64}$$

and for adsorption of organic ions

$$Bc = \frac{\theta}{1-\theta} \exp{(p_1\theta^{1/2} - q\theta^3)}. \tag{III.65}$$

In these equations $p > 0$, $p_1 > 0$, and $q > 0$ are certain constants, expressed in terms of the dipole moment (or charge), the area occupied by an adsorbed molecule or ion, certain other characteristics of the adsorbed particles, and the dielectric constant of the surface layer.

It is known from experimental data [34, 46] that the adsorption isotherms of organic substances may be either sigmoid or logarithmic, dependent on whether attractive or repulsive interaction predominates between the adsorbed particles. If attractive interaction predominates and the isotherm is S-shaped, its slope in the middle region when $\theta = \theta*$ should be greater than the initial slope at $\theta = 0$. In other words, the following condition must be satisfied in this case:

$$\frac{f'(\theta^*)}{f'(0)} < 1, \tag{III.66}$$

where $f'(\theta*)$ and $f'(0)$ represent the values of $\partial f(\theta)/\partial\theta$ when $\theta = \theta*$ and $\theta = 0$.

Analysis of Eqs. (III.51)-(III.56) and (III.60)-(III.62) shows that they do not satisfy the condition (III.66), and are therefore not applicable to experimental S-shaped adsorption isotherms and to the corresponding differential capacity curves.

It follows from the paper by Lorenz et al. [46] that the Lorenz isotherm (III.63) satisfactorily represents the sigmoid shape of experimental adsorption isotherms at $\nu > 1$ and K_{as} of the order of 10^4-10^5. When $K_{as} \to 0$, this isotherm becomes the Langmuir equation, which is the limiting case of the Lorenz isotherm when the dissociation rate greatly exceeds the association rate ($v_{dis} \gg v_{as}$).

However, in order to represent repulsive interaction between adsorbed particles with the aid of Eq. (III.63) it is necessary to postulate formally that $\nu < 1$; however, this is devoid of physical meaning, because ν is the number of particles in an associated group and is therefore always greater than or equal to unity.

Thus, it is impossible with the aid of the Lorenz isotherm to represent the logarithmic form of the adsorption isotherm with repulsive interaction between adsorbed organic particles, such as is observed, e.g., during adsorption of $[(CH_3)_4N]^+$ and $[(C_2H_5)_3HN]^+$ cations on mercury [46].

Another defect of the Lorenz isotherm is that two-dimensional interaction is represented with the aid of two arbitrary constants, K_{as} and ν. The Blomgren–Bockris isotherms (III.64) and (III.65) also suffer from this disadvantage in comparison with the isotherms (III.57)-(III.59). Moreover, as has been shown by Damaskin et al. [36], at not very high degrees of surface coverage by the organic substance ($\theta \ll 0.6$) it is impossible to establish experimentally the difference between the isotherms of Frumkin (III.57) and of Blomgren and Bockris (III.64) and (III.65), as this requires determination of θ to an accuracy of at least 0.01. Thus, the linear dependences of $\Delta\overline{G}_A$ on $\theta^{3/2}$ (for neutral molecules) or of $\Delta\overline{G}_A$ on $\theta^{1/2}$ (for ions) which have been found experimentally [79, 80-82] do not exclude, with the degree of experimental accuracy presently available, a linear dependence of $\Delta\overline{G}_A$ on θ, corresponding to Eq. (III.57). However, since the Blomgren–Bockris isotherms at not very high θ represent only repulsive interaction [the constants p and p_1 in Eqs. (III.64) and (III.65) are always positive by their physical meaning], Eq. (III.57) is preferable for representation of experimental data.

Thus, of the equations (III.51)-(III.65) considered above, the most suitable for representation of adsorption of organic substances on electrodes are the isotherms (III.57)-(III.59), in which one arbitrary constant (a) is used to represent two-dimensional interaction between the adsorbed particles; dependent on the sign of this attraction constant, Eqs. (III.57)-(III.59) may describe both attractive ($a > 0$) and repulsive ($a < 0$) interaction.

This conclusion is illustrated by Fig. 17, which shows curves of the dependence of θ on the relative concentration $c/c_{\theta=\theta^*}$ calculated from Eq. (III.57) for different values of a.

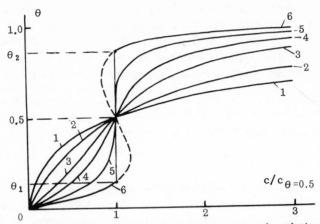

Fig. 17. Dependence of the surface coverage θ on the relative concentration $c/c_{\theta=\theta*}$, calculated from Eq. (III.57) for different values of a. 1) $a = -1$; 2) $a = 0$; 3) $a = 1$; 4) $a = 1.5$; 5) $a = 2.0$; 6) $a = 2.5$. Under equilibrium conditions with $a = 2.5$, θ changes abruptly from θ_1 to θ_2.

vs $c/c_{\theta=\theta*}$ of similar form are also obtained from Eqs. (III.58) and (III.59). Figure 17 shows that when $a < 0$ the isotherm is of logarithmic form, indicating repulsive interaction between the adsorbed particles, while at $a > 0$ the isotherm begins to assume a sigmoid shape characteristic of attractive interaction.

At a certain critical value a_{cr} of the attraction constant the tangent to the isotherm at $\theta = \theta*$ becomes parallel to the ordinate axis. It follows from Eq. (III.46) that in this case the peak on the C vs φ curve degenerates to a vertical line. If $a > a_{cr}$, a part of the theoretical isotherm corresponds to unstable states of the adsorption layer (dashed portion of curve 6 in Fig. 17).

The values of $\theta*$ and a_{cr} for Eqs. (III.57)-(III.59) can be found from the conditions

$$\frac{\partial^2 \ln f\,(\theta)}{\partial\theta^2} = 0 \tag{III.67}$$

and

$$\left[\frac{\partial f\,(\theta)}{\partial\theta}\right]_{\theta=\theta*} = 0. \tag{III.67a}$$

These values are given in Table 1.

TABLE 1

Isotherm	θ^*	a_{cr}
Frumkin (III.57)	0.500	2.000
Hill−De Boer (III.58)	$1/3 = 0.333$	3.375
Parsons (III.59)	$(\sqrt{7} - 2)/3 = 0.215$	5.841

Thus, all the three equations (III.57)-(III.59) satisfy the selection conditions for the function $f(\theta)$ and are in qualitative agreement with experimental data. We use the following criterion [13] for selecting one of these equations for quantitative interpretation of differential capacity curves in presence of organic compounds. It is easily seen that Eqs. (III.57)-(III.59) may be written in the form

$$Bc = f(\theta) = F(\theta)\exp(-2a\theta), \tag{III.68}$$

where $F(\theta)$ is a certain function of θ, independent of a. Converting Eq. (III.68) into logarithmic form and then differentiating with respect to θ at φ = const, we obtain

$$\left(\frac{\partial \ln c}{\partial \theta}\right)_\varphi = \frac{\partial \ln f(\theta)}{\partial \theta} = \frac{\partial \ln F(\theta)}{\partial \theta} - 2a. \tag{III.69}$$

With the condition (III.67) taken into account it follows that when $\theta = \theta^*$ the $(\partial \ln c/\partial \theta)_\varphi$ vs θ curve passes through a minimum, the position of which is determined only by the function $F(\theta)$ and is independent of the attraction constant a.

Plots of $(\partial \ln c/\partial \theta)_\varphi$ vs θ, calculated with the aid of Eq. (III.69) for the isotherms (III.57)-(III.59), are shown in Fig. 18a; for convenient comparison the attraction constant was so chosen that at the minimum $(\partial \ln c/\partial \theta)_\varphi = 0$. It follows from Table 1 that the position of the minimum on these curves corresponds to the following values of θ: 0.500, 0.333, and 0.215. Therefore the position of the minimum on the experimental $(\partial \ln c/\partial \theta)_\varphi$ vs θ curves is a convenient criterion for selection of one of the adsorption isotherms (III.57)-(III.59).

On the other hand, it is seen in Fig. 18a that the difference between the two values of θ under the condition $(\partial \ln c/\partial \theta) -$

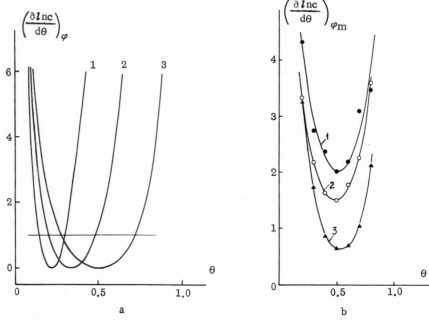

Fig. 18. Dependence of $(\partial \ln c / \partial \theta)_{\varphi}$ on the surface coverage θ. a) Calculated from the Parsons (1), Hill–De Boer (2), and Frumkin (3) isotherms; b) based on experimental isotherms for adsorption of aniline (1), n-amylamine (2), and tert-amyl alcohol (3). (From data in [13], Chap. III.)

$(\partial \ln c / \partial \theta)_{\min} = 1$, i.e., the breadth of the minimum $(\Delta \theta)$ on the plot of $(\partial \ln c / \partial \theta)_{\varphi}$ vs θ, can also serve as the criterion for selection of the adsorption isotherm equation. Our calculated values of $\Delta \theta$ for the isotherms (III.57)-(III.59) are 0.447, 0.281, and 0.158, respectively.

The proposed method was applied to adsorption of a number of organic compounds on mercury [13]. The experimental adsorption isotherms were determined at the potential of maximum adsorption $\varphi = \varphi_m$ when, in accordance with Eq. (III.36), regardless of the particular isotherm equation,

$$C = C_0 (1 - \theta) + C'\theta. \qquad \text{(III.70)}$$

With the aid of this equation it was possible to plot an experimental θ vs c curve and, with the aid of graphical differentiation, to

TABLE 2

System studied	θ^*	$\Delta\theta$
n-C_3H_7OH + $1N$ Na_2SO_4	0.48	0.48
n-$C_5H_{11}OH$ + $1N$ Na_2SO_4	0.53	0.45
tert-$C_5H_{11}OH$ + $1N$ KF	0.54	0.46
n-$C_5H_{11}NH_2$ + $1N$ Na_2SO_4	0.49	0.45
iso-$C_5H_{11}NH_2$ + $1N$ Na_2SO_4	0.49	0.46
[$(C_4H_9)_4N$]$_2SO_4$ + $1N$ Na_2SO_4	0.55	0.50
$C_6H_5NH_2$ + $1N$ KCl	0.52	0.44
C_6H_5OH + $1N$ Na_2SO_4	0.45	0.32

find values of $\dfrac{\partial \ln c}{\partial \theta} = \dfrac{1}{c} \cdot \dfrac{\partial c}{\partial \theta}$ at given values of θ. The curves of $(\partial \ln c / \partial \theta)_\varphi$ vs θ found in this way were used for determining ex-experimental values of θ^* and $\Delta\theta$. The results are presented in Table 2 and in Fig. 18b.

Figure 18 and Table 2 show that adsorption of all the organic compounds studied conforms with good approximation to the Frumkin isotherm and cannot be represented by Eqs. (III.58) and (III.59).

Thus, of all the isotherms examined here, the Frumkin isotherm (III.57) gives the best agreement with experimental data on adsorption of various organic compounds on mercury. Therefore, this is the isotherm which should be chosen for representation of differential capacity curves determined in presence of organic compounds.

Equation (III.57) was obtained by Frumkin [1] by inclusion in the equation of state of the surface layer

$$\Delta\sigma = -A \ln(1 - \theta) \qquad (III.71)$$

of a term representing attractive interaction between the adsorbed particles

$$\Delta\sigma = -A [\ln(1 - \theta) + a\theta^2]. \qquad (III.72)$$

In fact, combination of Eq. (III.72) with the fundamental electrocapillary equation (III.1) leads directly to the isotherm (III.57).

Equation (III.57) was later derived on statistical grounds by Fowler and Guggenheim [83] in relation to localized adsorption

of particles from a gaseous phase. Accordingly, objections were
raised by Parsons [49] to the application of Eq. (III.57) to non-
localized adsorption of ions and molecules at interfaces between
a liquid metal (mercury) and a solution. However, these objections are not justified, because lowering of interfacial tension
by an adsorbed layer should be interpreted as the two-dimensional analog of the osmotic pressure of concentrated solutions
rather than as the analog of a compressed gas [84]. In other
words, in the case of adsorption from solutions, when the molecules of adsorbate replace solvent molecules in the surface layer,
localized adsorption is not an essential condition for conformity
to the Langmuir equation (III.53) or the Frumkin equation (III.57).
For these conditions Eq. (III.57) can be derived statistically [85],
or from the osmotic pressure [84], but with the assumption that the
adsorbed particles of the solvent and the solute occupy equal
areas on the electrode surface. At first sight, comparison of the
area corresponding to a water molecule ($\sim$10 $\overset{\circ}{A}{}^2$) with the area
corresponding to maximum adsorption of many aliphatic com-
pounds ($\sim$20-30 $\overset{\circ}{A}{}^2$) appears to contradict this assumption.

It is of interest in this connection to consider an adsorption
isotherm in which it is taken into account that n solvent molecules
are displaced from the surface when one organic molecule is adsorbed. An isotherm of this type, with attractive interaction between the adsorbed particles ignored, was derived by Zhukhovitskii
[86]:

$$\frac{b_1}{b_2^n} = \frac{a_1}{a_2^n} \exp\left(\frac{\sigma_2 - \sigma_1}{RT\Gamma_m}\right), \qquad (III.73)$$

where the activities in the solution volume and in the surface layer
are represented by a_i and b_i; the subscripts 1 and 2 refer to the
solute and solvent respectively.

Assuming that $b_1 = \theta$ and $b_2 = 1 - \theta$, for low concentrations of
the organic substance, when $a_1 \approx nc/55.5$ and $a_2 \approx 1$, Eq. (III.73)
can be written in the form

$$Bc = \frac{\theta}{n(1-\theta)^n} \qquad (III.74)$$

or, with a correction for attractive interaction between the adsorbed molecules

$$Bc = \frac{\theta}{(1-\theta)^n} \exp(-2a\theta). \tag{III.75}$$

Equation (III.74) was used by Dahms and Green [87], and Eq. (III.75) by Damaskin [20] and Parsons [49].

On the other hand, the activities of the organic substance and water in the surface layer may be equated to the corresponding molar fractions, i.e.,

$$b_1 = \frac{\Gamma_1}{\Gamma_1 + \Gamma_2} = \frac{\theta}{\theta + n(1-\theta)} \text{ and } b_2 = \frac{\Gamma_2}{\Gamma_1 + \Gamma_2} = \frac{n(1-\theta)}{\theta + n(1-\theta)}. \tag{III.76}$$

In that case we have from Eq. (III.73)

$$Bc = \frac{\theta}{n(1-\theta)^n}\left(1 - \theta + \frac{\theta}{n}\right)^{n-1} \tag{III.77}$$

or, with a correction for attractive interaction

$$Bc = \frac{\theta}{n(1-\theta)^n}\left(1 - \theta + \frac{\theta}{n}\right)^{n-1} \cdot \exp(-2a\theta). \tag{III.78}$$

Equation (III.77) was obtained by Bockris and Swinkels [88], and Eq. (III.78) by Kastening and Holleck [89].

It is easily seen that when n = 1 the two equations (III.75) and (III.78) both become the Frumkin isotherm (III.57). When $n \to \infty$, Eq. (III.78) becomes the Hill−De Boer isotherm (see [89]) whereas Eq. (III.75) has no analogous limiting case. Comparison of various derivations of Eqs. (III.74) and (III.77), recently carried out by Damaskin [37], shows that a preferential choice between the isotherms (III.75) and (III.78) cannot be made at the present time.*

---

*Schuhmann's paper [90], in which yet another derivation of Eq. (III.77) was given recently, does not alter this conclusion. In fact, Schuhmann used Frumkin's method [84], based on the analogy between osmotic pressure and $\Delta\sigma$, for the derivation. According to Schuhmann

$$\Delta\sigma = -nA \ln(1-x) \tag{III.79}$$

where x is the molar fraction of the adsorbate in the surface layer, expressed by the first of the relations (III.76). However, the applicability of Eq. (III.79) in the case of athermal solutions is not evident, because it is again assumed that the activity in the surface layer is equal to the corresponding molar fraction (see [37, 91]).

TABLE 3

n	0.5	1	2	3	4	∞
θ^* from Eq. (III.82)	0.585	0.500	0.415	0.366	0.333	0
θ^* from Eq. (III.83)	0.577	0.500	0.423	0.394	0.378	0.333

From the general expression (III.39) for the function h, the isotherms (III.75) and (III.78) respectively give

$$h = \left[\frac{1 + (n-1)\theta}{\theta(1-\theta)} - 2a \right]^{-1} \tag{III.80}$$

and

$$h = \left[\frac{1}{\theta(1-\theta)(1-\theta+\theta/n)} - 2a \right]^{-1}. \tag{III.81}$$

The maxima of these functions, and the minima on the plots of $(\partial \ln c / \partial \theta)_\varphi$ vs θ, correspond to

$$\theta^* = \frac{1}{1 + \sqrt{n}} \tag{III.82}$$

$$\theta^* = \frac{2n - 1 - \sqrt{n^2 - n + 1}}{3(n-1)}. \tag{III.83}$$

Values of θ^* calculated by the formulas (III.82) and (III.83) for various values of n are given in Table 3.

Comparison of the data in Tables 2 and 3 shows that n ≈ 1 for all the compounds investigated, with the exception of phenol. The probable explanation of this result is that a group of bonded H_2O molecules rather than a single water molecule occupies one adsorption site on the mercury surface. From the thermodynamic standpoint this may mean that the work of transfer of such a group of molecules as a whole from the surface into the volume is less than the total work of transfer of all the molecules individually. This view is supported by the sigmoid shape of the isotherms of water vapor adsorption on mercury, obtained in several investigations [92-94],* indicating considerable attraction interaction between the adsorbed H_2O molecules.

---

*The data obtained on adsorption of water vapor on mercury [95-97] contradict this result; however, it has been pointed out [93, 94] that this discrepancy is probably due to inadequate experimental cleanness in the investigations cited [96, 98], and apparently also in [97].

The concept of association of water molecules adsorbed on the mercury surface theoretically justifies the use of the Frumkin isotherm (III.57) as the semiempirical basis for examination of the adsorption of organic molecules at the mercury–electrolyte interface.

4. Properties of Differential Capacity Curves Determined in Presence of Organic Substances the Adsorption of which Conforms to the Frumkin Isotherm

a. Properties of Equilibrium C vs φ Curves. It has already been noted that the surface coverage at the maximum of the C vs φ curve, i.e., θ^{max}, in general depends on the position of the maximum, i.e., on φ^{max}, and is not a constant equal to θ^*. Let us examine certain properties of equilibrium differential capacity curves, which follow from Eq. (III.42) with the use of the Frumkin isotherm (III.57) with a constant value of the attraction constant, a = const. Since Eq. (III.42) was derived with the assumption that C_0 = const and C' = const, these conditions will also be imposed on the results obtained.

In the case of the Frumkin isotherm (III.57)

$$h = \frac{\theta\,(1-\theta)}{1-2a\theta\,(1-\theta)} \tag{III.84}$$

[see, e.g., Eq. (III.81) when n = 1] and

$$\frac{\partial h}{\partial\theta} = \frac{1-2\theta}{[1-2a\theta\,(1-\theta)]^2}. \tag{III.85}$$

It follows from Eqs. (III.42) and (III.85) that the relation between the peak potential φ^{max} and the surface coverage at $\varphi = \varphi^{max}$, i.e., θ^{max}, is given by the following expression:

$$\varphi^{max} = \varphi_m \pm \sqrt{\frac{3}{2\alpha}} \cdot \frac{1-2a\theta^{max}\,(1-\theta^{max})}{\sqrt{1-2\theta^{max}}}. \tag{III.86}$$

Figure 19 shows the results obtained by the use of this equation for calculation of the dependence of θ^{max} on $|\varphi^{max} - \varphi_m|$

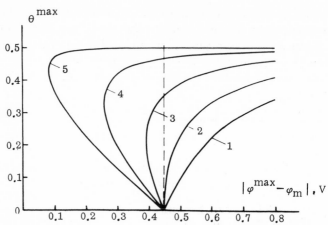

Fig. 19. Dependence of the surface coverage at the maximum of the equilibrium C vs φ curve (θ^{max}) on the peak potential for different values of the attraction constant a. 1) $a = 0$; 2) $a = 0.5$; 3) $a = 1.0$; 4) $a = 1.5$; 5) $a = 1.9$; the dash line represents the limiting value of $|\varphi^{max} - \varphi_m|$ when c $\to$ 0 (from data in [7], Chap. III).

when $\alpha = 7.5$ V^{-2} with different values of the attraction constant. It is seen that the deviations of θ^{max} from θ^* diminish with increasing distance between the peak potential and the potential of maximum adsorption, i.e., with increasing concentration of the organic substance. However, at $a < 1$, even when φ^{max} is relatively far from φ_m ($|\varphi^{max} - \varphi_m| \sim 0.7$-$0.8$ V), θ^{max} deviates considerably from the limiting value of $\theta^* = 0.5$. With decreasing concentration of the organic substance, when $\theta^{max} \to 0$, the position of the maxima on the differential curves at any value of a tends to $\varphi_m \pm \sqrt{3/2\alpha}$.

If we determine the derivative $[\partial (\varphi^{max} - \varphi_m)/\partial \theta^{max}]_{\theta^{max}=0}$ on the basis of Eq. (III.86) and equate it to zero, we find that the value of a satisfying this condition is 0.5. It follows that at $a \leq 0.5$ the dependence of θ^{max} on $|\varphi^{max} - \varphi_m|$ is monotonic, whereas at $a > 0.5$ certain values of $|\varphi^{max} - \varphi_m|$ may correspond to two values of θ^{max} (see Fig. 19). In other words, at $a \leq 0.5$ a definite position of the maximum on the C vs φ curve corresponds to a perfectly definite value of coverage of the surface by the organic substance. On the other hand, when $a > 0.5$ certain positions of the maximum on the C vs φ curve at which $|\varphi^{max} - \varphi_m| < \sqrt{3/2\alpha}$, may correspond to two values of the surface coverage, one at a higher and one at a lower concentration of the organic substance. In the latter case it may be expected that with decreasing concentration of the organic substance the peak potentials on the C vs φ curves will first come closer together and then diverge again, tending asymptotically to $\varphi_m \pm \sqrt{3/2\alpha}$.

To verify this conclusion, the dependence of $(\varphi^{max} - \varphi_m)$ on the logarithm of the concentration of the organic substance should be calculated. Equation (III.37) may be used for this purpose; in the case of the Frumkin isotherm (III.57) it follows from this equation that

$$K + \log c = \frac{\alpha}{2.3} (\varphi^{max} - \varphi_m)^2 + \log \frac{\theta^{max}}{1 - \theta^{max}} + \frac{a}{2.3} (1 - 2\theta^{max}), \quad \text{(III.87)}$$

where

$$K = \log [B_m / f(\theta^*)] = \log B_m + \frac{a}{2.3}.$$

It is easy to show that at $\theta^{max} \approx \theta^* = 0.5$, Eq. (III.87) passes into the previously derived Eq. (III.45). Taking definite values of $(\varphi^{max} - \varphi_m)$ from the graph in Fig. 19, we can find the corresponding values of θ^{max} and then calculate the K + log c values with the aid of Eq. (III.87). The calculated dependence of K + log c on $(\varphi^{max} - \varphi_m)$ at $a = 0, 1,$ and 1.9, together with the limiting parabolic relationship corresponding to $\theta^{max} \approx \theta^* = 0.5$ are shown in Fig. 20a.

It follows from Fig. 20 that at high concentrations of the organic substance the dependence of log c on φ^{max} approaches the limiting parabolic relationship determined by Eq. (III.45), whereas at $c \to 0$, when θ^{max} also tends to zero, $\varphi^{max} \to \varphi_m \pm \sqrt{3/2\alpha}$. These limiting relationships are indicated in Fig. 20a by dash lines.

However, it is evident from Fig. 20a that the transition from one limiting case to the other depends to a considerable extent on the magnitude of the attraction constant. Thus, as already noted, at $a \leq 0.5$ the dependence of θ^{max} on $|\varphi^{max} - \varphi_m|$ is monotonic and with decreasing concentration of the organic substance the peaks on the C vs φ curves gradually come closer together, tending to the limiting value of $\varphi_m \pm \sqrt{3/2\alpha}$. In partic-

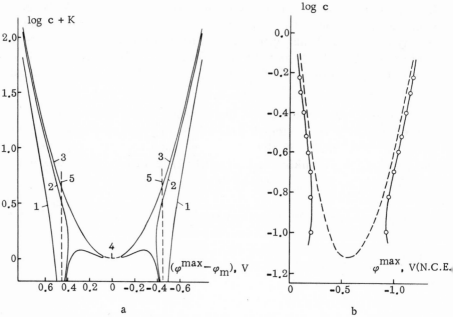

Fig. 20. Dependence of the peak potentials of the equilibrium C vs φ curves on the logarithm of the concentration of the organic substance. a) Calculated from Eqs. (III.86) and (III.87) with α = 7.5 V^{-2}; 1) a = 0; 2) a = 1.0; 3) a = 1.9; 4) limiting relationship, corresponding to $\theta^{max} \approx \theta * = 0.5$; 5) limiting relationship for c → 0 (from data in [7], Chap. III). b) Experimental data for the system 1 N Na$_2$SO$_4$ = n-C$_3$H$_7$OH; the dash line represents the limiting relationship corresponding to $\theta^{max} \approx$ 0.5 (from data in [22], Chap. III).

ular, when a = 0, this transition may be expressed with the aid of one equation, derived from Eqs. (III.86) and (III.87):

$$K + \log c = \frac{\alpha}{2.3} (\varphi^{max} - \varphi_m)^2 + \log \left[\frac{(\varphi^{max} - \varphi_m)^2 - 3/2\alpha}{(\varphi^{max} - \varphi_m)^2 + 3/2\alpha} \right]. \quad \text{(III.88)}$$

Calculation with the aid of this equation shows (see curve 1 in Fig. 20a) that a virtually linear relationship, first noted by Breyer and Hacobian [67] and by Doss [98], exists between log c and φ^{max} over a fairly wide range of concentrations of the organic substance. However, it follows from Eq. (III.88) that even for this special case the conclusion that φ^{max} is a linear function of log c [67, 68] over a wider range of concentrations is erroneous, to say nothing of systems with $a > 0$.

When $a > 0.5$, as already noted in relation to the dependence of θ^{max} on $|\varphi^{max} - \varphi_m|$, the peaks on the C vs φ curves first come closer together with decreasing concentration of the organic substance and then diverge again (curve 2 in Fig. 20a), tending to $\varphi_m \pm \sqrt{3/2a}$. The dependence of φ^{max} on log c found for the system $1\,N$ Na_2SO_4 + n-C_3H_7OH [22], shown in Fig. 20b, is a characteristic example of this case.

It is seen in Fig. 20a that when the attraction constant is close to 2 ($a = 1.9$) in a certain rather narrow concentration range three extremal points (two maxima separated by a minimum) correspond to one value of log c on each branch of the C vs φ curve. The cathodic branch of the C vs φ curve calculated for $a = 1.9$ and (log c + K) = 0.05, illustrating this, is shown in Fig. 21a. This rather rare effect was observed by Damaskin et al. [99] on the

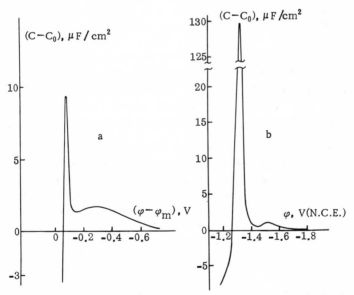

Fig. 21. Cathodic branches of differential capacity curves. a) Calculated for the conditions: $a = 1.9$; $\alpha = 7.5\ V^{-2}$; $A = 1\ \mu J/cm^2$; $B_m c = 0.168$; b) experimental data for $0.5\,N$ CsCl + $10^{-3}\ N$ $C_{12}H_{25}OSO_3Na$ solution (from data in [19], Chap. III).

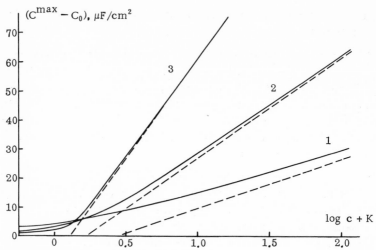

Fig. 22. Dependence of the peak height on the equilibrium C vs φ curves on the logarithm of the concentration of the organic substance. 1) $a = 0$; 2) $a = 1.0$; 3) $a = 1.5$; the dash lines are calculated with the condition $\theta^{max} \approx \theta* = 0.5$ for the corresponding values of a (from data in [7], Chap. III).

cathodic branch of the C vs φ curve for a solution of 0.5 N CsCl + $10^{-3}\,N$ $C_{12}H_{25}OSO_3Na$, shown in Fig. 21b. Similar results were later obtained by Tedoradze and Zolotovitskii [23] for $6 \cdot 10^{-3} M$ solution of 2,6-lutidine in 0.5 N KCl, and by Damaskin et al. [34] for 0.04 M solutions of n-butyl alcohol in 2 N and 4 N $MgSO_4$.

Taking a definite equation for the adsorption isotherm, we can calculate the dependence of the peak height on the C vs φ curve on log c for any value of the attraction constant and any concentration of the organic substance. Using the Frumkin isotherm (III.57), we obtain from Eq. (III.40) written for $\varphi = \varphi^{max}$:

$$C^{max} = C_0 - 2A\alpha\theta^{max} + 4A\alpha^2 \left(\varphi^{max} - \varphi_m\right)^2 \frac{\theta^{max}\left(1 - \theta^{max}\right)}{1 - 2a\theta^{max}\left(1 - \theta^{max}\right)}. \quad \text{(III.89)}$$

Taking definite values of θ^{max}, we can use Eq. (III.86) to calculate the corresponding values of $(\varphi^{max} - \varphi_m)$, and then use Eqs. (III.87) and (III.89) to find mutually corresponding values of

$(C^{max} - C_0)$ and $(\log c + K)$ at these values of θ^{max} and $(\varphi^{max} - \varphi_m)$. The relationship between $(C^{max} - C_0)$ and $(\log c + K)$ calculated in this way for three values of the attraction constants is compared in Fig. 22 with the limiting linear relationship corresponding to Eq. (III.47). It is not difficult to show that if the Frumkin isotherm (III.57) is used the constants K_1 and K_2 in this equation are

$$K_1 = \frac{2.3 \, (C_0 - C')}{2 - a}; \quad K_2 = \frac{2.3 \, (C_0 - C')}{2 - a} \left(\log B_m + \frac{a}{2.3}\right) + \frac{C_0 + C'}{2}.$$

Figure 22 shows that at $a \geq 1$ the exact dependence of the peak height on the C vs φ curve on $\log c$ is nearly linear over a considerable range of concentrations of the organic substance. This explains the experimentally observed linear relationship between C^{max} and $\log c$ (see Fig. 16).

It is of interest to find the relation between the degree of coverage by the organic substance at the potential of maximum adsorption, designated θ_m, and θ^{max}. This relation may be obtained as follows. From Eq. (III.37), written for $\varphi = \varphi^{max}$ and $\varphi = \varphi_m$, it follows that

$$f(\theta_m) = f(\theta^{max}) \exp [\alpha (\varphi^{max} - \varphi_m)^2]. \qquad \text{(III.90)}$$

Inserting the expression for $f(\theta)$ corresponding to the Frumkin isotherm (III.57) and $(\varphi^{max} - \varphi_m)$ from Eq. (III.86), we finally obtain [23]

$$\frac{\theta_m}{1 - \theta_m} \exp (- 2a\theta_m) =$$

$$= \frac{\theta^{max}}{1 - \theta^{max}} \exp \left\{ \frac{3 [1 - 2a\theta^{max} (1 - \theta^{max})]^2}{2(1 - 2\theta^{max})} - 2a\theta^{max} \right\}. \qquad \text{(III.91)}$$

Curves representing the dependence of θ^{max} on θ_m for different values of the attraction constant, calculated from Eq. (III.91), are given in Fig. 23. In agreement with our earlier conclusions, θ^{max} tends to its limiting value $\theta^* = 0.5$ at θ_m values which decrease with increase of the attraction constant a. At the same time, at low concentrations of the organic substance, corresponding to $\theta_m < 0.5$, $\theta^{max} \ll \theta^*$ at any value of a, and therefore under these conditions the square-law relationship between φ^{max} and $\log c$ and the linear relationship between C^{max} and $\log c$ break down. It also follows from Fig. 23 that at $1.7 \leq a \leq 2$ three

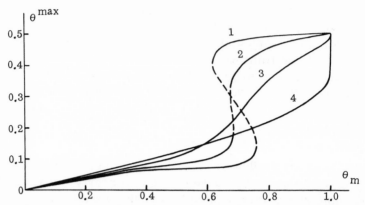

Fig. 23. Dependence of the extremal values of θ^{max} on the surface coverage at the potential of maximum adsorption (θ_m), calculated from Eq.
(III.91) for different values of the attraction constant. 1) $a = 1.9$; 2) $a = 1.7$; 3) $a = 1.3$; 4) $a = 0$; the dash line represents the portion of the curves
corresponding to the minimum on the differential capacity curves (from
data in [19], Chap. III).

extremal values of θ^{max} correspond to the same value of θ_m;
two correspond to maxima and one to a minimum on the C vs φ
curves. The characteristic form of the differential capacity curves
under these conditions has already been discussed (see Fig. 21).

  b. Properties of Nonequilibrium C vs φ
Curves. Since experimentally determined C vs φ curves are
always nonequilibrium to a certain extent, it is of interest to study
the relationships applicable to nonequilibrium differential capacity
curves in presence of organic substances. Some of these relationships have been examined in the first approximation by Tedoradze and Zolotovitskii [23]. A more exact solution of this
problem was given later by Damaskin [24].

  Suppose that the dependence of the capacity on the ac frequency is determined entirely by diffusion of the organic substance to the electrode surface, which is consistent with experimental data for not very high values of ω [44, 100]. Under these
conditions, in accordance with [100], [see Eq. (II.5)]:

$$C_{add(\omega)} = C_{add(\omega=0)} \frac{N+2}{(N+1)^2+1},$$ (III.92)

where

$$N = \sqrt{\frac{2\omega}{D}} \left(\frac{\partial \Gamma}{\partial c}\right)_{\varphi} = \frac{\Gamma_m}{c} \sqrt{\frac{2\omega}{D}} \, h \qquad (III.93)$$

and D is the diffusion coefficient of the organic substance.

It follows from Eqs. (III.40) and (III.92) that the total capacity at a given frequency ω is

$$C_{(\omega)} = C_0 - 2A\alpha\theta + 4A\alpha^2 (\varphi - \varphi_m)^2 h \frac{N+2}{(N+1)^2+1}. \qquad (III.94)$$

It is easy to see that the deviation of the capacity $C_{(\omega)}$ from the limiting value $C_{(\omega=0)}$ is determined by the factor N. In limiting cases, with N = 0, Eq. (III.94) becomes Eq. (III.40), and with $N \gg 2$, when $\frac{N+2}{(N+1)^2+1} \approx \frac{1}{N}$, from Eqs. (III.93) and (III.94) we have

$$C_{(\omega)} \approx C_0 - 2A\alpha\theta + \frac{4A\alpha^2 (\varphi - \varphi_m)^2 c}{\Gamma_m} \sqrt{\frac{D}{2\omega}}. \qquad (III.95)$$

It follows from Eq. (III.95) that when $N \gg 2$ the adsorption maxima are completely absent from the C vs φ curves. Indeed, differentiating Eq. (III.95) with respect to potential, we obtain

$$\frac{\partial C_{(\omega)}}{\partial \varphi} = 4A\alpha^2 (\varphi - \varphi_m) \left(h + \frac{2c}{\Gamma_m} \sqrt{\frac{D}{2\omega}}\right), \qquad (III.96)$$

which is equal to zero only at $\varphi = \varphi_m$, when the condition $\partial C/\partial \varphi = 0$ corresponds to the minimum of the C vs φ curve.

We will now find the general relationship between the peak potentials on the nonequilibrium C vs φ curves and θ^{max}, assuming that adsorption of the organic substance conforms to the Frumkin isotherm (III.57). Taking into account Eq. (III.38) and the fact that in this case

$$\frac{dN}{d\varphi} = -2\alpha (\varphi - \varphi_m) N \frac{1 - 2\theta}{[1 - 2a\theta (1 - \theta)]^2}, \qquad (III.97)$$

and differentiating Eq. (III.94) with respect to potential, we obtain

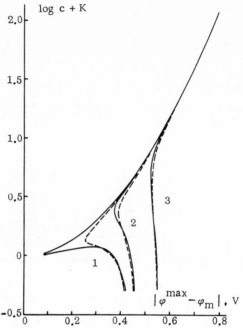

Fig. 24. Dependence of the peak potentials on
the C vs φ curves on the logarithm of the concentration of the organic substance, calculated
for $a = 1.9$ (continuous lines) and $a = 1.5$ (dash
line) with different values of the nonequilibrium
parameter k: 1) k = 0 (equilibrium curves); 2)
k = 1; 3) k = 5 (from data in [24], Chap. III).

$$\frac{\partial C_{(\omega)}}{\partial \varphi} = 4 A \alpha^2 (\varphi - \varphi_m) h \left\{ 1 + \frac{2(N+2)}{(N+1)^2 + 1} - \right.$$
$$\left. - 2\alpha (\varphi - \varphi_m)^2 \frac{1 - 2\theta}{[1 - 2a\theta(1-\theta)]^2} \frac{4(N+1)}{[(N+1)^2 + 1]^2} \right\}. \qquad \text{(III.98)}$$

It follows from Eq. (III.98) that the position of the minimum on the
C vs φ curves ($\varphi = \varphi_m$) does not alter with variation of ω. At the
same time, the expression for the peak potentials becomes

$$\varphi^{\max} - \varphi_m = \pm \sqrt{\frac{3}{2\alpha}} \cdot \frac{1 + 2a\theta^{\max}(1 - \theta^{\max})}{\sqrt{1 - 2\theta^{\max}}} \times$$
$$\times \sqrt{\frac{[(N^{\max} + 2)^2 + 2][(N^{\max} + 1)^2 + 1]}{12(N^{\max} + 1)}}. \qquad \text{(III.99)}$$

where N^{max} is the value of N at $\varphi = \varphi^{max}$, which is in its turn a certain function of φ^{max} and θ^{max}. In fact, putting the expression for c from the isotherm (III.37) and the expression for h from Eq. (III.84) into Eq. (III.93), we obtain, with the condition $\theta = \theta^{max}$ and $\varphi = \varphi^{max}$

$$N^{max} = k \; \frac{\exp\left[-\alpha\,(\varphi^{max} - \varphi_m)^2\right](1 - \theta^{max})^2 \exp(2a\theta^{max})}{1 - 2a\theta^{max}(1 - \theta^{max})} \; . \quad \text{(III.100)}$$

where

$$k = \Gamma_m B_m \sqrt{2\omega/D} \; .$$

Consequently, to find the dependence of θ^{max} on φ^{max} in the case of nonequilibrium differential capacity curves it is necessary to solve Eqs. (III.99) and (III.100) simultaneously. This problem was solved graphically by Damaskin [24], who calculated θ^{max} vs φ^{max} curves at $\alpha = 7.5$ V^{-2} for the following values of a and k: $a = 1.5$ and 1.9; k = 1 and 5. Assuming that for aliphatic compounds in the first approximation $\Gamma_m \approx 5 \cdot 10^{-10}$ M/cm^2 [14] and D $\approx 9 \cdot 10^{-6}$

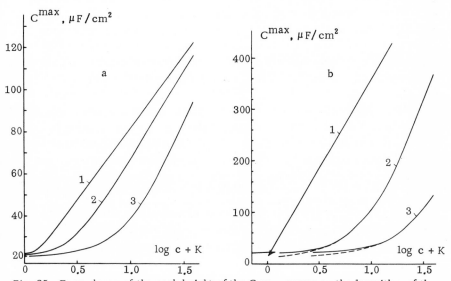

Fig. 25. Dependence of the peak height of the C vs φ curves on the logarithm of the concentration of the organic substance, calculated for $a = 1.5$ (a) and $a = 1.9$ (b) with different values of the nonequilibrium parameter k: 1) k = 0 (equilibrium curves); 2) k = 1; 3) k = 5 (from data in [24], Chap. III).

cm/sec [101], at an ac frequency of 400 Hz ($\omega = 2\pi \cdot 400$) the chosen values of k correspond to $B_m \approx 84.5$ liters/mole and $B_m \approx 421$ liters/mole; i.e., they are attainable when the length of the carbon chain in the organic molecule is sufficiently large [34]. The dependence of θ^{max} on φ^{max} found in this way was then used for calculating theoretical φ^{max} vs log c and C^{max} vs log c curves [24]. The results are presented in Figs. 24 and 25. Examination of Fig. 24 reveals the following: the nonequilibrium and equilibrium curves diverge appreciably at low concentrations of the organic substance; this divergence increases with increase of a and k; divergence occurs from the equilibrium curves, in the direction of apparently lower values of the effective attraction constant (cf. Fig. 20); curves for equal k and different values of the attraction constant are close together.

It is also clear from Fig. 24 that the range of potentials in which adsorption maxima can be observed on the nonequilibrium differential capacity curves narrows appreciably with increase of k. Finally, it follows from Fig. 24 that even when k = 1 the nonequilibrium differential capacity curve corresponding to $a = 1.9$ does not have the two adsorption maxima on each branch of the C vs φ curve, characteristic of equilibrium C vs φ curves (see Fig. 21); only one maximum remains, corresponding to higher values of $|\varphi^{max} - \varphi_m|$. A similar conclusion was reached by Tedoradze and Zolotovitskii [23] on the basis of direct calculation of the cathodic branch of the C vs φ curve with the aid of Eq. (III.94). Comparison of Figs. 24 and 25 shows that the nonequilibrium nature of the C vs φ curves has a much stronger influence on the height than on the position of the adsorption–desorption peaks. Thus, the nonequilibrium peak becomes noticeably the lower of the two even when their potentials are still almost the same.

It should also be noted that the approximation used by Tedoradze and Zolotovitskii [23] is more applicable in the case of $C_{(\omega)}^{max}$ vs log c curves. This is evident from Fig. 25b, where the dash lines represent $C_{(\omega)}^{max}$ vs log c curves calculated with the aid of this approximation.

It follows from numerous experimental data on adsorption of organic substances that the energy of adsorption of organic molecules, determined by B_m, and therefore the parameter k, increase with increasing length of the hydrocarbon chain. On the other hand, the attraction constant a may be expected to increase in the same direction. Therefore in the light of our results it is to be expected that at a given ac frequency the deviations from the properties of equilibrium C vs φ curves in a particular homologous series will increase with increasing molecular weight of the organic compound.

c. Properties of C vs φ Curves under Conditions of Two-Dimensional Condensation of the Organic Substance. It follows from [1, 102] that two-dimensional condensation of an organic substance in the adsorption layer can be represented, with some degree of approximation, by the equation of state (III.72), corresponding to the Frumkin isotherm (III.57), under the condition that the attraction constant $a > a_{cr} = 2$. In order to examine the properties of C vs φ curves under these conditions, it is convenient to write Eq. (III.57) in dimensionless coordinates:

$$\ln y = \ln \frac{\theta}{1 - \theta} + a(1 - 2\theta), \qquad (III.101)$$

where $y = c/c_{\theta = 0.5}$.

It is easy to see that the dependence of $\ln y$ on θ is symmetrical about a point with the coordinates ($\theta = 0.5$; $\ln y = 0$) (see [33]). If $a > 2$, three values of θ correspond to the condition $\ln y = 0$: 1) the value θ_1, corresponding to very low surface coverages; 2) $\theta = 0.5$; 3) the value $\theta_2 = 1 - \theta_1$, close to unity (see Fig. 17, curve 6). The region between θ_1 and θ_2 corresponds to unstable states of the adsorption layer, and under equilibrium conditions the surface coverage increases abruptly from θ_1 to θ_2 with increasing concentration of the organic substance. Under nonequilibrium conditions unstable states, for which $d\theta/dc > 0$ (see Fig. 17), are attainable in practice, whereas states for which $d\theta/dc < 0$ (in particular, when $\theta = 0.5$) are not.

The values of θ_1 and θ_2 can be found from the relation

$$a = \frac{\ln (1 - \theta) - \ln \theta}{1 - 2\theta}, \qquad (III.102)$$

which is obtained from Eq. (III.101) when $\ln y = 0$. Equation (III.102) is easily solved graphically [33].

An abrupt change of θ under the condition $a > 2$ occurs not only with increasing concentration of the organic substance when $\varphi = $ const but also with decrease of $|\varphi - \varphi_m|$ when c = const because, in accordance with Eqs. (III.30), (III.57), and (III.101),

$$\left(\frac{\partial\theta}{\partial\varphi}\right)_c = -2\alpha(\varphi - \varphi_n) \cdot \left(\frac{\partial\theta}{\partial \ln y}\right)_\varphi .$$ (III.103)

Under equilibrium condition this jump in the degree of surface coverage leads to a sudden change in the capacity of the double layer:

$$\Delta C = C(\theta_1) - C(\theta_2) = (\theta_2 - \theta_1)(C_o - C') .$$ (III.104)

The expression (III.104) is obtained from Eqs. (III.36) and (III.103) with the symmetry of the ln y vs θ curve about the point having the coordinates ($\theta = 0.5$; ln y = 0) taken into account, from which it follows that $\left(\frac{\partial\theta}{\partial \ln y}\right)_{\varphi_1\theta_1} = \left(\frac{\partial\theta}{\partial \ln y}\right)_{\varphi_1\theta_2}$. Thus, the degree of coverage diminishes with increasing distance from the potential of maximum adsorption and, in accordance with Eqs. (III.40) and (III.84), the capacity of the double layer begins to increase. However, at a certain potential which corresponds to the sudden decrease of coverage from θ_2 to θ_1, the capacity also increases abruptly by ΔC, and this is followed by a gradual decrease of capacity to C_0. The shape of the C vs φ curves obtained under these conditions depends to a considerable extent on the magnitude of the attraction constant. Several differential capacity curves calculated for conditions of two-dimensional condensation of the organic substance ($a > 2$) are given in Fig. 26. The following values [33] were taken in the calculation: $C_0 = 20~\mu F/cm^2$, C' = 5 $\mu F/cm^2$; A = 1 $\mu J/cm^2$; $a = 2.3, 2.5, 3.0$, and 3.5; the B_{mc} values were chosen so that in every case the sharp change of coverage and capacity corresponded to $|\varphi - \varphi_m|$ = 0.7 V.

It follows from Fig. 26 that the peak height of the C vs φ curves under conditions of two-dimensional condensation d e - c r e a s e s instead of increasing with increase of the attraction constant, so that at $a \geq 3.5$ the peak virtually vanishes. Differential capacity curves of this type were observed experimentally by Lorenz [45] in presence of nonylic acid, by Frumkin and Damaskin [103] for 1 N KI solutions with additions of various amounts of $[(C_4H_9)_4N]I$, and by various investigators [104-107] for camphor solutions. Qualitative explanations of the unusual form of the C vs φ curves were offered in all these publications; the explanation put forward by Sathyanarayana [107] is difficult to accept.

If the C vs φ curves are determined with a stationary electrode under conditions of two-dimensional condensation of the organic substance, when the potential is varied rapidly it is possible to achieve a part of the unstable states ($\theta_1 < \theta < \theta_2$) where the coverage decreases with increase of $|\varphi - \varphi_m|$ (see [33]). Under these conditions a hysteresis loop, as indicated by a dash line in Fig. 26d, should be observed when the C vs φ curve is recorded in different directions from positive to negative φ and vice versa). Such a result was observed by Lorenz [45] experimentally for nonylic acid solutions.

Figure 26 shows that the abrupt changes in the C vs φ curves at $a > 2$ become distinct only when $a \geq 3$, when the peaks become

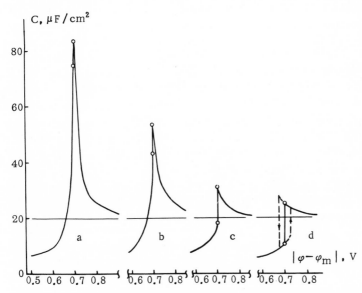

Fig. 26. Theoretically calculated dependence of differential capacity on the electrode potential with two-dimensional condensation of the organic substance: a) $a = 2.3$; b) $a = 2.5$; c) $a = 3.0$; d) $a = 3.5$ (from data in [33], Chap. III).

so low that ΔC is comparable to the peak height. Under these conditions it is possible to make an approximate estimate of a for a given system with the aid of Eqs. (III.104) and (III.102). On the other hand, if $2 \leq a \leq 2.5$, the peak height of the C vs φ curve is noticeably greater than the capacity change ΔC at the peak potential, and the experimental capacity curves corresponding to the condition $a > 2$ can be mistaken for the usual C vs φ curves when $a < 2$. This may lead to certain erroneous conclusions; e.g., with regard to the kinetics of adsorption of the organic molecules. In this connection the possibility is not excluded that the anomalous dependence of cot δ on $1/\sqrt{\omega}$, found by Lorenz [108] for n-octyl alcohol, does not in reality represent two-dimensional association (see Chapter II, Section 3), as the value of the attraction constant found by extrapolation from the data of Damaskin et al. [34] for this alcohol is $a \approx 2.1$.

5. Interpretation of the Complete Differential Capacity Curves Determined in Presence of Aliphatic Compounds

In our discussion of the general properties of differential capacity curves determined in presence of organic substances it has up to now been assumed that the capacities at $\theta = 0$ and $\theta = 1$ are independent of the electrode potential, i.e., that $C_0 = $ const and $C' = $ const. In reality, determinations of C vs φ curves in pure solutions of supporting electrolytes show that the condition $C_0 = $ const is not satisfied. Apparently, the condition $C' = $ const is not strictly satisfied either; however, since the capacity of the covered part of the surface is in most cases considerably less than C_0, we can put $C' = $ const in the next approximation and consider the dependence of C_0 only on the potential. In this case Eq. (III.15) remains valid for the charge ε' of the completely covered surface, and the surface charge ε_0 when $\theta = 0$ may be represented as the integral

$$\varepsilon_0 = \int_0^{\varphi} C_0 \, d\varphi, \tag{III.105}$$

which is easy to calculate graphically from experimental data.

Thus, in accordance with Eq. (III.13), the surface charge in presence of an organic substance is

$$\varepsilon = \varepsilon_0 (1 - \theta) + C' (\varphi - \varphi_N) \theta. \qquad \text{(III.106)}$$

Taking Eq. (III.12) into account, we obtain

$$\frac{d \ln B}{d\varphi} = - \frac{\varepsilon_0 - C' (\varphi - \varphi_N)}{A} , \qquad \text{(III.107)}$$

which, after integration and conversion to exponential form, gives

$$B = B_0 \exp \left[- \frac{E + C'\varphi (\varphi_N - \varphi/2)}{A} \right] , \qquad \text{(III.108)}$$

where B_0 is the value of B when $\varphi = 0$ and

$$E = \int_0^\varphi \varepsilon_0 \, d\varphi = \int_0^\varphi \int_0^\varphi C_0 \, d\varphi^2, \qquad \text{(III.109)}$$

i.e., the lowering of interfacial tension due to adsorption only of the ions of the supporting electrolyte.

Equation (III.108), in conjunction with the isotherm (III.57) for given constants A, B_0, C', φ_N, and a, and an experimental C_0 vs φ curve, makes it possible to calculate the dependence of surface coverage on the potential. Equation (III.108) is first used to calculate B at a given φ, and then θ is found from a plot of θ vs Bc based on Eq. (III.57) for a given a. Another value of φ is then taken and the corresponding θ is found, etc.

If the dependence of θ on φ is known Eq. (III.72) can be used for calculation of the lowering of interfacial tension due to adsorption of organic molecules, and thus to obtain an electrocapillary curve in presence of an organic substance. Calculations of this type were carried out by Frumkin [2, 3] for various concentrations of tertiary amyl alcohol in $1N$ NaCl, and showed satisfactory agreement between theory and experiment.

If the electrode charge in presence of an organic substance is an homogeneous function of the potential and surface coverage, i.e., $\varepsilon = \varepsilon (\varphi, \theta)$, we have the following expression for the differential capacity:

$$C = \left(\frac{\partial \varepsilon}{\partial \varphi}\right)_\theta + \left(\frac{\partial \varepsilon}{\partial \theta}\right)_\varphi \cdot \left(\frac{\partial \theta}{\partial \varphi}\right)_C \qquad \text{(III.110)}$$

or, if the condition (III.7) is satisfied,

$$C = \left(\frac{\partial \varepsilon}{\partial \varphi}\right)_\theta + \left(\frac{\partial \varepsilon}{\partial \theta}\right)_\varphi \frac{d \ln B}{d \varphi} h. \qquad \text{(III.110a)}$$

The derivatives $\left(\frac{\partial \varepsilon}{\partial \varphi}\right)_\theta$ and $\left(\frac{\partial \varepsilon}{\partial \theta}\right)_\varphi$ are found from Eq. (III.106), and the derivative $d \ln B / d\varphi$ from Eq. (III.108).

Putting the values of these derivatives and the expression for h [see Eq. (III.84)] into Eq. (III.110), we obtain

$$C = C_0 (1 - \theta) + C'\theta + \frac{[\varepsilon_0 + C'(\varphi_N - \varphi)]^2}{A} \frac{\theta(1-\theta)}{1 - 2a\theta(1-\theta)} \cdot \qquad \text{(III.111)}$$

A similar equation was obtained by Hansen et al. [42] and used by them for calculating the apparent degree of surface coverage by the organic substance:

$$\theta_{app} = (C_0 - C)/(C_0 - C').$$

With given constants A, C', φ_N, and a, the dependence of the differential capacity in presence of an organic substance on the electrode potential can be calculated with the aid of Eq. (III.111) from an experimental C_0 vs φ curve and a previously calculated θ vs φ curve. The results of such a calculation, for constants corresponding to adsorption of tert-amyl alcohol on mercury [3, 5] are shown in Fig. 27a. For comparison, Fig. 27b shows experimental C vs φ curves for $0.9\ N$ NaF with similar additions of tert-amyl alcohol.

It is seen that the theoretical and experimental curves agree satisfactorily in shape and position. However, the experimental anodic capacity peaks are noticeable higher and the cathodic peaks somewhat lower than the calculated peaks. A similar deviation was observed by Hansen et al. [42] in the case of adsorption of n-valeric acid on mercury in presence of $HClO_4$. Finally, a detailed comparison of experimental and theoretically calculated electrocapillary curves in presence of tert-$C_5H_{11}OH$ [2, 3] also indicates the existence of some discrepancies; however, these are not as great as in the case of the C vs φ curves (see Fig. 27).

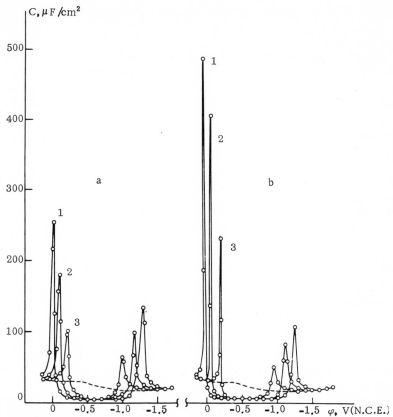

Fig. 27. Differential capacity curves for 0.9 N NaF solution (dash line) and in presence of various amounts of tert-amyl alcohol. 1) 0.3 M; 2) 0.1 M; 3) 0.03 M. a) Calculated from Eq. (III.111) with a = 1.6; b) experimental data determined at 400 Hz (from data in [5], Chap. III).

All these discrepancies between calculated and experimental results are due to small deviations of real systems from our model of two parallel condensers. As this model corresponds to isotherms which are strictly congruent with respect to the electrode potential, e.g., the isotherm (III.57) for a = const, certain deviations from the model of two parallel condensers can be formally expressed with the aid of the dependence of the attraction constant on the electrode potential, $a = a(\varphi)$. This necessitates revision of the derivation of the fundamental equations for calculating differential capacity in presence of an organic substance, found for the condition $a = $ const [29].

Suppose that the adsorption of the organic substance on the electrode surface conforms to the Frumkin isotherm (III.57), where a is now an arbitrary function of the electrode potential. Converting (III.57) to logarithmic form and differentiating with respect to potential at c = const we obtain after algebraic transformations

$$\frac{\partial \theta}{\partial \varphi c} = \left(\frac{d \ln B}{d\varphi} + 2\theta \frac{da}{d\varphi}\right) h, \qquad (III.112)$$

where h, as before, is given by Eq. (III.84). On the other hand, differentiation with respect to φ at $\theta = $ const gives

$$\left(\frac{\partial \ln c}{\partial \varphi}\right)_{\theta} = -\left(\frac{d \ln B}{\partial \varphi} + 2\theta \frac{da}{\partial \varphi}\right). \qquad (III.113)$$

Taking the thermodynamic relation (III.3) into account, we obtain

$$\frac{d \ln B}{\partial \varphi} + 2\theta \frac{da}{d\varphi} = \frac{1}{A} \left(\frac{\partial \varepsilon}{\partial \theta}\right)_{\varphi}. \qquad (III.114)$$

Therefore for the additional capacity in Eq. (III.110) we can write

$$C_{add} = \frac{1}{A} \left(\frac{\partial \varepsilon}{\partial \theta}\right)_{\varphi}^2 h = A \left(\frac{\partial \ln c}{\partial \varphi}\right)_{\theta}^2 h. \qquad (III.115)$$

Integration of Eq. (III.114) gives

$$\varepsilon = const + A \frac{d \ln B}{\partial \varphi} \theta + A\theta^2 \frac{da}{d\varphi}, \qquad (III.116)$$

where const is the integration constant, equal to ε_0, because when $\theta = 0$, $\varepsilon = \varepsilon_0$. Since when $\theta = 1$, $\varepsilon = \varepsilon' = C'(\varphi - \varphi_N)$, therefore

$$\varepsilon_0 + A \frac{d \ln B}{d\varphi} + A \frac{da}{d\varphi} = C'(\varphi - \varphi_N), \qquad (III.117)$$

hence

$$\frac{d \ln B}{d\varphi} = -\frac{\varepsilon_0 + C'(\varphi_N - \varphi)}{A} - \frac{da}{d\varphi}. \qquad (III.118)$$

Integration of this equation between 0 and φ with subsequent conversion to exponential form gives

$$B = B_0 \exp\left[-\frac{E + C'\varphi\,(\varphi_N - \varphi/2)}{A}\right] \exp(a_0 - a), \qquad \text{(III.119)}$$

where a_0 is the value of a when $\varphi = 0$, instead of Eq. (III.108), valid when $a = \text{const}$. On the other hand, by putting Eq. (III.118) into (III.116) we obtain the following equation for the surface charge:

$$\varepsilon = \varepsilon_0(1 - \theta) + C'\theta(\varphi - \varphi_N) - A\frac{da}{d\varphi}\,\theta(1 - \theta). \qquad \text{(III.120)}$$

It is evident that when $a = \text{const}$ Eq. (III.120) becomes Eq. (III.106). In the general case, the third term in Eq. (III.120) represents deviation of real systems from the model of two parallel condensers. From Eq. (III.120) we have

$$\left(\frac{\partial\varepsilon}{\partial\varphi}\right)_\theta = C_0(1 - \theta) + C'\theta - A\frac{d^2a}{d\varphi^2}\,\theta(1 - \theta) \qquad \text{(III.121)}$$

and

$$\left(\frac{\partial\varepsilon}{\partial\theta}\right)_\varphi = -\varepsilon_0 + C'(\varphi - \varphi_N) - A\frac{da}{d\varphi}\,(1 - 2\theta), \qquad \text{(III.122)}$$

hence, in accordance with Eqs. (III.110) and (III.115), the total differential capacity in presence of an organic substance is

$$C = C_0(1 - \theta) + C'\theta - A\frac{d^2a}{d\varphi^2}\,\theta(1 - \theta) +$$

$$+ \frac{\left[\varepsilon_0 + C'(\varphi_N - \varphi) + A\dfrac{da}{d\varphi}\,(1 - 2\theta)\right]^2}{A} \cdot \frac{\theta(1 - \theta)}{1 - 2a\theta(1 - \theta)}. \qquad \text{(III.123)}$$

When $a = \text{const}$, Eq. (III.123) becomes Eq. (III.111).

Equations (III.119), (III.120), and (III.123), in conjunction with the isotherm (III.57), can be used for calculating the dependence of surface coverage, electrode charge, and differential capacity on the potential for any given $a = a\,(\varphi)$ function if the constants A, C', B_0, and φ_N are known. For calculations of electrocapillary curves in presence of an organic substance, after the θ vs φ curve has been found with the dependence of a on φ taken into account, it is possible as before to use Eq. (III.72), which remains valid under the condition $a = a(\varphi)$.

Thus, the first requirement is to calculate the dependence of surface coverage on the electrode potential. For this purpose it is convenient, first, to calculate the dependence of log B on φ with the aid of Eq. (III.119), and then to determine the dependence of θ on φ for each of the chosen concentrations of the organic substance from log (Bc) vs θ curves plotted for various values of a on the basis of Eq. (III.57) (Fig. 28). When the dependence of θ

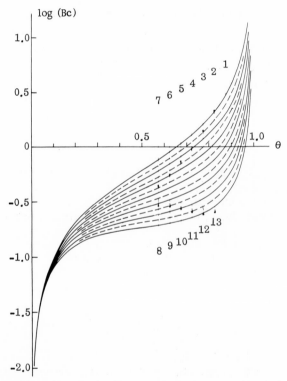

Fig. 28. Adsorption isotherms, calculated in the form of log (Bc) vs θ curves from Eq. (III.57) for various values of the attraction constant a: 1) 0.5; 2) 0.6; 3) 0.7; 4) 0.8; 5) 0.9; 6) 1.0; 7) 1.1; 8) 1.2; 9) 1.3; 10) 1.4; 11) 1.5; 12) 1.6; 13) 1.7.

on φ has been found, the use of Eqs. (III.72) and (III.123) for calculation of the complete σ vs φ and C vs φ curves presents no
difficulties.

For many aliphatic compounds the deviation of the behavior
of the system from the model of two parallel condensers is such
that the attraction constant is a linear function of the potential
(see the next section):

$$a = a_0 + \beta\varphi \qquad \text{(III.124)}$$

(where a_0 and β are constants). Under these conditions the calculations are simplified somewhat. Since in this case $da/d\varphi = \beta$
and $d^2a/d\varphi^2 = 0$, we obtain from Eqs. (III.119), (III.120), and (III.123),
respectively:

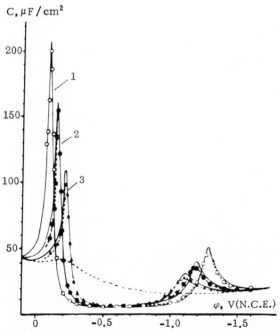

Fig. 29. Differential capacity curves for a mercury electrode with the system 1 N Na_2SO_4 + iso-$C_5H_{11}NH_2$. The
points and dash lines represent experimental data; the
continuous lines are calculated for a linear dependence
of a on φ. Concentration of iso-$C_5H_{11}NH_2$: 1) 0.04; 2)
0.02; 3) 0.01 M (from data in [16], Chap. III).

$$B = B_0 \exp\left[-\frac{E + C'\varphi(\varphi_N - \varphi/2)}{A} \right] \exp(-\beta\varphi), \qquad \text{(III.125)}$$

$$\varepsilon = \varepsilon_0(1 - \theta) + C'\theta(\varphi - \varphi_N) - \beta A\theta(1 - \theta), \qquad \text{(III.126)}$$

and

$$C = C_0(1 - \theta) + C'\theta + \frac{[\varepsilon_0 + C'(\varphi_N - \varphi) + \beta A(1 - 2\theta)]^2}{A} \frac{\theta(1 - \theta)}{1 - 2a\theta(1 - \theta)}.$$
$$\text{(III.127)}$$

The results of calculations with the use of these equations for a number of aliphatic compounds (n-C_3H_7OH; n-C_4H_9OH; n-, iso-, and tert-$C_5H_{11}OH$; n-$C_4H_9NH_2$; n- and iso-$C_5H_{11}NH_2$; n-C_4H_9COOH; $C_2H_5COC_2H_5$ [5, 10, 16, 21, 22] are in quantiative agreement with experimental data. Agreement between the experimental and theoretically calculated C vs φ curves is especially convincing confirmation of the foregoing theory because, as was shown above, the differential capacity is the most sensitive function of the structure of the electric double layer in presence of organic substances. In Fig. 29 the agreement between calculated and experimental data is illustrated by differential capacity curves for $1\,N$ Na_2SO_4 solutions with various additions of isoamylamine.

6. Certain Causes of the Deviation
of Real Systems from the Model of
Two Parallel Condensers

In 1926 Frumkin [2] showed that one of the causes of deviations of real systems from the model of two parallel condensers may be variation of the area S per adsorbed molecule with increase of the amount Γ of the organic substance adsorbed. Suppose in the first approximation that S decreases linearly with Γ, and therefore with $\theta = \Gamma/\Gamma_m$, which under these conditions is no longer the true surface coverage $\theta_r = \Gamma_S$. Since $S_{\theta=1} = 1/\Gamma_m$, the equation of the linear S vs Γ plot may be written as

$$S = \frac{k}{\Gamma_m} - \frac{k-1}{\Gamma_m^2}\Gamma, \qquad \text{(III.128)}$$

where $k = S_{\theta=0}/S_{\theta=1}$. From Eq. (III.128) we have

$$\theta_r = \Gamma S = k\theta_S - (k - 1)\theta_S^2. \qquad \text{(III.129)}$$

After substituting the true surface coverage θ_r into Eq. (III.106) we should obtain the general expression (III.120) for the surface charge. It can be easily shown that this is possible only under the condition that

$$\frac{da}{d\varphi} = (k-1)\frac{\varepsilon_0 + C'(\varphi_N - \varphi)}{A}, \qquad \text{(III.130)}$$

hence

$$a = a_0 + (k-1)\frac{E + C'\varphi\left(\varphi_N - \frac{\varphi}{2}\right)}{A}. \qquad \text{(III.131)}$$

Therefore, if linear decrease of S with increase of surface coverage is the cause of deviation of the real system from the model of two parallel condensers, application of the Frumkin equation (III.57) to a system of this kind should give an approximately parabolic dependence of the attraction constant on the electrode potential. A dependence of this type is observed in the case of adsorption of tetrabutylammonium cations on mercury (Fig. 30). It is seen that in this case there is satisfactory agreement between

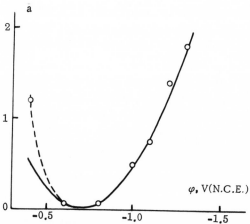

Fig. 30. Dependence of the attraction constant on the potential for the system 1 N Na_2SO_4 + $[(C_4H_9)_4N]_2SO_4$. The continuous line was calculated from Eq. (III.131) with a_0 = 0.3; k = 1.38; C' = 4.5 $\mu F/cm^2$; φ_N = 0.7 V; A = 0.51 $\mu J/cm^2$; the points and the dash line represent experimental data (from data in [6], Chap. III).

the experimental dependence of a on φ and the results of calculations based on Eq. (III.131); all the constants in this equation (with
the exception of a_0) were found from the results of independent
measurements [6].

As was noted in Section 1 of this chapter, for exact conformity to the model of two parallel condensers [which we designate
model (I)] both phases (metal and solution) must have infinitely
large dielectric constants. If only one of the two contiguous phases
has a high dielectric constant, Eq. (III.17), the equation for model
(I), is inapplicable to the system. Under such conditions the equation proposed by Hansen et al. [43]

$$\varepsilon = [C_0(1-\theta) + C'\theta](\varphi - \varphi_N\theta) \qquad \text{(III.132)}$$

may be used instead of Eq. (III.17). The model corresponding to
Eq. (III.132) we will designate model (II).

Models (I) and (II) give the same expression $(\partial\varepsilon/\partial\varphi)_\theta = C_0(1-\theta) + C'\theta$ for the true capacity, but the dependence of the
potential shift $\varphi_{\varepsilon=0}$ due to adsorption on θ (or Γ) is different:
nonlinear for model (I) [see Eq. (III.19)] and linear for model (II).
If the orientation of the adsorbed dipoles does not change with increase of θ, it is therefore to be expected that model (I) is approximately applicable to the mercury−aqueous solution interface, and model (II) to the solution−air interface. Figure 13
shows that this conclusion is confirmed experimentally.

However, since the dielectric constant of water is nevertheless considerably lower than that of the metal, the surface of the
double-layer condenser on the solution side cannot be regarded
as a strictly equipotential surface (see [109]). In other words, as
the result of the discrete nature of the adsorbed organic dipoles
some deviation of the real interface between the electrode and the
solution from model (I) in the direction of model (II) is to be expected.

If model (I) is realized then, as already noted, a = const. In
the case of model (II) it follows from a comparison of Eqs. (III.120)
and (III.132) that

$$\frac{da}{d\varphi} = \frac{\varphi_N(C_0 - C')}{A}, \qquad \text{(III.133)}$$

hence

$$a = a_0 + \frac{\varphi_N (C_0 - C')}{A} \varphi \; . \qquad (III.134)$$

Thus, transition from model (I) to model (II) leads to a linear dependence of the attraction constant on the potential, with a positive value of the coefficient $\beta = \varphi_N (C_0 - C')/A$. Putting, in accordance with experimental data, $\varphi_N \approx 0.3$, $A \approx 1$, and $C_0 - C' \approx 15$, we obtain $\beta \approx 4.5$, which is greater by an order of magnitude than the values of β found experimentally for a number of aliphatic compounds [5, 16, 34]. In other words, the behavior of the real interface between the electrode and the solution deviates but little from the behavior of model (I) toward the behavior of model (II). This explains the linear dependence, with a small positive value of the coefficient β, of the attraction constant on the potential.

Another cause of deviation of real systems from the model of two parallel condensers is π-electron interaction between the organic molecules and the surface of the mercury electrode. As the result of π-electron interaction, organic molecules can be adsorbed in two different positions, flat and vertical, on the electrode. As the number of molecules adsorbed in either position depends simultaneously on the electrode potential and on the concentration of the organic substance, deviations from model (I) due to this cause cannot in general be interpreted with the aid of the $a = a(\varphi)$ relationship. The behavior of systems of this type will be discussed in the next section. Here we merely note that even the deviations from the model of two parallel condensers which can be represented with the aid of the $a = a(\varphi)$ relationship introduce additional capacity and electrocapillary curves, discussed above. For example, in the case of a linear dependence of a on φ with the condition $\beta < 0$ [see Eq. (III.124)] the minimum on the C vs φ curves corresponding to the potential of the maximum adsorption φ_m should shift in the negative direction with increasing concentration of the organic substance, while the potential of maximum lowering of the interfacial tension, $\Delta\sigma$, is somewhat on the positive side of φ_m. These points have been discussed more fully by Damaskin and Dyatkina [35].

7. Interpretation of the Adsorption Behavior of Aromatic and Heterocyclic Compounds

Adsorption of aromatic or heterocyclic compounds at a mercury electrode is influenced in a varying degree by π-electron interaction with the charges of the electrode surface. If, as the result of π-electron interaction, the organic molecules can be adsorbed in two different positions the model of two parallel condensers becomes inapplicable in the general case. For example, owing to variation of the nature of adsorption of aniline molecules with the potential the calculated differential capacity curves can reproduce only the cathodic branch of the experimental C vs φ curves. At $\varepsilon > 0$ the calculated and experimental C vs φ curves no longer agree (Fig. 31).

To represent systems in which the organic substance is adsorbed at the electrode in two different positions, the two-condenser model must be replaced by a model consisting of three parallel condensers, with water molecules between the plates of the first, organic molecules in the first (e.g., vertical) position (1) between the plates of the second, and organic molecules in the second (e.g., flat) position (2) between the plates of the third. The equation for the electrode charge ε corresponding to this model is

$$\varepsilon = \varepsilon_o(1 - \theta_1 - \theta_2) + C_1(\varphi - \varphi_{N1})\theta_1 + C_2(\varphi - \varphi_{N2})\theta_2 , \quad \text{(III.135)}$$

where θ_1 and θ_2 represent the coverage by organic molecules in positions 1 and 2 respectively; C_1 and C_2 are the capacities of the double layer with complete surface coverage in positions 1 and 2; φ_{N1} and φ_{N2} are the shifts of p.z.c. in transition from the pure electrolyte solution to $\theta_1 = 1$ or $\theta_2 = 1$, respectively.

For practical application of Eq. (III.135) we must additionally choose two adsorption isotherms connecting θ_1 and θ_2 with the volume concentration c of the organic substance. This choice cannot be arbitrary, as the isotherms are linked with Eq. (III.135) by the fundamental electrocapillary equation (III.1), in which the value of Γ for an organic substance adsorbed in two positions should be written in the form

$$\Gamma = \Gamma_{m1}\theta_1 + \Gamma_{m2}\theta_2 , \quad \text{(III.136)}$$

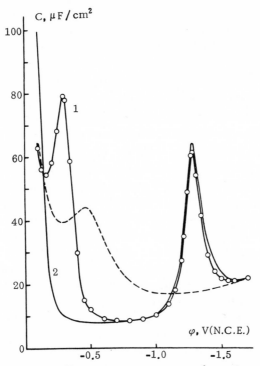

Fig. 31. Differential capacity curves of pure 1 N
KCl solution (dash line) and with addition of 0.1 M
aniline: 1) Experimental data, 400 Hz; 2) calculated from Eq. (III.127) (from data in [17],
Chap. III).

where Γ_{m1} is the value of Γ when $\theta_1 = 1$ and $\theta_2 = 0$, and Γ_{m2} is the
value of Γ when $\theta_2 = 1$ and $\theta_1 = 0$.

We will show that the model of three parallel condensers is
satisfied by the following system of isotherms:

$$\left. \begin{array}{l} B_1 c = \dfrac{\theta_1}{n_1(1 - \theta_1 - \theta_2)^{n_1}} \exp\left(- 2a_1 n_1 \theta_1 - 2a_3 n_1 \theta_2\right) \\[3mm] B_2 c = \dfrac{\theta_2}{n_2(1 - \theta_1 - \theta_2)^{n_2}} \exp\left(- 2a_2 n_2 \theta_2 - 2a_3 n_2 \theta_1\right) \end{array} \right\} \quad \text{(III.137)}$$

based on the Flory–Huggins model [91, 110], but with molecular interaction in the adsorption layer taken into account. In Eqs. (III.137) the adsorption interaction constants B_1 and B_2 are certain functions of the electrode potential; the attraction constants a_1, a_2, and a_3 represent interaction between vertically adsorbed molecules, molecules in flat orientation, and molecules adsorbed in different positions respectively; n_1 is the ratio of the area occupied by an organic molecule in vertical orientation to the area per associated group of adsorbed water molecules; n_2 is the corresponding ratio for flat orientation of adsorbate molecules.

It should be noted that an analogous system of equations was used by Parry and Parsons [51], but they assumed that B_1 and B_2 are certain functions of the electrode charge and not of the potential. Analysis of this assumption shows that it is equivalent to a model of three condensers in series, containing water molecules and adsorbate molecules in two different positions respectively. Certain difficulties arise in physical substantiation of this model, although it does not contradict the experimental dependence of Γ on ε for sodium p-toluene-sulfonate solutions [51]. Moreover, the isotherms of Parry and Parsons do not contain the factors n_1 and n_2 in the exponents; these isotherms therefore cannot be correlated with the equation of state of the adsorption layer.

At a constant electrode potential, it follows from Eqs. (III.1) and (III.136) that

$$d\sigma = -\frac{A}{n_1}\theta_1 d\ln c - \frac{A}{n_2}\theta_2 d\ln c , \qquad \text{(III.138)}$$

where $A = n_1 RT\Gamma_{m1} = n_2 RT\Gamma_{m2}$. On the other hand, it is easy to obtain from Eqs. (III.137) at $\varphi = \text{const}$

$$\left. \begin{aligned} d\ln c &= \frac{d\theta_1}{\theta_1} + \frac{n_1 d\theta}{1-\theta} - 2n_1 a_1 d\theta_1 - 2n_1 a_3 d\theta_2 \\ d\ln c &= \frac{d\theta_2}{\theta_2} + \frac{n_2 d\theta}{1-\theta} - 2n_2 a_2 d\theta_2 - 2n_2 a_3 d\theta_1 \end{aligned} \right\} , \qquad \text{(III.139)}$$

where $\theta = \theta_1 + \theta_2$.

Introducing (III.139) into Eq. (III.138) we find after integration:

$$\sigma = \sigma_o + A\left[\ln(1-\theta) + \frac{n_1 - 1}{n_1}\theta_1 + \right.$$

$$\left. + \frac{n_2 - 1}{n_2}\theta_2 + a_1\theta_1^2 + a_2\theta_2^2 + 2a_3\theta_1\theta_2\right], \qquad \text{(III.140)}$$

where σ_0 is the integration constant, equal to the interfacial tension in a solution of the pure supporting electrolyte, because $\sigma = \sigma_0$ when $\theta = \theta_1 = \theta_2 = 0$. It is easily seen that with the conditions $n_1 = n_2 = 1$ and $a_1 = a_2 = a_3 = a$, Eq. (III.140) becomes the equation of state (III.72).

Suppose that Eq. (III.140), like (III.72), is valid at all potentials. The differentiation of Eq. (III.140) with respect to potential gives:

$$\varepsilon = -\frac{d\sigma}{d\varphi} = \varepsilon_o + A\left[\frac{1}{1-\theta} \cdot \frac{d\theta}{d\varphi} - \frac{n_1-1}{n_1} \cdot \frac{d\theta_1}{d\varphi} - \frac{n_2-1}{n_2} \cdot \frac{d\theta_2}{d\varphi} - \right.$$
$$\left. - 2a_1\theta_1 \frac{d\theta_1}{d\varphi} - 2a_2\theta_2 \frac{d\theta_2}{d\varphi} - 2a_3\theta_1 \frac{d\theta_2}{d\varphi} - 2a_3\theta_2 \frac{d\theta_1}{d\varphi}\right]. \qquad \text{(III.141)}$$

It is easy to show that the expression in square brackets in Eq. (III.141) is equal to $(P_1/n_1)\theta_1 + (P_2/n_2)\theta_2$, where $P_1 \equiv d\ln B_1/d\varphi$ and $P_2 \equiv d\ln B_2/d\varphi$ which, in accordance with the system of isotherms (III.137), are expressed as follows:

$$\left.\begin{array}{l} P_1 = \dfrac{1}{\theta_1} \cdot \dfrac{d\theta_1}{d\varphi} + \dfrac{n_1}{1-\theta} \dfrac{d\theta}{d\varphi} - 2n_1a_1 \dfrac{d\theta_1}{d\varphi} - 2n_1a_3 \dfrac{d\theta_2}{d\varphi} \\[3mm] P_2 = \dfrac{1}{\theta_2} \cdot \dfrac{d\theta_2}{d\varphi} + \dfrac{n_2}{1-\theta} \dfrac{d\theta}{d\varphi} - 2n_2a_2 \dfrac{d\theta_2}{d\varphi} - 2n_2a_3 \dfrac{d\theta_1}{d\varphi} \end{array}\right\}, \quad \text{(III.142)}$$

Thus, the expression for the charge may be written as

$$\varepsilon = \varepsilon_o + \frac{A}{n_1}\theta_1 \cdot \frac{d\ln B_1}{d\varphi} + \frac{A}{n_2}\theta_2 \frac{d\ln B_2}{d\varphi}, \qquad \text{(III.143)}$$

which is equivalent to the model of three parallel condensers [see Eq. (III.135)] under the conditions:

$$\left.\begin{array}{l} \dfrac{d\ln B_1}{d\varphi} = -\dfrac{n_1[\varepsilon_o - C_1(\varphi - \varphi_{N1})]}{A} \\[3mm] \dfrac{d\ln B_2}{d\varphi} = -\dfrac{n_2[\varepsilon_o - C_2(\varphi - \varphi_{N2})]}{A} \end{array}\right\}, \qquad \text{(III.144)}$$

and therefore

$$B_1 = B_{01} \cdot \exp\left\{-\frac{n_1\left[\int_0^\varphi \varepsilon_o \, d\varphi + C_1\varphi(\varphi_{N1} - \varphi/2)\right]}{A}\right\}$$

$$B_2 = B_{02} \cdot \exp\left\{-\frac{n_2\left[\int_0^\varphi \varepsilon_o \, d\varphi + C_2\varphi(\varphi_{N2} - \varphi/2)\right]}{A}\right\} . \quad \text{(III.145)}$$

Equations (III.145) specify the dependence of the adsorption equilibrium constants in the isotherms (III.137) on potential.

Finally, differentiation of Eq. (III.143) with respect to potential gives the expression for the differential capacity of the double layer in presence of organic substances adsorbed in two different positions:

$$C = \frac{d\varepsilon}{d\varphi} = C_0 + \frac{A}{n_1} \cdot \frac{d^2 \ln B_1}{d\varphi^2}\,\theta_1 + \frac{A}{n_2} \cdot \frac{d^2 \ln B_2}{d\varphi^2}\,\theta_2 +$$

$$+ \frac{A}{n_1} \cdot \frac{d \ln B_1}{d\varphi} \cdot \frac{d\theta_1}{d\varphi} + \frac{A}{n_2} \cdot \frac{d \ln B_2}{d\varphi} \cdot \frac{d\theta_2}{d\varphi} . \quad \text{(III.146)}$$

The derivatives $d\theta_1/d\varphi$ and $d\theta_2/d\varphi$ in this equation can be found from the linear system of equations (III.142) if $d\theta/d\varphi$ in it is replaced by the sum $(d\theta_1/d\varphi) + (d\theta_2/d\varphi)$. We thus obtain:

$$\frac{d\theta_1}{d\varphi} = \frac{\dfrac{P_1}{n_1}\left(\dfrac{1}{n_2\theta_2} + \dfrac{1}{1-\theta} - 2a_2\right) - \dfrac{P_2}{n_2}\left(\dfrac{1}{1-\theta} - 2a_3\right)}{\left(\dfrac{1}{n_1\theta_1} + \dfrac{1}{1-\theta} - 2a_1\right)\left(\dfrac{1}{n_2\theta_2} + \dfrac{1}{1-\theta} - 2a_2\right) - \left(\dfrac{1}{1-\theta} - 2a_3\right)^2}$$

$$\frac{d\theta_2}{d\varphi} = \frac{\dfrac{P_2}{n_2}\left(\dfrac{1}{n_1\theta_1} + \dfrac{1}{1-\theta} - 2a_1\right) - \dfrac{P_1}{n_1}\left(\dfrac{1}{1-\theta} - 2a_3\right)}{\left(\dfrac{1}{n_1\theta_1} + \dfrac{1}{1-\theta} - 2a_1\right)\left(\dfrac{1}{n_2\theta_2} + \dfrac{1}{1-\theta} - 2a_2\right) - \left(\dfrac{1}{1-\theta} - 2a_3\right)^2} .$$

$$\text{(III.147)}$$

With the aid of Eqs. (III.135)-(III.137), (III.140), and (III.145)-(III.147) we can calculate the adsorption isotherms, dependence of adsorption on the potential, dependence of θ_1, θ_2, and the charge on the potential, and the electrocapillary and differential capacity curves for various concentrations of an organic substance adsorbed in two different positions.

Such calculations are exceedingly cumbersome. To diminish this difficulty, Damaskin et al. [38] introduced the following simplifying assumptions: $a_1 = a_2 = a_3 = 0$; $C_2 = C_0 = \text{const}$; $n_1 = 1$ and $n_2 = 2$; this greatly simplified the formulas. For example, the final expression for the differential capacity becomes:

$$C = C_o - 2A\alpha\theta_1 + \frac{A}{2(1 + \theta_2)}\{8\alpha^2(\varphi - \varphi_m)^2\theta_1(1 - \theta_1) +$$
$$+ \rho^2\theta_2(1 - \theta_2) + 8\theta_1\theta_2[\alpha^2(\varphi - \varphi_m)^2 + \alpha\rho(\varphi - \varphi_m)]\} , \quad \text{(III.148)}$$

where

$$\alpha = \frac{C_o - C_1}{2A}, \qquad \varphi_m = -\frac{C_1\varphi_{N1}}{C_o - C_1} \qquad \text{and} \qquad \rho = -\frac{2C_o\varphi_{N2}}{A} .$$

In calculations of the Γ vs φ, σ vs φ, and C vs φ curves Damaskin et al. [38] chose the following values of the double-layer parameters: $C_0 = 20\ \mu F/cm^2$; $C_1 = 7\ \mu F/cm^2$; $A = 1.6\ \mu J/cm^2$; $B_{01}c = 2$; $B_{02}c = 10$; $\varphi_{N1} = +0.5$ V and 0; $\varphi_{N2} = -0.25$ and -0.5 V. The calculated results were compared with experimental data on adsorption of o- and p-phenylenediamine from solutions in 1 M KCl on mercury. Some of the results are given in Figs. 32-35.

Figure 32 shows that the theoretical Γ vs φ curves are in qualitative agreement with the shape of the Γ vs φ curve in presence of p-phenylenediamine. Similar Γ vs φ curves were obtained earlier for solutions of aniline in 1M KCl and 1 M KI [17]. It also follows from Fig. 32 that in the case of an organic substance adsorbed in two different positions the shape of the Γ vs φ curve is virtually independent of φ_{N2}, whereas variation of φ_{N1} leads to a shift of the Γ vs φ curve along the abscissa axis.

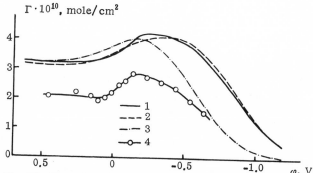

Fig. 32. Dependence of the amount adsorbed on the potential. 1) Calculated for $\varphi_{N1} = 0.5$ V and $\varphi_{N2} = -0.25$ V; 2) calculated for $\varphi_{N1} = 0.5$ V and $\varphi_{N2} = -0.5$ V; 3) calculated for $\varphi_{N1} = 0$ and $\varphi_{N2} = -0.5$ V; 4) experimental curve for 1 M KCl + 0.02 M p-phenylenediamine (from data in [38], Chap. III).

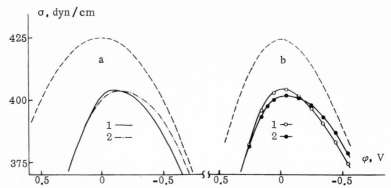

Fig. 33. Electrocapillary curves: calculated theoretically (a); experimental (b). a: 1) $\varphi_{N1} = 0.5$ and $\varphi_{N2} = -0.5$ V; 2) $\varphi_{N1} = 0$ and $\varphi_{N2} = -0.5$ V. b: 1) 1 M KCl + 0.03 M p-phenylenediamine; 2) 1 M KCl + 0.03 M o-phenylenediamine. Electrocapillary curves for the pure electrolyte solution are indicated by dash lines (from data in [38], Chap. III).

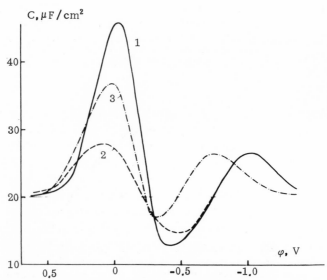

Fig. 34. Theoretically calculated differential capacity curves: 1) $\varphi_{N1} = 0.5$ and $\varphi_{N2} = -0.5$ V; 2) $\varphi_{N1} = 0.5$ and $\varphi_{N2} = -0.25$ V; 3) $\varphi_{N1} = 0$ and $\varphi_{N2} = -0.5$ V (from data in [38], Chap. III).

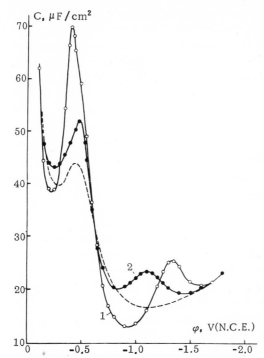

Fig. 35. Differential capacity curves for 1 M KCl
(dash line) and for 1 M KCl with additions of 0.02 M
p-phenylenediamine (1) and o-phenylenediamine
(2). Frequency 400 Hz (from data in [38], Chap.III).

Electrocapillary curves theoretically calculated for two values of φ_{N1}, +0.5 V and 0, at the same $\varphi_{N2} = -0.5$ V, are shown in Fig. 33a. It is seen that these curves intersect near the potential of zero charge and merge at $\varepsilon > 0$. In distinction from these curves, the σ vs φ curve calculated for $\varphi_{N2} = -0.25$ V is located at $\varepsilon > 0$ between the curve for the supporting electrolyte and curves 1 and 2 in Fig. 33a. Comparison of Figs. 33a and 33b shows that when o-phenylenediamine is replaced by p-phenylenediamine the positive value of φ_{N1} increases, whereas the value of φ_{N2}, determined by π-electron interaction of these molecules with the mercury surface remains unchanged.

This result is confirmed by a comparison of theoretical and experimental differential capacity curves (Figs. 34 and 35). When

o-phenylenediamine is replaced by the para-isomer the anodic
peak increases appreciably, the lowering of capacity in the middle
region of the C vs φ curve is greater in comparison with the curve
for the supporting electrolyte, and the cathodic peak is shifted toward more negative potentials (Fig. 35). All these effects can be
observed in Fig. 34 as we pass from curve 3, theoretically calculated for $\varphi_{N1} = 0$, to curve 1 calculated for $\varphi_{N1} = +0.5$ V. At
the same time, Fig. 34 shows that increase of the negative value of
φ_{N2} at constant φ_{N1} (curves 2 and 3) does not affect the position of
the cathodic peak, although it is again accompanied by substantial
increase of the anodic peak and some decrease of capacity at the
minimum of the C vs φ curve.

It follows from Fig. 34 that the theoretical C vs φ curves also
reproduce satisfactorily the experimentally observed agreement
between the capacities at $\varepsilon > 0$ in solutions of the pure electrolyte
and in presence of the organic additive (see Figs. 7 and 35), although under these conditions desorption of the organic molecules
from the electrode surface does not occur (Figs. 32 and 33). The
consequence of this effect, detected earlier and explained qualitatively in the case of adsorption of pyridine and aniline on mercury [8, 17] is that the area under the differential capacity curve
in presence of an organic substance may differ greatly from the
area under the capacity curve for the pure electrolyte solution.

In the general case, positions 1 and 2 of the adsorbed molecules may differ in the values of the limiting capacity and the
limiting adsorption potential difference, in the area per adsorbate
molecule, and in the values of the standard free energy of adsorption at $\varphi = 0$ and $\theta \to 0$. However, among aromatic and heterocyclic compounds adsorbed on mercury, systems can be found for
which positions 1 and 2 of the adsorbate molecules differ, in the
first approximation, only in the adsorption potential difference.
Suppose that in these systems interaction between the adsorbed
molecules leads to change of their orientation from flat to vertical
with increase of the adsorbate concentration, such that the part
of the adsorption potential difference due to vertically oriented
molecules, i.e., $\theta_1 \varphi_{N1}/\theta$ increases linearly with θ while the part
of the adsorption difference due to molecules oriented flat, i.e.,
$\theta_2 \varphi_{N2}/\theta$, decreases linearly with the surface coverage:

$$\frac{\theta_1 \varphi_{N1}}{\theta} = k_1 \theta; \qquad \frac{\theta_2 \varphi_{N2}}{\theta} = k_2 (1 - \theta) , \qquad \text{(III.149)}$$

where k_1 and k_2 are proportionality factors.

If the relations (III.149) and the condition $C_1 = C_2 = C'$ are satisfied, it follows from the model of three parallel condensers (III.135) that

$$\varepsilon = \varepsilon_0 (1 - \theta) + C'\theta\{\varphi - [k_1\theta + k_2(1 - \theta)]\} . \qquad \text{(III.150)}$$

Comparing Eq. (III.150) with the equation for the model of two parallel condensers (III.106), we find

$$\varphi_N(\theta) = k_1\theta + k_2(1 - \theta) , \qquad \text{(III.151)}$$

where φ_N (θ) represents the shift of potential in the transition from $\theta = 0$ to $\theta = 1$ which would be observed if the orientation of the adsorbed molecules at $\theta = 1$ remained the same as at the given coverage θ. It is easy to see that $\varphi_N (0) = k_2$ while $\varphi_N (1) = k_1$.

According to our assumption, the organic substance is adsorbed only in position 2 (flat) when $\theta \to 0$, and only in position 1 (vertical or inclined) when $\theta = 1$. Therefore $k_2 = \varphi_N(0) = \varphi_{N2}$, while $k_1 = \varphi_N (1) = \varphi_{N1}$. Hence it follows from Eq. (III.149) that

$$\theta_1 = \theta^2 \qquad \text{and} \qquad \theta_2 = \theta(1 - \theta) \qquad \text{(III.152)}$$

and Eq. (III.150) can now be written as follows:

$$\varepsilon = \varepsilon_0 (1 - \theta) + C'\theta\{\varphi - [\varphi_{N1}\theta + \varphi_{N2}(1 - \theta)]\} . \qquad \text{(III.153)}$$

The relations (III.152) show that for the systems under consideration the number of molecules adsorbed in the two positions is not simultaneously a function of θ and φ, but may be expressed either in terms of θ or of φ and the concentration c, since $\theta = f(\varphi, c)$. Thus, for this special case the deviation of a real system from the model of two parallel condensers due to π-electron interaction can be expressed with the aid of the dependence of the attraction constant on the potential, $a = a(\varphi)$.

Comparing Eq. (III.153) with Eq. (III.120), in which $\varphi_N = \varphi_{N1}$ by physical meaning, we find

$$\frac{da}{d\varphi} = - \frac{C'(\varphi_{N1} - \varphi_{N2})}{A} \qquad \text{and} \qquad a = a_o - \frac{C'(\varphi_{N1} - \varphi_{N2})}{A} \cdot \varphi \qquad \text{(III.154)}$$

According to Eq. (III.154), the dependence of a on φ in these systems should be linear in character, of the type of (III.124), with a negative value of the coefficient $\beta = -C'(\varphi_{N1} - \varphi_{N2})/A$, since by the physical meaning φ_{N1} is always more positive than φ_{N2}. It has been shown [41] that a_0 in Eq. (III.154) represents the sum of two effects: 1) increase of the free energy of adsorption $\Delta\bar{G}_A$ with increase of θ as the result of attractive interaction between the adsorbate molecules (positive component of a_0); 2) decrease of $\Delta\bar{G}_A$ with increase of θ as the result of transition from flat orientation, when $\Delta\bar{G}_A$ is generally greater (e.g., as the result of π-electron interaction), to vertical or inclined orientation (negative component of a_0). Dependent on the ratio between these two effects, a_0 in Eq. (III.154) may be either positive or negative.

Thus, if the two positions in which the organic molecules can be adsorbed essentially differ only in the values of the adsorption potential difference, while the proportions of these potential differences for the two positions and the total energy of adsorption at $\varphi = 0$ are linear functions of the coverage, the behavior of such systems can be represented by the model of two parallel condensers with linear dependence of a on φ ($\beta < 0$) taken into account. These results were found experimentally for adsorption of pyridine [8, 36], phenol [18], and toluidine [111] on mercury. They can be illustrated by the example of adsorption of phenol on mercury from fairly dilute solutions, when adsorption is not yet complicated by formation of multimolecular layers [18, 112]. The calculated and experimental data for this case are compared in Fig. 36. It is seen that the calculated Γ vs φ and C vs φ curves are in satisfactory agreement with experimental data. Some degree of deviation between the calculated and experimental values is probably due to the approximate nature of the assumptions (III.149).

We note in conclusion that the model of three parallel condensers (III.135) in conjunction with the system of isotherms (III.137) can also serve as the basis for representing simultaneous adsorption of two surface-active substances at an electrode. However, in this case in the first of Eqs. (III.137) the total concentration c should be replaced by c_1, the concentration of the first substance, while in the second of these equations c_2, the volume concentration of the second organic substance should be put instead of c. The physical meaning of the other quantities in Eqs. (III.135) and (III.137) alters correspondingly; all quantities with subscript 1

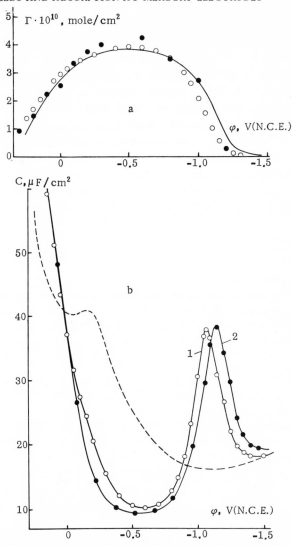

Fig. 36. Comparison of theoretically calculated and experimental data on the dependence of adsorption (a) and differential capacity (b) on the potential of the mercury electrode for the system 1 N Na_2SO_4 + 0.05 M phenol. a) Continuous line calculated from Eqs. (III.57) and (III.125); the points represent data obtained with the aid of the Gibbs equation from electrocapillary curves: 1) from experimental σ vs φ curves; 2) from σ vs φ curves obtained by twofold "back integration" of the capacity curves. b) Differential capacity: 1) experimental values; 2) calculated from Eq. (III.127): the dash line is the experimental C vs φ curve for 1 N Na_2SO_4 solution (from data in [18], Chap. III).

refer to the first substance, and those with subscript 2 to the second.

Exact solution of the system of equations (III.137) in the case of simultaneous adsorption of two organic substances on the electrode presents the same difficulties as the case of one substance adsorbed in two different positions. In some special cases of joint adsorption of two organic substances, when these difficulties could be avoided, the model of three parallel condensers was used successfully by Tedoradze et al. [31, 32]. An essentially similar theoretical approach to systems of this type was also developed by Kastening and Holleck [89].

References

1. A. N. Frumkin, Z. Phys. Chem., 116:466 (1925).
2. A. N. Frumkin, Z. Phys., 35:792 (1926).
3. A. N. Frumkin and B. B. Damaskin, in: Modern Aspects of Electrochemistry, Vol. 3, Plenum Press, New York (1964), p. 149.
4. B. B. Damaskin, Dokl. Akad. Nauk SSSR, 144:1073 (1962).
5. B. B. Damaskin and N. B. Grigor'ev, Dokl. Akad. Nauk SSSR, 147:135 (1962).
6. B. B. Damaskin, S. Vavrzhichkba, and N. B. Grigor'ev, Zh. Fiz. Khim., 36:2530 (1962).
7. B. B. Damaskin and G. A. Tedoradze, Dokl. Akad. Nauk SSSR, 152:1151 (1963).
8. L. D. Klyukina and B. B. Damaskin, Izv. Akad. Nauk SSSR, Otd. Khim. Nauk, 1022 (1963).
9. B. B. Damaskin, Zh. Fiz. Khim., 37:2483 (1963).
10. B. B. Damaskin, Electrochim. Acta, 9:231 (1964).
11. B. B. Damaskin, J. Electroanal. Chem., 7:155 (1964).
12. G. A. Tedoradze, Dokl. Akad. Nauk SSSR, 155:1423 (1964).
13. B. B. Damaskin, Dokl. Akad. Nauk SSSR, 156:128 (1964).
14. G. A. Tedoradze and R. A. Arakelyan, Dokl. Akad. Nauk SSSR, 156:1170 (1964).
15. A. N. Frumkin, B. B. Damaskin, V. M. Gerovich, and R. I. Kaganovich, Dokl. Akad. Nauk SSSR, 158:706 (1964).
16. R. Lerkkh and B. B. Damaskin, Zh. Fiz. Khim., 38:1154 (1964); 39:211 (1965).
17. B. B. Damaskin, I. P. Mishutushkina, V. M. Gerovich, and R. I. Kaganovich, Zh. Fiz. Khim., 38:1797 (1964).
18. B. B. Damaskin, V. M. Gerovich, I. P. Gladkikh, and R. I. Kaganovich, Zh. Fiz. Khim., 38:2495 (1964).
19. B. B. Damaskin and G. A. Tedoradze, Electrochim. Acta, 10:529 (1965).
20. B. B. Damaskin, Élektrokhimiya, 1:63 (1965).
21. V. K. Venkatesan, B. B. Damaskin, and N. V. Nikolaeva-Fedorovich, Zh. Fiz. Khim., 39:129 (1965).
22. B. B. Damaskin and R. Lerkkh, Zh. Fiz. Khim., 39:495 (1965).
23. G. A. Tedoradze and Ya. M. Zolotovitskii, Élektrokhimiya, 1:201 (1965).

24.   B. B. Damaskin, Élektrokhimiya, 1:255 (1965).
25.   B. B. Damaskin and S. L. Dyatkina, Élektrokhimiya, 1:706 (1965).
26.   A. N. Frumkin, B. B. Damaskin, and A. A. Survila, Élektrokhimiya, 1:738 (1965).
27.   B. B. Damaskin, Zh. Fiz. Khim., 39:1636 (1965).
28.   B. B. Damaskin, Élektrokhimiya, 1:1123 (1965).
29.   B. B. Damaskin, Usp. Khim., 34:1764 (1965).
30.   Ya. M. Zolotovitskii and G. A. Tedoradze, Élektrokhimiya, 1:1339 (1965).
31.   G. A. Tedoradze, R. A. Arakelyan, and E. D. Belokolos, Élektrokhimiya, 2:563 (1966).
32.   R. A. Arakelyan and G. A. Tedoradze, Élektrokhimiya, 4:144 (1968).
33.   B. B. Damaskin and S. L. Dyatkina, Élektrokhimiya, 2:981 (1966).
34.   B. B. Damaskin, A. A. Survila, and L. E. Rybalka, Élektrokhimiya, 3:146, 927, 1138 (1967).
35.   B. B. Damaskin and S. L. Dyatkina, Élektrokhimiya, 3:531 (1967).
36.   B. B. Damaskin, A. A. Survila, S. Ya. Vasina, and A. I. Fedorova, Élektrokhimiya, 3:825 (1967).
37.   B. B. Damaskin, Élektrokhimiya, 3:1390 (1967).
38.   B. B. Damaskin, A. N. Frumkin, and S. L. Dyatkina, Izv. Akad. Nauk SSSR, Ser. Khim., No. 10, 2171 (1967).
39.   A. A. Survila and B. B. Damaskin, in: Electrochemical Processes Involving Organic Substances [in Russian], Izd. Nauka, Moscow (1969).
40.   A. N. Frumkin, B. B. Damaskin, and A. A. Survila, J. Electroanal. Chem., 16:493 (1968).
41.   B. B. Damaskin, Élektrokhimiya, 4:6 (1968).
42.   R. S. Hansen, R. E. Minturn, and D. A. Hickson, J. Phys. Chem., 60:1185 (1956); 61:953 (1957).
43.   R. S. Hansen, D. J. Kelsh, and D. H. Grantham, J. Phys. Chem., 67:2316 (1963).
44.   W. Lorenz and F. Möckel, Z. Elektrochem., 60:507 (1956).
45.   W. Lorenz, Z. Elektrochem., 62:192 (1958).
46.   W. Lorenz, F. Möckel, and W. Müller, Z. Phys. Chem. (N.F.), 25:145 (1960).
47.   W. Lorenz and W. Müller, Z. Phys. Chem. (N.F.), 25:161 (1960).
48.   R. Parsons, Trans. Faraday Soc., 51:1518 (1955); 55:999 (1959).
49.   R. Parsons, J. Electroanal. Chem., 7:136 (1964); 8:93 (1964).
50.   J. M. Parry and R. Parsons, Trans. Faraday Soc., 59:241 (1963).
51.   J. M. Parry and R. Parsons, J. Electrochem. Soc., 113:992 (1966).
52.   R. Parsons and F. G. R. Zobel, Trans. Faraday Soc., 62:3511 (1966).
53.   E. Dutkievicz, J. D. Garnish, and R. Parsons, J. Electroanal. Chem., 16:505 (1968).
54.   M. A. V. Devanathan, Proc. Roy. Soc., A264:133 (1961); A267:256 (1962).
55.   M. A. V. Devanathan and B. V. K. S. R. A. Tilak, Chem. Rev., 65:635 (1965).
56.   J. O'M. Bockris, M. A. V. Devanathan, and K. Müller, Proc. Roy. Soc., A274:55 (1963).
57.   J. O'M. Bockris, E. Gileadi, and K. Müller, Electrochim. Acta, 12:1301 (1967).
58.   G. Gouy, Ann. Phys., 7:129 (1917).
59.   A. G. Stromberg and L. S. Zagainova, Zh. Fiz. Khim., 31:1042 (1957).

60. I. Zwierzykowska, Roczniki Chem., 39:101 (1965).

61. R. G. Barradas, P. G. Hamilton, and B. E. Conway, J. Phys. Chem., 69:3411 (1965).

62. B. B. Damaskin, R. I. Kaganovich, V. M. Gerovich, and S. L. Dyatkina, Élektrokhimiya (in press).

63. M. Breiter and P. Delahay, J. Am. Chem. Soc., 81:2938 (1959).

64. N. F. Mott, R. Parsons, and R. J. Watts-Tobin, Phil. Mag., (75), 7:483 (1962).

65. D. C. Grahame, Chem. Rev., 41:441 (1947).

66. J. A. V. Butler, Proc. Roy. Soc., A122:399 (1929).

67. B. Breyer and S. Hacobian, Australian J. Sci. Res., A5:500 (1952).

68. I. R. Miller, Electrochim. Acta, 9:1453 (1964).

69. M. Senda and J. Tachi, Rev. Polarogr. (Japan), 10:79 (1962).

70. R. Parsons, J. Electroanal. Chem., 5:397 (1963).

71. R. Parsons, in: Proceedings of the Fourth Conference on Electrochemistry [in Russian], Izd. Akad. Nauk SSSR, Moscow (1959), p. 42.

72. H. Freundlich, Colloid and Capillary Chemistry, Methuen, London (1926).

73. I. Langmuir, J. Am. Chem. Soc., 40:1369 (1918).

74. M. Volmer, Z. Phys. Chem., 115:253 (1925).

75. E. Helfand, H. L. Frisch, and J. L. Lebowitz, J. Chem. Phys., 34:1037 (1961).

76. T. L. Hill, J. Chem. Phys., 20:141 (1952).

77. J. H. de Boer, The Dynamical Character of Adsorption, Oxford University Press (1953).

78. M. I. Temkin, Zh. Fiz. Khim., 15:296 (1941).

79. E. Blomgren and J. O'M. Bockris, J. Phys. Chem., 63:1475 (1959).

80. B. E. Conway and R. G. Barradas, Electrochim. Acta, 5:319, 349 (1961).

81. B. E. Conway, R. G. Barradas, P. G. Hamilton, and J. M. Parry, J. Electroanal. Chem., 10:485 (1965).

82. I. Zwierzykowska, Roczniki Chem., 38:663, 1169, 1195, 1367 (1964).

83. R. Fowler and E. A. Guggenheim, Statistical Thermodynamics, University Press, Cambridge (1939), p. 431.

84. A. N. Frumkin, J. Electroanal. Chem., 7:152 (1964).

85. E. A. Guggenheim, Trans. Faraday Soc., 41:150 (1945).

86. A. A. Zhukhovitskii, Zh. Fiz. Khim., 18:214 (1944).

87. H. Dahms and M. Green, J. Electrochem. Soc., 110:1075 (1963).

88. J. O'M. Bockris and D. A. J. Swinkels, J. Electrochem. Soc., 111:736 (1964).

89. B. Kastening and L. Holleck, Talanta, 12:1259 (1965).

90. D. Schuhmann, J. Chim. Phys., 64:1399 (1967).

91. M. L. Huggins, J. Phys. Chem., 46:151 (1942).

92. H. M. Cassel and F. Salditt, Z. Phys. Chem., A155:321 (1931).

93. B. P. Bering and K. A. Ioileva, Izv. Akad. Nauk SSSR, Otd. Khim. Nauk, 9 (1955).

94. M. E. Nicholas, P. A. Joyner, B. M. Tessem, and M. D. Olson, J. Phys. Chem., 65:1373 (1961).

95. C. Kemball, Proc. Roy. Soc., A187:73 (1946); A190:117 (1947).

96. S. V. Karpachev, M. V. Smirnov, and Z. S. Volchenkova, Zh. Fiz. Khim., 27:1228 (1953).

97. N. K. Roberts, J. Chem. Soc., 1907 (1964).
98. K. S. G. Doss and A. Kalyanasundaram, Current Sci., 20:199 (1951); Proc. Indian Acad. Sci., A35:27 (1952).
99. B. B. Damaskin, N. V. Nikolaeva-Fedorovich, and R. V. Ivanova, Zh. Fiz. Khim., 34:894 (1960).
100. A. N. Frumkin and V. I. Melik-Gaikazyan, Dokl. Akad. Nauk SSSR, 77:855 (1951).
101. J. Thovert, Compt. Rend., 135:579 (1901).
102. L. Saraga and I. Prigogine, Compt. Rend. de la 2-e Reunion de Chim. Phys. Paris (1952).
103. A. N. Frumkin and B. B. Damaskin, Dokl. Akad. Nauk SSSR, 129:862 (1959).
104. M. A. Loshkarev and M. A. Sevryugina, Dokl. Akad. Nauk SSSR, 108:111 (1956).
105. A. A. Moussa, H. M. Sammour, and H. A. Ghaly, Egypt. J. Chem., 2:169 (1959).
106. H. Jehring, Z. Phys. Chem., 226:59 (1964).
107. S. Sathyanarayana, J. Electroanal. Chem., 10:56 (1965).
108. W. Lorenz, Z. Phys. Chem. (N.F.), 26:424 (1958).
109. J. R. Macdonald and C. A. Barlow, J. Electrochem. Soc., 113:978 (1966).
110. P. J. Flory, J. Chem. Phys., 10:51 (1942).
111. B. B. Damaskin, S. L. Dyatkina, and V. K. Venkatesan, Élektrokhimiya (in press).
112. A. N. Frumkin, A. V. Gorodetskaya, and P. S. Chugunov, Acta Physicochim. URSS, 1:12 (1934).

Chapter 4

Nonthermodynamic Methods for Calculating Adsorption of Organic Substances on Electrodes

The thermodynamic approach to the study of adsorption of organic compounds, considered in Chapter I, Section 3 and based on experimental capacity or interfacial tension data, involves graphical differentiation, which greatly lowers the calculation accuracy. Investigators therefore turned their attention to development of nonthermodynamic methods, depending on selection of a particular equation for the adsorption isotherm and comparison of experimentally determined values (interfacial tension or the capacity of the double layer) with results which follow from the fundamental electrocapillary equation incorporating the chosen adsorption isotherm.

1. The Surface Pressure Method

The method proposed by Parsons [1] involves comparison, at various concentrations of the adsorbate, of the experimental values of surface pressure at constant charge $(\Delta\xi)$ [or at constant potential $(\Delta\sigma)$] with $\Delta\xi$ (or $\Delta\sigma$) values calculated with the aid of a particular adsorption isotherm. In practice, $\Delta\xi$ is first plotted against log c at various constant ε (or $\Delta\sigma$ vs log c at various constant φ). If all the experimental points can be fitted on one common curve by shifting these curves along the abscissa, this means that the form of the isotherm governing adsorption is congruent with respect to the corresponding electrical variable. In this case a plot

141

of log ($\Delta \xi$) vs log c [or log ($\Delta\sigma$) vs log c] is compared with corresponding plots calculated from various adsorption isotherms. If
the equation of state contains only one variable parameter (e.g.,
the Langmuir or Volmer equations, or the isotherm with virial
coefficients), it is possible by such comparison to determine the
maximum adsorption Γ_m and the energy of adsorption $\Delta\overline{G}_A$. This
can be illustrated by the example of the Langmuir isotherm

$$Bc = \frac{\theta}{1-\theta} \qquad (IV.1)$$

for which the equation of state is of the form

$$\Delta\sigma = -A \ln(1-\theta). \qquad (IV.2)$$

In this case

$$\log c = \log\left(\frac{\theta}{1-\theta}\right) - \log B \qquad (IV.3)$$

and

$$\log(\Delta\sigma) = \log[-\ln(1-\theta)] + \log A. \qquad (IV.4)$$

Thus, the form of the log ($\Delta\sigma$) vs log c plot is determined by
the functional dependence of log $[-\ln(1-\theta)]$ on $\log[\theta/(1-\theta)]$, and
comparison of experimental and calculated data permits determination of log A and log B, which are themselves connected with
the maximum adsorption of the organic substance (A = $RT\,\Gamma_m$) and
with the free energy of adsorption log B = $-1.74 - (\Delta\overline{G}_A/2.3RT)$
[see Eq. (III.31)].

Parsons [2] applied this method to the data of Schapink et al.
[3], who determined differential capacities in $0.1\,N$ NaF solutions
with addition of various amounts of thiourea. The experimental
data on adsorption of thiourea on mercury proved to give the best
agreement with the isotherm with virial coefficients (III.60) rather
than with the Langmuir isotherm as was supposed by Schapink et al.
[3]. A similar method was used by Parry and Parsons [4] for
studying adsorption of m-phenylene-disulfonate ions on mercury,
by Dutkiewicz and Parsons [5] for studying adsorption of thiourea
in formamide solutions on mercury, and by Parsons and Zobel [6]
for studying adsorption of acetanilide on mercury from aqueous
solutions of NaH_2PO_4 (see also [46], relating to adsorption of

thiourea on mercury in presence of KNO_3). The isotherms of Temkin (III.62) and Frumkin (III.57), and the equation with virial coefficients (III.60) were used in these investigations for comparing calculated and experimental $\Delta\xi$ vs log c curves.

Application of the surface pressure method to these systems showed that adsorption of thiourea on mercury from aqueous solutions is similar in many respects to adsorption from solutions in formamide. In both cases the adsorbed thiourea dipoles are oriented perpendicularly to the mercury surface and appreciable repulsive interaction occurs between them. Moreover, the relation between the free energy of adsorption $\Delta\overline{G}_A$ and the electrode charge ε is approximately linear in both cases. A linear dependence of $\Delta\overline{G}_A$ on ε and repulsive interaction between the adsorbed particles are also characteristic of m-phenylenedisulfonate anions; however, in contrast to thiourea molecules, these anions are oriented parallel to the mercury surface.

The surface pressure method has also been used for studying adsorption of inorganic anions (I^-, NO_3^-, ClO_4^-, and SCN^-) on mercury [7-9, 46, 47].

Unfortunately, as was noted by Parsons [10], the essentially simple surface pressure method suffers from several serious defects. First, in the case of isotherms whose equations of state contain two variable parameters, A and the attraction constant a [e.g., the Frumkin (III.57), Hill—De Boer (III.58), Parsons (III.59), and Temkin (III.62) isotherms], experimental determination of the adsorption parameters by comparison of calculated and experimental plots of log ($\Delta\xi$) vs log c at constant ε [or log ($\Delta\sigma$) vs log c at constant φ] leads to ambiguity. By variation of the parameter a it is possible to make the calculated and experimental log ($\Delta\xi$) vs log c curves coincide with the use of different adsorption isotherms; this, in turn, leads to substantially different values of $A = RT\Gamma_m$ for the same compound. This conclusion may be illustrated by the example of adsorption of m-phenylenedisulfonate anions on mercury. Parsons [10] showed with the aid of a more accurate method based on measurements of the differential capacity (this method will be discussed in Section 4) that adsorption of these anions conforms to the modified Helfand—Frisch—Lebowitz isotherm (III.59), and the maximum adsorption (expressed in electrical units) is 38 $\mu C/cm^2$. However, in an earlier investigation [4] the experimental $\Delta\xi$ vs log c curves for adsorption of m-phenylenedisulfonate ions were compared with calculated curves based on the Temkin isotherm (III.62) with a correspondingly chosen value of the attraction constant. This gave a value of 26.8 $\mu C/cm^2$, which is lower than the true value by a factor of 1.4, for the maximum adsorption.

If the equation of state contains two variable parameters, arbitrary choice of the adsorption isotherm naturally makes it impossible to use the surface pressure method for studying variations of the attraction constant with the potential or electrode charge. This limits the information obtainable on adsorption of organic compounds by the surface pressure method.

Finally, the low accuracy of the surface pressure method should be noted. This defect of the method may be illustrated by the following example.

Equations (III.30) and (III.57) can be used for calculating the dependence of θ on φ for various concentrations of the organic substance if all the parameters in these equations are known. Further, a series of ε vs φ curves can be calculated from Eq. (III.18) and a series of electrocapillary curves from Eq. (III.72), in presence of various concentrations of the organic substance. These data are sufficient for plotting $\Delta\sigma$ vs log c curves for different φ, which, by the calculation conditions, should coincide when shifted along the abscissa axis, and $\Delta\xi$ vs log c curves for different ε, which should not coincide in the same way under the condition $C_0 \neq C'$. This last conclusion follows from the fact that isotherms congruent with respect to potential are compatible with isotherms congruent with respect to charge only when $C_0 = C'$, when the expressions (III.19) and (III.27) for $\varphi_{\varepsilon=0}$, which follows from Eqs. (III.7) and (III.8) respectively, coincide (see Chapter III, Section 1).

Calculations of this type were carried out by Frumkin et al. [11]; the adsorption parameters required for the calculation were chosen close to the values corresponding to the system 0.1 N NaF + n-C_4H_9OH [12]. As was to be expected, the $\Delta\sigma$ vs log c curves calculated for different φ coincide completely when moved along the abscissa axis. As regards the $\Delta\xi$ vs log c curves, in the charge range where adsorption of the usual organic compounds leads to appreciable lowering of σ the deviations between the calculated values of $\Delta\xi$ after displacement of these curves along the abscissa axis does not exceed 3-4 dyn/cm, which is within the limits of experimental error. This calculation explains why, despite the evident incompatibility of Eqs. (III.7) and (III.8) for the system 0.1 N NaF + n-C_4H_9OH, where $C_0/C' \approx 4.5$ [12], Parsons [13] found that the $\Delta\sigma$ vs log c and $\Delta\xi$ vs log c curves for this system coincided simultaneously within about 3-4 dyn/cm.

We must therefore conclude that the sensitivity of the surface pressure method is low and that this method cannot be used for solving the problem of the choice of the independent electrical variable.

Model methods based on measurements of differential capacity in presence of organic compounds are more accurate. It was shown in Chapter III, Section 3 that of the 14 equations for the various isotherms the best agreement with experimental data on adsorption of a number of organic compounds on mercury is given by the Frumkin isotherm (III.57). Accordingly, in our discussion of model methods for determination of the adsorption parameters of organic substances, based on measurements of the double layer capacity, we will use the equation of the Frumkin isotherm

$$Bc = \frac{\theta}{1-\theta} \exp(-2a\theta), \qquad (IV.5)$$

in which, in accordance with the data in Chapter III, Section, 1 the adsorption equilibrium constant B is a function of the electrode potential and not of the electrode charge.

2. Methods for Determination of the Attraction Constant at Various Electrode Potentials

a. Determination of the Attraction Constant at the Potential of Maximum Adsorption from the Shape of the Adsorption Isotherm. It follows from Eq. (III.110) that at the potential of maximum adsorption, when $(\partial\theta/\partial\varphi)_c = 0$, the additional capacity is zero and consequently the total differential capacity is $(\partial\varepsilon/\partial\varphi)_\theta$, i.e., is represented by Eq. (III.121). It is found experimentally that $d^2a/d\varphi^2$ is usually close to zero, so that the third term in Eq. (III.121) may be neglected. Therefore at the potential of maximum adsorption we can, in the first approximation, find the surface coverage from the expression

$$\theta = \frac{C_0 - C}{C_0 - C'} \qquad (IV.6)$$

if C' is known. The value of C' can usually be found to be a sufficient degree of accuracy by extrapolation of the $1/C$ vs $1/c$ curve
at $\varphi = \varphi_m$ to $1/c = 0$. Although Eq. (IV.6) is restricted to the region of potentials of maximum adsorption, where the differential
capacity curves have a minimum, it is extremely convenient for
determination of the adsorption isotherms of organic substances.

Laitinen and Mosier [14] used Eq. (IV.6) for studying adsorption isotherms of 30 different organic compounds on mercury, and
found that they approximately corresponded in form to the Langmuir isotherm. However, in a number of cases, when the concentration of highly adsorbable substances was low, adsorption
equilibrium was probably not reached when a dropping electrode
with a relatively short drop time (8-10 sec) was used [15-17].
This limits the significance of the conclusions reached in [14]
with regard to the form of the adsorption isotherm.

Frumkin and Damaskin [18] found from the results of differential capacity measurements in 1 N KI solutions with various
additions of $[(C_4H_9)_4N]I$ that the adsorption isotherm of tetrabutylammonium cations on mercury is S-shaped, which corresponds to
Eq. (IV.5) and indicates strong attractive interaction between the
adsorbed particles. At the same time, according to several investigators [15-17], when a dropping electrode was used at low
$[(C_4H_9)_4N]I$ concentrations the total amount of adsorbed particles
was determined by the diffusion rate and could, in the first approximation, be calculated with the aid of the Ilkovic equation [19].
Biegler and Laitinen [20] subsequently also found that the Ilkovic
equation is applicable to adsorption of leucoriboflavine from dilute
solutions on a mercury electrode.

Lorenz et al. [21, 22] determined the adsorption isotherms of
various amines, alcohols, and organic acids on mercury from
differential capacity data with the aid of Eq. (IV.6). It was shown
that, if attractive forces predominate between the adsorbed particles (higher alcohols and amines) the adsorption isotherm is
S-shaped; when repulsive forces predominate, as in the case of
$[(CH_3)_4N]^+$; $[(C_2H_5)_3HN]^+$, the adsorption isotherm lies below the

Langmuir isotherm. In the light of these results, the data of Laitinen and Morinaga [23], according to whom the adsorption isotherm of n-nonylic acid conforms to the Langmuir equation, appear to be highly improbable.

Dobren'kov and Bankovskii [24] used the dependence of capacity at the potential of maximum adsorption on the degree of surface coverage for studying the simultaneous adsorption of two organic substances on the electrode surface.

The following procedure is used for determining the attraction constant a in the Frumkin isotherm (IV.5) from the form of the adsorption isotherm calculated with the aid of Eq. (IV.6). The concentration of the organic substance corresponding to $\theta = 0.5$ is first found from a plot of θ vs c, and the experimental isotherm is then plotted in θ vs $y = c/c_{\theta=0.5}$ coordinates.

Relative concentrations are used because in this case the Eq. (IV.5) does not contain B. From Eq. (IV.5) at $\theta = 0.5$ it follows that

$$Bc_{\theta=0.5} = \exp(-a). \tag{IV.7}$$

Dividing Eq. (IV.5) by (IV.7), we obtain

$$y = \frac{c}{c_{\theta=0.5}} = \frac{\theta}{1-\theta}\exp[a(1-2\theta)]. \tag{IV.8}$$

Solving Eq. (IV.8) for a, we find

$$a = \frac{2.3}{1-2\theta}\log\left[\frac{y(1-\theta)}{\theta}\right]. \tag{IV.9}$$

Substitution of various values of θ and y from the experimental θ vs y curve into this equation gives a series of values of the attraction constant, from which the arithmetic mean value can be found.

Moreover, from Eq. (IV.8) it follows that

$$\frac{dy}{d\theta} = \exp[a(1-2\theta)]\frac{1-2a\theta(1-\theta)}{(1-\theta)^2}, \tag{IV.10}$$

$$\text{when } \theta = 0 \qquad (dy/d\theta)_{\theta=0} = e^a, \tag{IV.10a}$$

$$\text{when } \theta = 0.5 \quad (dy/d\theta)_{\theta=0.5} = 4-2a. \tag{IV.10b}$$

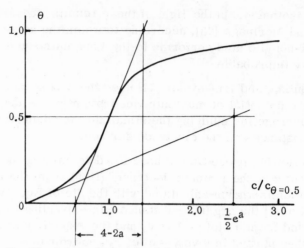

Fig. 37. Scheme for graphical determination of the attraction constant from the slope of the experimental adsorption isotherm.

Thus, the relations (IV.10a) and (IV.10b) give two more graphical methods for determination of a from the shape of the experimental isotherm; the principle of these methods is clear from the schematic diagram in Fig. 37.

b. Determination of the Attraction Constant at the Peak Potentials from Equilibrium C vs φ Curves. We first suppose that attractive interaction between the adsorbed particles is considerable and the concentration of the organic substance is not too low, so that the condition (III.43) is satisfied, when $\theta^{\max} \approx \theta^* = 0.5$. Under these conditions the peak height of the C vs φ curve is determined by the function h, which is of the following form for the Frumkin isotherm (IV.5):

$$h = \frac{\theta(1-\theta)}{1 - 2a\theta(1-\theta)} .$$
(IV.11)

Solving this equation for θ, we find

$$\theta = \frac{1}{2}(1 \pm r),$$
(IV.12)

where

$$r = \sqrt{\frac{2 - 2h(2-a)}{1 + 2ah}} .$$
(IV.13)

When $\theta = \theta^* = 0.5$, the function h passes through a maximum

$$h^* = \frac{1}{4 - 2a}. \tag{IV.14}$$

Let h be a certain fraction m of h^*. In this case, Eq. (IV.12) gives two values of θ, θ_1 and θ_2, corresponding at $\theta^{\max} \approx 0.5$ to the given relative peak height of the C vs φ curve: $m = C_{add}/C_{add}^{max}$. It is found that r is a monotonic function of a. Putting $h = m/(4 - 2a)$ into (IV.13), we obtain

$$r = \sqrt{\frac{(2 - a)(1 - m)}{2 - a(1 - m)}}. \tag{IV.15}$$

In the special cases of $m = 1/2$ or $m = 3/4$

$$r_{1/2} = \sqrt{\frac{2 - a}{4 - a}} \text{ and } r_{3/4} = \sqrt{\frac{2 - a}{8 - a}}. \tag{IV.15a}$$

Thus, if the surface coverages θ_1 and θ_2 are connected with the corresponding potentials (φ_1 and φ_2), the width of the peak on the C vs φ curve at a definite relative peak height becomes a function of the attraction constant a. For this we use Eqs. (III.112) and (III.113), from which it follows that

$$\left(\frac{\partial \theta}{\partial \varphi}\right)_c = -\left(\frac{\partial \ln c}{\partial \varphi}\right)_\theta h = -\left(\frac{\partial \ln c}{\partial \varphi}\right)_\theta \cdot \frac{\theta(1 - \theta)}{1 - 2a\theta(1 - \theta)}. \tag{IV.16}$$

Since the condition $\theta^{\max} \approx 0.5$ is equivalent to the relation

$$\frac{\partial \ln c}{\partial \varphi^{\max}} \approx \left(\frac{\partial \ln c}{\partial \varphi}\right)_{\theta = 0,5}, \tag{IV.17}$$

we find from Eq. (IV.16)

$$-\left(\frac{\partial \ln c}{\partial \varphi^{\max}}\right) d\varphi \approx \frac{1 - 2a\theta(1 - \theta)}{\theta(1 - \theta)} d\theta. \tag{IV.18}$$

We integrate the left-hand side of this equation between φ_1 and φ_2, which determine the peak width at a given relative peak height (Fig. 38), and the right-hand side between θ_1 and θ_2, which correspond to the values of φ_1 and φ_2 and are given by Eq. (IV.12). Taking into account the Lagrange mean value theorem, we obtain

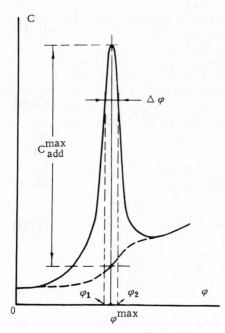

Fig. 38. Determination of peak width at a given relative peak height.

$$\left| \frac{\partial \ln c}{\partial \varphi^{\max}} \right|_{av} \Delta\varphi \approx 2 \left(\ln \frac{1+r}{1-r} - ar \right), \qquad \text{(IV.19)}$$

where r is defined by Eq. (IV.15) while $\left| \frac{\partial \ln c}{\partial \varphi^{\max}} \right|_{av}$ is the average slope of the $\ln c$ vs $\varphi^{\max}$ curve in the potential range from φ_1 to φ_2, approximately corresponding to $\varphi = \varphi^{\max}$. If we express the peak width $\Delta\varphi$ in mV and convert from natural to common logarithms, Eq. (IV.19) can be rewritten in the form

$$f(a) = 1000 \left(\log \frac{1+r}{1-r} - r \frac{a}{2.3} \right) \approx \frac{\Delta\varphi}{2} \left| \frac{\partial \log c}{\partial \varphi^{\max}} \right|_{av}, \qquad \text{(IV.20)}$$

where $f(a)$ is a monotonic function of a, which can be calculated with the aid of Eqs. (IV.15) and (IV.20). Calculated values of the function $f(a)$ for m = 1/2 and m = 3/4 are given in Table 4; the dependence of the function $f(a)$ on the attraction constant is shown graphically in Fig. 39.

As the right-hand side of Eq. (IV.20), equal to the function $f(a)$, is easy to determine experimentally, values of the attraction con-

TABLE 4

a	$f(a)$		a	$f(a)$	
	$m = \frac{1}{2}$	$m = \frac{3}{4}$		$m = \frac{1}{2}$	$m = \frac{3}{4}$
2.0	0	0	1.4	162.64	86.96
1.9	12.60	6.16	1.3	200.35	108.71
1.8	34.59	17.30	1.2	239.50	131.76
1.7	61.88	31.53	1.1	279.87	156.00
1.6	92.83	48.13	1.0	321.21	181.27
1.5	126.65	66.72	0.8	406.30	234.71

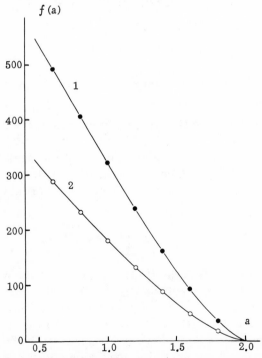

Fig. 39. Dependence of the function $f(a)$ on the attraction constant. 1) With m = 1/2; 2) with m = 3/4.

stant at the peak potentials of the C vs φ curves can be found without difficulty from Fig. 39.

The value of $\left| \dfrac{\partial \lg c}{\partial \varphi^{max}} \right|$ can be determined more exactly from the formula

$$\frac{\partial \log c}{\partial \varphi^{max}} \approx \left(\frac{\partial \log c}{\partial \varphi} \right)_{\theta=0.5} = \frac{\varepsilon_0 + C'\,(\varphi_N - \varphi^{max})}{2.3A}. \qquad (IV.21)$$

which follows from Eqs. (III.113) and (III.118). The constants φ_N and A in this formula are determined from the experimental plot of log c vs φ^{max} as described in the next section.

Another method for determining the attraction constant at $\varphi = \varphi^{max}$ is based on the dependence of the peak height on a. If we again assume that the condition (III.43) is satisfied and that therefore $\theta^{max} \approx 0.5$, we have from Eq. (III.123), neglecting the term $d^2a/d\varphi^2$, the following expression for the peak height:

$$C^{max} \approx \frac{C_0 + C'}{2} + \frac{[\varepsilon_0 + C'\,(\varphi_N - \varphi^{max})]^2}{2A\,(2-a)}, \qquad (IV.22)$$

from which a is easily determined, φ_N and A being known (see Section 3 of this chapter).

Finally, if the condition (III.43) is satisfied, the attraction constant in the region of the peak potentials of the C vs φ curves can be found from the slope of the linear plot of the peak height vs log c [see Eq. (III.47)], since K_1 in Eq. (III.47), i.e., the slope of the C^{max} vs log c plot in the case of the Frumkin isotherm is given by the expression

$$K_1 = \frac{2.3\,(C_0 - C')}{2-a}. \qquad (IV.23)$$

In order to establish the limits of applicability of these methods, restricted by the condition (III.43), for determination of the attraction constant, Damaskin and Tedoradze [25] used Eqs. (III.40) and (IV.11) to calculate C vs φ curves for the following conditions: $C_0 = 20\ \mu F/cm^2$; $C' = 5\ \mu F/cm^2$; $A = 1\ \mu J/cm^2$; $|\varphi^{max} - \varphi_m| = 0.7$ V, for $a = 1$ and $a = 0.5$. The attraction constants were then determined by the methods described above from the theoretical

TABLE 5

Assigned value of a	Calculated values of a			
	From peak widths		From peak height	From slope of C^{max} vs log c plots
	$m = 1/2$	$m = 3/4$		
1.0	0.93	0.94	0.95	0.98
0.5	0.23	0.29	0.27	0.37

curves. The results are given in Table 5; they show that the average error in determination of the attraction constant is 5% when $a = 1$, whereas when $a = 0.5$ it is as much as $\sim 42\%$.

Thus, the methods described above for determination of the attraction constant in the region of the peaks on the C vs φ curves can be used with sufficient accuracy only when $a \geq 1$; it follows from [26] that if a is a linear function of φ this condition refers only to the value of the attraction constant at the potential of maximum adsorption. At the same time, in the region of the peaks on the C vs φ curves the values of a may be considerably less than unity, but this does not affect fulfillment of the condition $\theta^{max} \approx 0.5$ and therefore the applicability of the methods described for determination of the attraction constant.

When $a < 1$, the relation between the peak height of the C vs φ curve and the value of the attraction constant is given by Eqs. (III.86) and (III.89). Eliminating a from these equations, we easily obtain [27] after algebraic transformations

$$\frac{(K^{max} + \theta^{max})^2 (1 - 2\theta^{max})}{(\theta^{max})^2 (1 - \theta^{max})^2} = 6\alpha (\varphi^{max} - \varphi_m)^2, \qquad (IV.24)$$

where $K^{max} = (C^{max} - C_0)/(C_0 - C')$.

With a known value of $\alpha = (C_0 - C')/2A$ (see Section 3 of this chapter), the surface coverage θ^{max} at the peak potential can be found with the aid of Eq. (IV.24). Equation (IV.24) is solved most simply by numerical trial and error with the condition $0 < \theta^{max} < 0.5$.

Subsequently, with θ^{max} known, we can finally use Eq. (III.89) to find the required value of a at the peak potential of the C vs φ

154

CHAPTER 4

curve. Solving Eq. (III.89) for a, we obtain after algebraic transformations

$$a = \frac{1}{2\theta^{max}(1-\theta^{max})} - \frac{\alpha(\varphi^{max}-\varphi_m)^2}{K^{max}+\theta^{max}}. \qquad (IV.25)$$

If $a \geq 1$ for the given system, then $\theta^{max} \approx 0.5$, and it is easy to show that if the dependence of C_0 on φ is taken into account Eq. (IV.25) becomes the relation (IV.22), already discussed.

c. Determination of the Attraction Constant at the Peak Potentials from Nonequilibrium C vs φ Curves. In Chapter III, Section 4b, we discussed certain properties of nonequilibrium differential capacity curves determined in presence of organic compounds; it was noted that the deviation from equilibrium increases with increasing energy of adsorption of the organic substance and with increase of attraction interaction between the adsorbed molecules. On the other hand, weak attractive interaction between adsorbed organic molecules is usually characteristic of weakly adsorbed substances having low molecular weights. Under these conditions appreciable adsorption effects occur only at fairly high concentrations of the organic substance, when deviations of capacity from the equilibrium values can be neglected at ac frequencies of the order of 50-400 Hz. Therefore the methods for determination of the attraction constant from nonequilibrium C vs φ curves can be fully limited by the condition $a \geq 1$, when the approximate relation (IV.18) is satisfied. With the Lagrange mean value theorem taken into account, integrating the left-hand side of Eq. (IV.18) between φ_1 and φ_2 we again obtain $\Delta\varphi \left| \frac{\partial \ln c}{\partial \varphi^{max}} \right|_{av}$, where $\Delta\varphi = |\varphi_2 - \varphi_1|$ is the peak width at the given relative height. However, in distribution from equilibrium C vs φ curves, the values of θ_1 and θ_2 for the potentials φ_1 and φ_2 no longer correspond to h = 3h*/4, and therefore cannot be calculated from Eqs. (IV.11) and (IV.15). The values θ_1 and θ_2 corresponding to the condition

$$C_{add(\omega)} = \frac{3}{4} C_{add(\omega)}^{max}$$

have been determined by Damaskin and Dyatkina [28].

According to [28],

$$\theta_{1,2} = \frac{1 \pm \sqrt{1-4x}}{2}, \qquad (IV.26)$$

where $x = [B - (B^2 - AC)^{\frac{1}{2}}]/A$, such that $A = (4a - n)(8a^2 - 4an + n^2) - 8(40a^2 - 20an + n^2) + 128(4a - n)$; $B = 128 - 8(4a - n) - (16a^2 - 8an - n^2)$; $C = 6(8 - 4a + n)$; a is the attraction constant; $n = (\Gamma_m/c)(2\omega/D)^{\frac{1}{2}}$ [see Eq. (III.93)].

The values of θ_1 and θ_2 found from Eq. (IV.26) are taken as the limits in integration of the right-hand side of Eq. (IV.18) in the case of nonequilibrium C vs φ curves; this gives, instead of Eq. (IV.20),

$$f(a, n) = 1000 \left(\log\frac{1 + \sqrt{1 - 4x}}{1 - \sqrt{1 - 4x}} - \frac{a}{2.3}\sqrt{1 - 4x}\right) \approx \frac{\Delta\varphi}{2}\left|\frac{\partial \log c}{\partial \varphi^{max}}\right|_{av}, \quad \text{(IV.27)}$$

where $f(a, n)$ is a function of a and n, which can be calculated

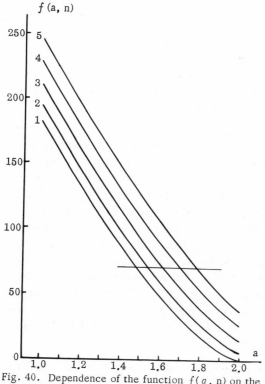

Fig. 40. Dependence of the function $f(a, n)$ on the attraction constant with the condition m = 3/4 for various values of n: 1) 0; 2) 0.5; 3) 1.0; 4) 1.5; 5) 2.0.

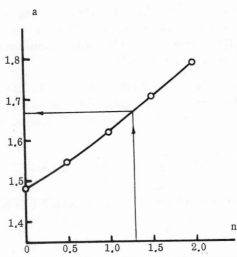

Fig. 41. Dependence of the attraction constant
on n for the condition $f(a, n) = 70$.

with the aid of Eqs. (IV.26) and (IV.27). Calculated values of the
function $f(a, n)$ for n = 0, 0.5, 1.0, 1.5, and 2.0 in the range from
$a = 1$ to $a = 2$ are given in Fig. 40.

These data can be used for approximate determination of the
attraction constant from experimental nonequilibrium differential
capacity curves. This is done as follows: values of the function
$f(a, n)$ are found from experimental values of $\Delta\varphi$ and
$|\partial \log c / \partial \varphi^{max}|_{av}$ in accordance with Eq. (IV.27), and a is then
plotted against n from points taken from Fig. 40. A plot of this
type for $f(a, n) = 70$ is given in Fig. 41. It is then necessary to
find the value of $n = (\Gamma_m /c)(2\omega/D)^{\frac{1}{2}}$ in order to determine a. The
values of c and ω are known from experimental data. Methods for
determination of Γ_m are described in the following section. As
regards the diffusion coefficient D, in view of the fact that the D
values for different organic molecules should not differ too much,
while n depends on $D^{\frac{1}{2}}$, we can take $D \approx 9 \cdot 10^{-6}$ cm^2/sec as a
first approximation [29].

When n is known, we can finally find the attraction constant a
from the plot of a vs n. However, it may be seen from Fig. 41
that in the first approximation the dependence of a on n may be
regarded as linear:

$$a \approx a_{\text{eff}} + bn = a_{\text{eff}} + b\frac{\Gamma_m}{c}\sqrt{\frac{2\omega}{D}}, \qquad \text{(IV.28)}$$

where a_{eff} is the effective value of the attraction constant, obtained with the nonequilibrium character of the C vs φ curves disregarded, and b is a certain coefficient, approximately equal to 0.15. The use of Eq. (IV.28) is all the more justified because a fairly approximate value of the diffusion coefficient must be used for finding n.

It is easy to see that the proposed method for calculating the attraction constant from nonequilibrium C vs φ curves involves a number of approximate assumptions (in particular $\theta^{\max} \approx 0.5$) which by their nature must lead to somewhat low values of a. To verify this, Eqs. (III.94) and (III.100) were used in conjunction with Eq. (III.30) and the Frumkin isotherm (IV.5) for calculating portions of nonequilibrium C vs φ curves, with the use of the following constants: $C_0 = 20 \ \mu\text{F/cm}^2$; $\alpha = 7.5 \ \text{V}^{-2}$; $A = 1 \ \mu\text{J/cm}^2$; $k = 5$; $|\varphi^{\max} - \varphi_m| = 0.65$ V, for two values of the attraction constant: $a = 1.5$ and $a = 2.9$ [30]. The method described above was then used for calculating the values of a, given in Table 6, from the theoretically calculated C vs φ curves.

Table 6 shows that correct values of the attraction constant are obtained if b = 0.2 is used.

Thus, the attraction constant can be determined with satisfactory accuracy from nonequilibrium C vs φ curves with the aid of the following semiempirical expression:

$$a \approx a_{\text{eff}} + 0.2 \frac{\Gamma_m}{c}\sqrt{\frac{2\omega}{D}} \approx a_{\text{eff}} + 236 \frac{\Gamma_m\sqrt{\nu}}{c}, \qquad \text{(IV.29)}$$

where ν is the ac frequency in Hz, Γ_m is in mole/cm^2, and c is in mole/cm^3.

TABLE 6

Assigned value of a	Calculated values of a		
	Nonequilibrium nature of C vs φ curves disregarded	Nonequilibrium nature of C vs φ curves taken into account	
		$b = 0.15$	$b = 0.2$
1.5	1.30	1.45	1.50
1.9	1.60	1.81	1.89

The use of formula (IV.29) may be illustrated by the example of adsorption of aliphatic alcohols on mercury [12]. Calculations showed that in the case of n-propyl and n-butyl alcohols corrections for the nonequilibrium nature of the C vs φ curves could be ignored. However, for n-amyl and especially n-hexyl alcohols the differences between the true values of the attraction constant and a_{eff} become appreciable (Table 7).

Tedoradze and Arakelyan [31] showed that reliable values of the attraction constant can also be obtained from the height of nonequilibrium C vs φ curves if the equilibrium peak height is first determined with aid of formula (II.12) (see Chapter II, Section 3), and then the relation (IV.22), discussed above, is applied. However, in addition to determinations of capacity, this method also requires accurate determination of the ohmic component of the adsorption impedance, which gives rise to additional experimental difficulties.

The methods described above can be used for determination of attraction constant in various regions of the C vs φ curve from experimental data, and hence for finding the dependence of a on the electrode potential.

In many cases a linear relationship is found between a and φ. Values of a obtained by different methods for the system $1 N$ $Na_2SO_4 + n\text{-}C_4H_9OH$, illustrating the linear dependence of the attraction constant on the electrode potential, are presented in Fig. 42. The constants a_0 and β in Eq. (III.124) can be determined without difficulty from this graph. A parabolic relationship between the attraction constant and the potential is much rarer (see Fig. 30). The causes of these relationships between a and φ have already been discussed in Chapter III.

TABLE 7

c, M	n-C_5H_{11}OH				c, M	n-C_6H_{13}OH			
	cathodic peak		anodic peak			cathodic peak		anodic peak	
	a_{eff}	a	a_{eff}	a		a_{eff}	a	a_{eff}	a
0.15	1.30	1.31	1.66	1.67	0.03	1.46	1.53	1.74	1.81
0.10	1.26	1.28	1.60	1.62	0.016	1.32	1.46	1.67	1.81
0.05	1.21	1.26	1.59	1.64	0.008	1.23	1.50	1.53	1.80

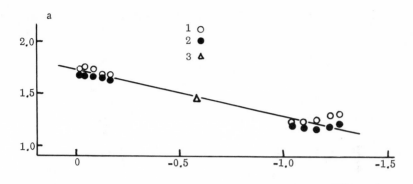

φ, V(N.C.E.)

Fig. 42. Dependence of the attraction constant on the potential of the mercury electrode in the system 1 N Na_2SO_4 + n-C_4H_9OH. Data obtained from:
1) peak heights of C vs φ curves; 2) peak widths of c vs φ curves; 3) form of
the adsorption isotherm.

3. Methods for Determination of the Adsorption Potential Difference, the Maximum Amount of Organic Substance Adsorbed, and the Free Energy of Adsorption

The adsorption potential difference is determined by the value
of φ_N; the maximum amount Γ_m of organic substance adsorbed, by

$$A = RT\Gamma_m \qquad\qquad (IV.30)$$

and the free energy of adsorption at $\varphi = 0$ and $\theta = 0$ by B_0:

$$-\Delta\overline{G}_A^0 = RT \ln(55.5B_0). \qquad\qquad (IV.31)$$

Therefore the problem is reduced to finding the values of φ_N, A,
and B_0 from experimental differential capacity curves, and in part,
from electrocapillary curves.

When the condition $a \geq 1$ is satisfied, φ_N, A, and B_0 are found
most simply from regions of the log c vs φ^{max} curve corresponding
to fairly high concentrations of the organic substance. Under such
conditions $\theta^{max} \approx 0.5$ may be assumed with sufficient accuracy.
Thus, from the isotherm (IV.5) and Eq. (III.119) we obtain the approximate equation of this curve

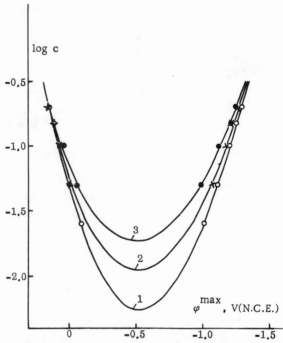

Fig. 43. Dependence of the peak potentials of the differential capacity curves on the logarithm of the concentration of amyl alcohol in the system $0.1\ N$ NaF +
n-C_5H_{11}OH at different temperatures: 1) 25°; 2) 50°;
3) 75°C. The continuous lines are calculated from Eq.
(IV.32); the points represent experimental data ([12] in
Chap. IV).

$$\log c \approx -\log B_0 - \frac{a_0}{2.3} + \frac{E + C'\varphi^{max}\,(\varphi_N - \varphi^{max}/2)}{2.3A}\ . \qquad \text{(IV.32)}$$

Figure 43 demonstrates the good agreement of this equation
with experimental data for the system $0.1\ N$ NaF + n-C_5H_{11}OH at
three different temperatures: 25, 50, and 75 °C [12]; accordingly,
Eq. (IV.32) may be used as the basis for determination of the constants φ_N, A, and B_0 from experimental log c vs φ^{max} curves.
Using the points of intersection of the log c vs φ^{max} curve with
the straight line corresponding to log c = const (see Fig. 44), we
obtain, in accordance with Eq. (IV.32)

$$\varphi_N \approx \frac{E_2 - E_1}{C'\,(\varphi_1 - \varphi_2)} + \frac{\varphi_1 + \varphi_2}{2}\ . \qquad \text{(IV.33)}$$

The choice of the potentials φ_1 and φ_2 in formula (IV.33) is clear from Fig. 44a, while E_1 and E_2 are values of the function E [see Eq. (III.109)] at the potentials φ_1 and φ_2.

It is convenient to use intersections of the log c vs φ^{max} curve with several log c = const lines, find several values of φ_N from Eq. (IV.33), and take the average.

To find A, we use Eq. (IV.32) for two concentrations of the organic substance: high (c_h) and low (c_l). It is seen in Fig. 44b that on each of the branches of the log c vs φ^{max} curve these concentrations correspond to the potentials φ_h and φ_l. Each potential, in its turn, has its own value of the function E [see Eq. (III.109)]. Writing Eq. (IV.32) for c_h and c_l and subtracting one from the other, we obtain after algebraic transformations

$$A \approx \frac{\Delta E + C' \varphi_N (\varphi_h - \varphi_l) - \frac{C'}{2} (\varphi_h^2 - \varphi_l^2)}{2.3 \Delta \log c} , \qquad \text{(IV.34)}$$

where $\Delta E = E(\varphi_h) - E(\varphi_l)$ and $\Delta \log c = \log c_h - \log c_l$. Formula (IV.34) is applied to both branches of the log c vs φ^{max} curve, and the average value of the constant A is then found.

When all the constants in Eq. (IV.32) with the exception of B_0 are known, it can be used directly for finding B_0. For this purpose it is convenient to calculate the dependence of the following function on the potential:

$$P(\varphi) = -\frac{a_0}{2.3} + \frac{E + C' \varphi (\varphi_N - \varphi/2)}{2.3 A} . \qquad \text{(IV.35)}$$

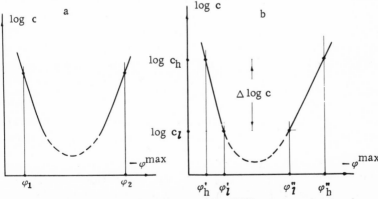

Fig. 44. Schematic log c vs peak potential (φ^{max}) curve, explaining determination of φ_N (a) and A (b) (explanation in text).

It follows from Eqs. (IV.32) and (IV.35) that the mean deviation of this function from experimental regions of the log c vs φ^{max} curves at fairly high concentrations of the organic substance is equal to $-\log B$. The last unknown constant, B_0, is thus found.

If, in consequence of the strong dependence of the attraction constant on the electrode potential, adsorption–desorption peaks are observed on only the cathodic (or anodic) branches of the C vs φ curves (see Fig. 36), or if the condition $a \geq 1$ is not satisfied, φ_N can be determined from the curve representing the dependence of the charge on the potential at the highest concentrations of the organic substance. It follows from Eq. (III.120) that under these conditions the equation of the tangent to the ε vs φ curve at the point of least slope (i.e., at $\theta \approx 1$) is $\varepsilon' = C'(\varphi - \varphi_N)$. Consequently this tangent cuts the abscissa axis at the point $\varphi = \varphi_N$ (see Fig. 12).

Finally, Frumkin et al. [32] showed that φ_N can be found by extrapolation of the plot of the point of zero charge $\varphi_{\varepsilon=0}$ vs the concentration to $1/c = 0$; by this method, the $\varphi_{\varepsilon=0}$ values are determined from the position of the minimum on the C vs φ curves for very dilute solutions. The considerable experimental difficulties in work with very dilute (10^{-4}-10^{-3} N) solutions of the supporting electrolyte are a disadvantage of this method.

The method of Damaskin and Lerkkh [27] is convenient for determination of A under the condition $a < 1$. An adsorption isotherm is first plotted with the aid of Eq. (IV.6) in θ vs $y = c/c_{\theta=0.5}$ coordinates for the potential of maximum adsorption $\varphi = \varphi_m$. The value of a at $\varphi = \varphi_m$ and the value of θ at a certain chosen concentration of the organic substance, at which the electrocapillary curve is determined, are found from this isotherm with the aid of Eq. (IV.9). Thus, at $\varphi = \varphi_m$ all the quantities with the exception of A are known in the equation of state corresponding to the Frumkin isotherm:

$$\Delta\sigma = -A\left[\ln(1-\theta) + a\theta^2\right]. \tag{IV.36}$$

Tedoradze and Arakelyan [31] have described a method of determining the maximum adsorption Γ_m, based on measurements of the components $R_{p(\omega)}$ and $C_{add(\omega)}$ of the adsorption impedance and on the use of Eq. (III.111). It is easy to show that in the case of the isotherm (IV.5)

$$\left(\frac{\partial \Gamma}{\partial c}\right)_\varphi = \frac{\Gamma_m}{c} \cdot \frac{\theta (1-\theta)}{1 - 2a\theta (1-\theta)}. \tag{IV.37}$$

Combination of this equation with Eq. (III.111) leads, after algebraic transformations, to the following expression for Γ_m^2:

$$\Gamma_m^2 = \frac{[\varepsilon_0 + C' (\varphi_N - \varphi)]^2 c}{RT C_{add\,(\omega=0)}} \cdot \left(\frac{\partial \Gamma}{\partial c}\right)_\varphi. \tag{IV.38}$$

From Eqs. (II.6) and (II.8) it follows that

$$\left(\frac{\partial \Gamma}{\partial c}\right)_\varphi = \sqrt{\frac{2D}{\omega}} \cdot \frac{1}{C_{add(\omega)}\, R_{p(\omega)}\, \omega - 1}. \tag{IV.39}$$

Dividing Eq. (IV.39) by (II.12), we obtain

$$\left(\frac{\partial \Gamma}{\partial c}\right)_\varphi \bigg/ C_{add(\omega=0)} = \frac{\sqrt{2D\omega} R_{p(\omega)}}{C_{add(\omega)}^2 R_{p(\omega)}^2 \omega^2 + 1}. \tag{IV.40}$$

Inserting (IV.40) into (IV.38), we obtain the final expression for Γ^2:

$$\Gamma_m^2 = \frac{[\varepsilon_0 + C' (\varphi_N - \varphi)]^2 c R_{p(\omega)} \sqrt{2D\omega}}{RT\, (C_{add(\omega)}^2 R_{p(\omega)}^2 \omega^2 + 1)}. \tag{IV.41}$$

The following method may be used for determining B_0 when $a < 1$. Equation (IV.6) is first used for plotting the adsorption isotherm at the potentials of maximum adsorption φ_m [if $a = a(\varphi)$, then φ_m depends on c], and the concentration at which $\theta = 0.5$ is found. Thus, B and a in Eq. (IV.7) are referred to $\varphi = \varphi_m$ at $\theta_m = 0.5$, and consequently we obtain from Eqs. (III.119) and (IV.7)

$$B_0 c_{\theta_m=0.5} \exp\left[-\frac{E_m + C'\varphi_m\left(\varphi_N - \frac{\varphi_m}{2}\right)}{A}\right] \exp(a_0 - a_m) = \exp(-a_m), \tag{IV.42}$$

where E_m is the value of the function E at $\varphi = \varphi_m$ for $\theta_m = 0.5$.

From Eq. (IV.42) it follows that

$$B_0 = \frac{\exp(-a_0)}{c_{\theta_m=0.5}} \exp\left[\frac{E_m + C'\varphi_m\left(\varphi_N - \frac{\varphi_m}{2}\right)}{A}\right], \tag{IV.43}$$

from which B_0 can be easily calculated if φ_m at $\theta_m = 0.5$ is known. The value of φ_m can be determined experimentally from the position of the minimum on the C vs φ curve corresponding to concentration $c = c_{\theta m = 0.5}$ of the organic substance.

The methods described in this and the preceding sections were used by Damaskin et al. [12] for systematic study of effects of the length of the hydrocarbon chain and of temperature on adsorption of aliphatic alcohols on mercury. The results are presented in Table 8, where the constants characterizing the adsorption state of the organic compounds have the following dimensions: a_0, dimensionless; β, V^{-1}; C', $\mu F/cm^2$; φ_N, V; A, $\mu J/cm^2$; Γ_m, M/cm^2; S, $\mathring{A}^2$/molecule; B_0, liters/mole; $-\Delta G_A^0$, kcal/mole.

It is seen from this table that attractive interaction between the adsorbed alcohol molecules clearly increases in the sequence from ethyl to n-hexyl alcohol, as was to be expected on simple physical grounds. In addition to increasing attractive interaction between the adsorbed molecules, increase of the chain length also leads to a slight increase of the slope of the a vs φ relationship, i.e., of β. As was shown in Chapter III, Section 6, the value $\beta > 0$ reflects the discrete nature of the adsorbed organic dipoles, which is therefore somewhat more pronounced in the case of longer organic molecules.

As was to be expected from the formula for a flat condenser, the limiting value of the capacity C' decreases with increasing chain length. Quantitative application of this formula for the capacity C' shows that under these conditions the dielectric constant in the electric double layer is considerably greater than the dielectric constant of the medium filled with hydrocarbon chains [33]. This indicates that at saturation adsorption, when the experimental $\theta = 1$, the electric double layer contains water molecules which increase its dielectric constant. It is also possible that owing to this effect the areas per adsorbed molecule at $\theta = 1$ (s) are considerably greater than the area corresponding to unbranched hydrocarbon chains in condensed films of higher alcohols or acids ($\sim$20-21 $\mathring{A}^2$). Table 8 shows that the maximum adsorption potential difference clearly diminishes as we pass from C_2H_5OH to $n-C_6H_{13}OH$. This effect can be attributed to a certain extent to the observed increase of the area occupied by one alcohol molecule with increasing length of the hydrocarbon chain (see Table 8), since the probability of deviation of the alcohol dipoles from vertical orientation at the interface increases with increase of S. It is evident that increase of φ_N is to be expected in the case of higher

TABLE 8

alcohol	t, °C	a_0	β	C'	φ_N	A	$\Gamma_m \cdot 10^{10}$	S	B_0	$-\Delta\bar{G}_A^0$
C_2H_5OH	25	0.80	—	6.40	0.36	—	—	—	0.41	1.86
n-C_3H_7OH	25	1.11	0.18	4.85	0.33	1.40	5.66	29.5	3.00	3.05
n-C_4H_9OH	25	1.28	0.20	4.80	0.24	1.28	5.18	32.2	11.2	3.83
	50	1.33	0.17	5.07	0.18	1.41	5.25	31.6	6.61	3.81
	75	1.35	0.06	5.15	0.16	1.50	5.19	32.0	4.52	3.83
n-$C_5H_{11}OH$	25	1.48	0.27	4.20	0.21	1.20	4.85	34.2	39.9	4.58
	50	1.49	0.24	4.45	0.18	1.32	4.90	33.9	20.0	4.52
	75	1.50	0.17	4.55	0.17	1.45	5.00	33.3	11.8	4.51
n-$C_6H_{13}OH$	25	1.68	0.26	3.85	0.20	1.14	4.61	36.2	154	5.39

alcohols forming condensed films at the interface ($S \approx 21$ Å^2). Thus, φ_N should pass through a minimum with increasing length of the hydrocarbon chain.

If the value of B_0 for ethyl alcohol is disregarded, the other alcohols, according to literature data, obey the Traube rule with the average value $k = 3.7$.

Rise of temperature leads to a very slight increase of the attraction constant and to decrease of the coefficient representing the deviation of the system from the model of two condensers in parallel, due to the discrete character of the adsorbed organic dipoles. In accordance with simple physical considerations, this last effect is somewhat more pronounced for alcohols of shorter chain length. Table 8 shows that Γ_m and $-\Delta\bar{G}_A^0$ are virtually independent of the temperature. Accordingly, with formulas (IV.30) and (IV.31) taken into account, variations of A and B_0 with temperature should be regarded as trivial.

4. Method Based on the Dependence of the Differential Capacity on the Concentration of the Organic Substance

Parsons [10] developed a method based on the assumption that the adsorption isotherms are congruent with respect to the electrode charge, i.e., that Eq. (III.8) is valid. However, it was shown in Chapter III, Section 1 that experimental data on the dependence of p.z.c. ($\varphi_{\varepsilon=0}$) on the adsorption of a number of organic compounds indicate that this assumption is not justified when $C' < C_0$ (see also [11]). We will therefore examine a modification of the Parsons method where the adsorption isotherms of the organic substance are in the first approximation congruent with respect to the electrode potential, i.e., conform to Eq. (III.7). As was shown in Chapter III, Section 1, under these conditions the relation (III.11), which may be written in the form

$$\varepsilon - \varepsilon_0 = A\theta \frac{d \ln B}{d\varphi} , \qquad (IV.44)$$

is satisfied. Differentiating (IV.44) with respect to potential, we obtain

$$C - C_0 = A\theta \frac{d^2 \ln B}{d\varphi^2} + A \frac{\partial \theta}{\partial \ln B} \cdot \left(\frac{d \ln B}{d\varphi}\right)^2 =$$
$$= A\theta \frac{d^2 \ln B}{d\varphi^2} + A \left(\frac{d \ln B}{d\varphi}\right)^2 h. \qquad (IV.45)$$

We will first suppose that the energy of adsorption is a linear function of the electrode potential. In this case $d \ln B / d\varphi = $ const [see Eq. (III.31)] and consequently $d^2 \ln B / d\varphi^2 = 0$. Equation (IV.45) is the simplified to

$$C - C_0 = A \left(\frac{d \ln B}{d\varphi}\right)^2 h, \qquad (IV.46)$$

i.e., $(C - C_0)$ turns out to be proportional to the function h.

We will now find the dependence of the function h on ln c under the condition that adsorption of the organic substance conforms to the Frumkin isotherm (IV.5). Taking logarithms of (IV.5), we obtain

$$\ln B + \ln c = \ln \frac{\theta}{1-\theta} - 2a\theta. \qquad (IV.47)$$

Inserting the values of θ expressed in terms of the function h in accordance with Eqs. (IV.12) and (IV.13), we obtain after algebraic transformations

$$\ln c = -a - \ln B \pm \left(\ln \frac{1+r}{1-r} - ar\right), \qquad \text{(IV.48)}$$

where r is given by the expression (IV.13). Analysis of Eq. (IV.48) shows that the dependence of h and also, in accordance with (IV.46), of $(C - C_0)$ at φ = const on ln c must pass through a maximum corresponding to r = 0. Under this condition h = h* and ln c = $-a - \ln B$. If $(C - C_0)_\varphi$ is a fraction of m (e.g., 1/2 or 3/4) of the maximum value $(C - C_0)_\varphi^{max}$, then h = mh*, r is represented by Eq. (IV.15), and the width of the $(C - C_0)_\varphi$ vs log c curve at a given relative height (Δ log c) is a monotonic function of the attraction constant

$$\Delta \log c = 2\left(\log\frac{1+r}{1-r} - \frac{a}{2.3}r\right). \qquad \text{(IV.49)}$$

Comparison of Eqs. (IV.20) and (IV.49) shows that the function $f(a)$, data on which are presented in Fig. 39 and Table 4, is connected with Δ log c by the simple relation

$$f(a) = 500\Delta\log c. \qquad \text{(IV.50)}$$

This makes it possible to determine attractive interaction between the adsorbed particles from a plot of $f(a)$ vs a, using the width of experimental plots of $(C - C_0)_\varphi$ vs log c.

Further, knowing the concentration corresponding to the maximum of the $(C - C_0)_\varphi$ vs log c curve (c*), and the value already found for a, we can use the formula

$$\log B = -\frac{a}{2.3} - \log c^* \qquad \text{(IV.51)}$$

for finding the adsorption equilibrium constant B at various electrode potentials.

When log B and the value of the attraction constant are known, it is possible to find θ for each concentration from the log (Bc) vs θ plot calculated with the aid of the adsorption isotherm equation (see Fig. 28), and consequently to find the function h with the aid of Eq. (IV.11). According to Eq. (IV.46), the plot of C vs h at φ = const at various concentrations of the organic substance should be a straight line having slope $A(d\ln B/d\varphi)^2$.

On the other hand, the slope of the linear plot of ε vs θ at φ = const, according to Eq. (IV.44), is $A \ln B/d\varphi$. This permits separate determination of $A = RT\Gamma_m$ and $d\ln B/d\varphi$. In addition, $d\ln B/d\varphi$ can be found from the slope of the linear log B vs φ plot, since the values of B at different potentials are already known from Eq. (IV.51).

It follows from Eq. (III.34) that a linear dependence of the energy of adsorption on the potential corresponds to fulfillment of the condition $C_0 = C'$. Then

$$\frac{d\ln B}{d\varphi} = -\frac{C'\varphi_N}{A} \tag{IV.52}$$

and it is therefore possible to determine the shift of potential due to adsorption, φ_N, from the value of the derivative $d\ln B/d\varphi$, while the formula

$$B = B_0 \exp\left(-\frac{C'\varphi_N \varphi}{A}\right) \tag{IV.53}$$

can be used for calculating B_0, all the other quantities being known.

The condition $C_0 \approx C'$ is satisfied, in the first approximation, in adsorption of thiourea [2], m-phenylenedisulfonate ions [4], and inorganic anions [6-8] on mercury. It follows from Chapter III, Section 1, that if this condition is satisfied the results should be independent of the assumption on which the nonthermodynamic method is based $B(\varphi)c = f(\theta)$ or $G(\varepsilon)c = f(\theta)$. This conclusion was confirmed experimentally by Payne [8]. Thus, certain results given by Parsons [10] for adsorption of thiourea, m-phenylenedisulfonate ions, and iodide ions may serve as an illustration of this method when the condition $d\ln B/d\varphi^2 \approx 0$ is satisfied.

We will now examine the case when the energy of adsorption is a quadratic function of the electrode potential, and the adsorption equilibrium constant is therefore given by Eq. (III.30). Under these conditions Eq. (III.40) for the differential capacity is valid; it is convenient to write it in the form

$$C - C_0 = -2A\alpha\theta + 4A\alpha^2(\varphi - \varphi_m)^2 h. \tag{IV.54}$$

Further, from Eq. (III.18), after simple transformations, we obtain

$$(\varepsilon - \varepsilon_m) - (\varepsilon_0 - \varepsilon_m) = -2A\alpha\theta(\varphi - \varphi_m). \tag{IV.55}$$

From Eq. (IV.55) we have

$$\frac{\varepsilon - \varepsilon_m}{\varphi - \varphi_m} - \frac{\varepsilon_0 - \varepsilon_m}{\varphi - \varphi_m} = -2A\alpha\theta, \tag{IV.56}$$

or

$$K - K_0 = -2A\alpha\theta, \tag{IV.57}$$

where K and K_0 are the integral capacities in presence of the organic substance and in a pure solution of the supporting electrolyte; however, they are referred to the potential of maximum adsorption and not to the p.z.c.

Introducing Eq. (IV.57) into (IV.54), we obtain

$$(C - C_0)_\varphi - (K - K_0)_\varphi = 4A\alpha^2 (\varphi - \varphi_m)^2 h, \tag{IV.58}$$

or, in more general form

$$(C - C_0)_\varphi - (K - K_0)_\varphi = A \left(\frac{d \ln B}{d\varphi}\right)^2 h. \tag{IV.59}$$

Comparison of Eqs. (IV.46) and (IV.59) leads to the conclusion that in the case of a quadratic dependence of the energy of adsorption on the electrode potential, instead of the dependence of $(C_0 - C)_\varphi$ on log c we must plot and analyze the dependence of $[(C - C_0)_\varphi - (K - K_0)_\varphi]$ on log c. The subsequent calculation of the adsorption parameters is exactly the same as in the case of a linear dependence of the energy of adsorption on the potential, discussed above.

Survila and Damaskin [34], with the system $0.1\,N$ Na_2SO_4 + n-C_4H_9OH taken as the example, compared two methods for calculating the adsorption parameters: the method described in Sections 2 and 3 (method I), and the modified Parsons method, described above for the condition $B(\varphi)c = f(\theta)$ and $C' < C_0$ (method II). The differential capacity curves of a mercury electrode in 16 different solutions (from 0.02 to 0.8 M) of n-butyl alcohol were determined, and the corresponding calculations were carried out. The dependences of log B and a on the potential calculated by the two methods virtually coincided. The constants A and B_0 determined by method I were 1.28 and 9.1, whereas the corresponding values found by method II were 1.18 and 7.8.

The observed agreement between the adsorption parameters found by methods I and II should be regarded as satisfactory in view of the considerable scatter of the points around the presumed linear dependences of $[(C - C_0)_\varphi - (K - K_0)_\varphi]$ on h and, to a lesser degree, of $(\varepsilon - \varepsilon_0)_\varphi$ on θ. This could not fail to affect the accuracy in determination of these linear plots and consequently of the values of A and d ln $B/d\varphi$ found by method II.

5. Devanathan's Method

Devanathan's nonthermodynamic method [35, 36] is based on the assumption that the potential differences dependent on the surface charge, specifically adsorbed ions, and organic molecules are additive at a constant electrode charge. On the basis of this assumption in the case when introduction of an organic substance into the surface layer does not affect its dielectric constant, Devananthan [35] derived the following expression for the dependence of adsorption on the electrode charge:

$$\frac{1}{C} = \frac{1}{C_o} + \frac{d\Gamma}{d\varepsilon} \cdot \frac{2\pi\mu}{D} , \qquad (IV.60)$$

where D is the dielectric constant in the layer between the metal surface and the external Helmholtz plane; μ is the dipole moment of the organic molecules.

In the derivation of Eq. (IV.60), Devanathan [35] used a relation according to which

$$\varphi_N = 2\pi\Gamma_m\mu/D. \qquad (IV.61)$$

The right-hand side of Eq. (IV.61) contains the coefficient 2 instead of the generally accepted value of 4. However, the arguments put forward in support of this change [35, 38] can hardly be accepted [39-41].

An expression analogous to Eq. (IV.60) can be derived from the model of two parallel condensers if it is assumed that $C_0 = C' = const$. If this condition is satisfied, it follows from Eq. (III.17) that

$$\varepsilon = C_0 \left(\varphi - \varphi_N \frac{\Gamma}{\Gamma_m}\right). \qquad (IV.62)$$

Differentiating (IV.62) with respect to charge, we obtain

$$\frac{1}{C} = \frac{1}{C_o} + \frac{d\Gamma}{d\varepsilon} \cdot \frac{\varphi_N}{\Gamma_m} . \qquad (IV.63)$$

It is easy to see that combination of Eqs. (IV-63) and (IV-61) gives Eq. (IV-60).

When D and μ are known, Eq. (IV.60) can be used for finding the dependence of $d\Gamma/d\varepsilon$ on ε, and hence Γ itself, because at sufficiently negative ε the value of Γ tends to zero. Devanathan [35] used Eq. (IV.60) for calculating the adsorption of thiourea from experimental C vs φ curves obtained by Schapink et al. [3]. Devanathan used the experimental value μ = 4.89 for thiourea, and the value D = 7.19, which he regards as the most probable on the basis of capacity measurements in solutions of potassium halides. The values of Γ calculated in this way are in approximate agreement with thermodynamic data [3], but this agreement disappears if the coefficient 2 is replaced by 4 in Eqs. (IV.60) and (IV.61). It should also be pointed out that, to judge by the values of capacity at different ε given in [3], the assumption that $C_0 = C'$ is a very crude approximation in the case of adsorption of thiourea (see also [42]).

Devanathan's method for calculating adsorption has the merit of simplicity. However, the arbitrary nature of a number of assumptions in the derivation of Eq. (IV.60), the arbitrary choice of D within the double layer, and the assumptions that all the adsorbed dipoles are perpendicular to the surface and that the values of μ are maintained in the field of the double layer cast doubt on the results obtained by this method.

In a later paper [36] Devanathan applied his method to adsorption of aliphatic compounds containing a polar group and a hydrocarbon chain in the molecule. Pointing out that in this case the assumption that D remains constant during adsorption is no longer tenable, Devanathan introduced a correction for the change of D into the capacity given by Eq. (IV.60). This was based on a relation somewhat similar to Eq. (III.13), namely:

$$D = D_0 (1 - \theta) + D'\theta = D_0 (1 - \alpha\theta), \qquad (IV.64)$$

where D_0 is the dielectric constant in absence of organic molecules, D' is the dielectric constant of the hydrocarbon chain, and $\alpha = (D_0 - D')/D_0$.

It was concluded from Eq. (IV.64) that the influence of the transition from D_0 to D' on the capacity can be taken into account by the correction factor $(1 - \alpha\theta)$:

$$\frac{1 - \alpha\theta}{C} = \frac{1}{C_o} + \frac{d\Gamma}{d\varepsilon} \cdot \frac{2\pi\mu}{D} . \qquad (IV.65)$$

Equation (IV.65) was used by Devanathan [36] for calculating θ from values obtained by Breiter and Delahay [43] for the capacity in n-amyl alcohol solutions. However, the results give rise to doubt because of additional errors made by Devanathan in derivation of Eq. (IV.65) (see [39]).

A correct expression for $d\Gamma/d\varepsilon$, taking into account both the dipole effect and the change of capacity upon adsorption of an organic substance, can be obtained from Eq. (III.17) if it is approximately assumed that C_0 = const and C' = const, and, following Devanathan, if it is assumed that these quantities are proportional to the dielectric constants, i.e.,

$$\frac{C'}{C_0} = \frac{D'}{D_0} = 1 - \alpha. \tag{IV.66}$$

Writing Eq. (III.17) in the form

$$\varepsilon = C_o[\varphi(1 - \theta) + (1 - \alpha)(\varphi - \varphi_N)\theta] = C_o[\varphi(1 - \alpha\theta) - C'\varphi_N\theta/C_o] \tag{IV.67}$$

and differentiating with respect to ε, we obtain after simple transformations

$$\frac{1 - \alpha\theta}{C} = \frac{1}{C_o} + \frac{d\Gamma}{d\varepsilon}\left(\frac{\varphi_N C'}{\Gamma_m C_o} + \frac{\alpha\varphi}{\Gamma_m}\right). \tag{IV.68}$$

It is easy to see that Eq. (IV.65) can be obtained from Eqs. (IV.68) and (IV.61) only if the condition $C' = C_0$ is satisfied, i.e., $\alpha = 0$. This condition is not fulfilled in the case of aliphatic compounds and Eq. (IV.65) is therefore inapplicable.

Equation (IV.68) explains why the values of θ calculated by Devanathan for n-amyl alcohol from Eq. (IV.65) are in good agreement at the potential of maximum adsorption with the values calculated by Breiter and Delahay [43] from Eq. (III.13). When $\varphi = \varphi_m$, $d\Gamma/d\varepsilon = 0$ and Eqs. (IV.65) and (IV.68) coincide.

The use of the dielectric constant calculated from Eq. (IV.64) for determination of capacity is based on the assumption that adsorption of the organic substance does not influence the thickness of the electric double layer. However, this assumption is not confirmed experimentally either [33, 39, 44, 45]. For all these reasons Devanathan's nonthermodynamic method has not gained acceptance.

References

1. R. Parsons, Trans. Faraday Soc., 51:1518 (1955).
2. R. Parsons, Proc. Roy. Soc., A261:79 (1961).
3. F. W. Schapink, M. Oudeman, K. W. Leu, and J. N. Helle, Trans. Faraday Soc., 56:415 (1960).
4. J. M. Parry and R. Parsons, Trans. Faraday Soc., 59:241 (1963).
5. E. Dutkiewicz and R. Parsons, J. Electroanal. Chem., 11:196 (1966).
6. R. Parsons and F. G. R. Zobel, Trans. Faraday Soc., 62:3511 (1966).
7. R. Payne, J. Chem. Phys., 42:3371 (1965).
8. R. Payne, J. Phys. Chem., 69:4113 (1965); 70:204 (1966).
9. E. Dutkiewicz and R. Parsons, J. Electroanal. Chem., 11:100 (1966).
10. R. Parsons, J. Electroanal. Chem., 7:136 (1964).
11. A. N. Frumkin, B. B. Damaskin, and A. A. Survila, J. Electroanal. Chem., 16:493 (1968).
12. B. B. Damaskin, A. A. Survila, and L. E. Rybalka, Élektrokhimiya, 3:146, 927, 1138 (1967).
13. R. Parsons, J. Electroanal. Chem., 8:93 (1964).
14. H. A. Laitinen and B. Mosier, J. Am. Chem. Soc., 80:2363 (1958).
15. P. Delahay and P. Trachtenberg, J. Am. Chem. Soc., 79:2355 (1957); 80:2094 (1958).
16. V. G. Levich, B. I. Khaikin, and E. D. Belokolos, Élektrokhimiya, 1:1273 (1965).
17. L. Němec, Collection Czech. Chem. Commun., 31:1162 (1966).
18. A. N. Frumkin and B. B. Damaskin, Dokl. Akad. Nauk SSSR, 129:862 (1959).
19. J. Koryta, Collection Czech. Chem. Commun., 18:206 (1953).
20. T. Biegler and H. A. Laitinen, J. Phys. Chem., 68:374 (1964).
21. W. Lorenz, F. Möckel, and W. Müller, Z. Phys. Chem. (N.F.), 25:145 (1960).
22. W. Lorenz and W. Müller, Z. Phys. Chem. (N. F.), 25:161 (1960).
23. H. A. Laitinen and K. Morinaga, J. Colloid. Sci., 17:628 (1962).
24. G. A. Dobren'kov and R. K. Bankovskii, Proceedings of Electrochemical Conference [in Russian], Kazan' (1959), p. 74; Izv. Vysshikh Uchebn. Zavedenii, Khim. i Khim. Tekhnol., 5:75 (1962).
25. B. B. Damaskin and G. A. Tedoradze, Dokl. Akad. Nauk SSSR, 152:1151 (1963).
26. B. B. Damaskin and S. L. Dyatkina, Élektrokhimiya, 3:531 (1967).
27. B. B. Damaskin and R. Lerkkh, Zh. Fiz. Khim., 39:495 (1965).
28. B. B. Damaskin and S. L. Dyatkina, Élektrokhimiya, 1:706 (1965).
29. J. Thovert, Compt. Rend., 135:579 (1901).
30. B. B. Damaskin, Élektrokhimiya, 1:255 (1965).
31. G. A. Tedoradze and R. A. Arakelyan, Dokl. Akad. Nauk SSSR, 156:1170 (1964).
32. A. N. Frumkin, B. B. Damaskin, and A. A. Survila, Élektrokhimiya, 1:738 (1965).
33. A. N. Frumkin and A. V. Gorodetskaya, Dokl. Akad. Nauk SSSR, 18:649 (1938).

34.   A. A. Survila and B. B. Damaskin, in: Electrochemical Processes Involving Organic Substances [in Russian], Izd. "Nauka," Moscow (1969).

35.   M. A. V. Devanathan, Proc. Roy. Soc., A264:133 (1961).

36.   M. A. V. Devanathan, Proc. Roy. Soc., A267:256 (1962).

37.   J. O'M. Bockris, M. A. V. Devanathan, and K. Müller, Proc. Roy. Soc., A274:55 (1963).

38.   M. A. V. Devanathan and B. V. K. S. R. A. Tilak, Chem. Rev., 65:635 (1965).

39.   A. N. Frumkin and B. B. Damaskin, in: Modern Aspects of Electrochemistry, Vol. 3, Plenum Press, New York (1964), p. 149.

40.   J. R. Macdonald and C. A. Barlow, J. Chem. Phys., 39:412 (1963).

41.   A. N. Frumkin, B. B. Damaskin, and Yu. A. Chizmadzhev, Élektrokhimiya, 2:875 (1966).

42.   A. M. Morozov, N. B. Grigor'ev, and I. A. Bagotskaya, Élektrokhimiya, 3:585 (1967).

43.   M. Breiter and P. Delahay, J. Am. Chem. Soc., 81:2938 (1959).

44.   M. A. Proskurnin and A. N. Frumkin, Trans. Faraday Soc., 31:110 (1935).

45.   W. Lorenz, Z. Elektrochem., 62:192 (1958).

46.   R. Parsons and P. C. Symons, Trans. Faraday Soc., 64:1077 (1968).

47.   S. Minc and J. Andrzeijczak, J. Electroanal. Chem., 17:101 (1968).

Part II

Adsorption of Organic Compounds
on Solid Electrodes

Introduction

The first investigations of adsorption processes on solid electrodes were carried out by Frumkin et al. [1-10], who studied adsorption of hydrogen, oxygen, and halide ions on metals of the platinum group and on gold, and by Butler [11]. Quantitative measurements of adsorption of organic substances on solid metals have been started only during the last 10-15 years. At the present time adsorption processes are attracting special attention in relation to investigations of electrochemical oxidation of organic substances, the action mechanism of corrosion inhibitors, electrochemical synthesis of organic compounds, and the part played by organic additives in electrodeposition of metals.

Studies of adsorption of organic substances on solid electrodes involve a number of additional experimental difficulties in comparison with measurements at liquid electrodes. For example, the capillary electrometer, which is one of the main instruments for investigating adsorption on liquid metals, cannot be used for adsorption measurements at solid surfaces. Measurements of interfacial impedance, which provide quantitative information on the surface state of liquid electrodes, are for a number of reasons of limited applicability to certain solid electrodes (e.g., metals of the platinum group). Quantitative studies of adsorption of organic compounds on solid metals became possible after the radioactive tracer method, the pulse potentiodynamic and galvanostatic methods, and other techniques, had been developed. However, even now the methods available for studying adsorption from solutions are less numerous than methods for study of adsorption from the gaseous phase [12], while adsorption measurements on solid metals are less accurate than in the case of the mercury electrode. This is due both to the limitations of the methods used and to the specific nature of solid surfaces.

Difficulties arise even in determinations of the surface areas of solid electrodes. In contrast to liquids, the surface of a solid electrode has various irregularities (cracks, projections, etc.), so that the true area may be considerably greater than the geometric area. The electrode areas are usually determined by the BET method, measurements of double-layer capacity, and determinations of adsorption of hydrogen, oxygen, or organic compounds on metals. The BET method, based on low-temperature adsorption of inert gases [13], is used for measurements of relatively large areas (some tens of cm^2 and more) [14]. The precision of the method is not better than ~10-15%. A defect of this method is that the surface area is measured under conditions differing significantly from those of the electrochemical experiment. Determinations of surface area from the double-layer capacity involve the assumption that the capacities of the double layer per cm^2 of true surface area are equal for different metals.* It is usually assumed that the double-layer capacity is 17-20 $\mu F/cm^2$ in the region of cation adsorption and 40 $\mu F/cm^2$ for adsorption of anions. This assumption has not yet been verified experimentally for all metals. In cases where fairly accurate determinations of the double-layer capacity were carried out, some differences were found between the capacities of different metals at the same potentials relative to the point of zero charge [15]. It must also be pointed out that for a number of metals the measured electrode capacity is not the true double-layer capacity but is determined to some extent by various electrochemical processes. In such cases determinations of surface area by this method may lead to considerable errors. In some investigations [16] the surface area of powder electrodes was found from the increase of the powder weight after adsorption of an organic substance having high molecular weight. The area was calculated with certain assumptions regarding the orientation and cross-sectional area of the adsorbed molecules in a monomolecular layer. Small areas can be cal-

---

*This method was first used by Frumkin et al. [2, 4].

ulated similarly from adsorption of organic substances con-
aining radioactive isotopes [17-19]. True electrode areas found
rom the adsorption of organic substances may differ appreciably
rom areas found by the BET method [20]. This appears to be due
o the presence of microcrevices and microcracks in the electrode
surface which are not accessible to large organic molecules.
Surface determinations by this method are also complicated if
multimolecular adsorption occurs.

In adsorption experiments on platinum, the true electrode
area is estimated from the amount of hydrogen adsorbed, with the
assumption that at 1 atm hydrogen pressure one atom of adsorbed
hydrogen corresponds to each surface metal atom [21].* The
amount of hydrogen adsorbed is found from the length of the hydrogen region of the charging curve, or by integration of the portion of the potentiodynamic I vs φ curve corresponding to the adsorption of hydrogen, with a correction for the quantity of elec-
ricity expended for charging the double layer (see [23] for de-
ails of the application of this correction). The results of determinations of the surface area of platinum by this method are in
agreement with the results given by the BET method [24-26].
There are contradictory reports regarding the possibility of determining the surface area of rhodium by this method (cf. [25-27]).
The most reliable results appear to be those of Tarasevich et al.
[25], indicating that the surface area of rhodium can be determined
from hydrogen adsorption with the assumption that one hydrogen
atom corresponds to each rhodium atom. An analogous method
has been proposed for determination of the surface area of ruthenium [28] and iridium [29]. The possibility of determining the
area of a palladium electrode from sorption of hydrogen has been
discussed by Burshtein et al. [30].

In some investigations [31] the areas of platinum metals were
determined from adsorption of oxygen, with certain assumptions
regarding the composition of the surface oxide layer.

It is usually assumed that there are $1.31 \cdot 10^{15}$ platinum atoms per cm^2 of true surface area of platinum; i.e., the amount of hydrogen adsorbed corresponds to 210 μC.
These values correspond to the (100) crystallographic face, but are used for determinations of surface areas of polycrycrystalline specimens. The numbers of platinum
atoms on the (110) and (111) faces are $0.93 \cdot 10^{15}$ and $1.51 \cdot 10^{15}$ respectively. Other
values for the number of atoms per cm^2 of polycrystalline platinum surface are used
in some publications (e.g., see [22]).

The difficulties met in studies of adsorption on solid electrode include the strong influence of preparatory treatment and surface state of the electrode on its adsorption characteristics. The absence of ideal polarizability, surface oxidation, and dissolution of the metals also introduce substantial complications into adsorption measurements and into interpretation of the results.

Measurements with the use of stationary electrodes require thorough purification of the reagents in order to remove surface-active impurities [12, 32]. The problem of solution purification which confronts electrochemists can be compared with the problem of attaining ultrahigh vacuum for preparation of clean catalyst surfaces.

Adsorption of organic compounds on solid electrodes is not only due to physical forces but is often accompanied by chemical interaction between the metal surface and the adsorbate; in some cases total destruction of the molecules occurs. Thus, chemisorption processes are more pronounced on many solid electrodes than on mercury. Because of this, the heats of adsorption of organic substances on solid metals reach considerable values and the adsorption process is characterized by high activation energies. In distinction from physical adsorption, chemisorption usually results in monolayer formation. In some cases the determining step in adsorption on solid metals becomes direct adsorptive interaction between the organic substance and the electrode surface.

Nonuniformity of surface energy is an important feature of solid electrodes. In adsorption studies it is necessary to take into account the texture of the solid surface, i.e., predominant orientation of particular crystal faces, and also lattice defects. Changes of texture and defect concentration after mechanical or heat treatment lead to considerable changes in the adsorption properties of metals.

The ability of many solid electrodes to adsorb hydrogen and oxygen must make the dependence of adsorption of organic substances on the electrode potential substantially different in character from that found in the case of mercury [33].

Chapter 5

Methods for Studying Adsorption of Organic Substances on Solid Electrodes

1. Methods for Determination of Adsorption from Changes in Interfacial Tension

There are as yet no methods for determining the absolute values of interfacial tension between a solid electrode and a solution. However, the dependence of interfacial tension on the electrode potential and solution composition can be investigated by a number of methods based on study of mechanical properties and wetting of metals by solutions. Therefore these methods will be discussed first, although as yet they have not been widely used for investigation of adsorption of organic substances on solid electrodes.

In the case of liquid metals, the work done in increasing the surface by 1 cm^2 under isothermal conditions in a reversible process is a measure of interfacial tension. Increase of the surface area of a solid metal (mechanical subdivision) is accompanied by irreversible changes, because destruction of the crystal lattice occurs. Therefore the work of subdivision of solids is greater than the interfacial tension. Moreover, the surfaces of solids are heterogeneous, with regions differing in energy and hence in interfacial tension (different crystallographic orientation of individual crystallites, lattice defects).

Despite these complications, several mechanical properties of solid metals (hardness, creep, coefficient of friction, etc.) are determined mainly by the interfacial tension. This dependence is used for determinations of the points of zero charge of metals. The region of adsorption of an organic substance can be determined, and the degree of its adsorbability can be assessed qualitatively, from variations in the dependence of mechanical properties on the potential [34-38].

Rebinder et al. [35, 39, 40] have shown that the hardness of metals depends on the potential and that this dependence is of the form of the electrocapillary curves obtained for liquid metals. The hardness of a metal reflects the degree of its dispersibility. Subdivision (dispersion) is accompanied by formation and widening of microcracks. The rate of these processes increases with decreasing surface tension of the metal. Therefore the hardness of a metal decreases when the electrode potential is shifted in the positive or negative direction relative to its potential of zero charge and when organic substances are adsorbed at electrode — solution interfaces. The pendulum method [41-43] is often used for hardness determinations. The determination is performed as follows. The pendulum rests on the metal surface by two tungsten carbide points. When the pendulum oscillates, these points deform and destroy the metal surface, and the amplitude of the pendulum decreases. The hardness (H) of the metal can be calculated from the damping of the oscillations:

$$H = -\frac{1}{\left(\frac{d \ln a}{dt}\right)_{t=0}}, \qquad (V.1)$$

where a is the oscillation amplitude.

Hardness curves of thallium and electrocapillary curves for 41.5% thallium amalgam in $1\,N$ Na_2SO_4 with addition of isoamyl alcohol [44] are shown in Fig. 45. The maxima of the hardness curves and of the electrocapillary curves coincide. The regions of adsorption of the organic substance determined by these methods also coincide.

In certain investigations [45] metals were dispersed by ultrasonic action or with the aid of abrasive suspensions [45]. The degree of dispersion was determined from the weight loss of the

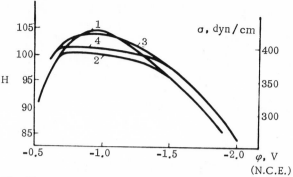

Fig. 45. Dependence of the hardness of thallium on the
potential in solutions of: 1) 1 N Na_2SO_4; 2) 1 N Na_2SO_4 +
0.185 M iso-$C_5H_{11}OH$ [35], and electrocapillary curves
for 41.5% thallium amalgam in solutions of: 3) 1 N
Na_2SO_4; 4) 1 N Na_2SO_4 + 0.175 M iso-$C_5H_{11}OH$ [44].

specimen. The minimum decrease of the electrode weight corresponds to the potential of zero charge.

Measurements of the tensile deformation of metals under
constant load in relation to the electrode potential show that the
relative elongation (Δl) of the specimen is least in the region of
the potential of zero charge [46, 47].

Data obtained by Venstrem and Rebinder [46] for a lead single
crystal in 0.1 N Na_2SO_4 are shown in Fig. 46. The minimum of the
curve corresponds to a potential of -0.6 V (vs a normal hydrogen
electrode). The potential of zero charge of lead (-0.64 V) [48, 49],
found from differential capacity curves determined in dilute solutions, is in fairly good agreement with this experimental result.
Addition of sodium hexadecylbenzene sulfonate (0.07 M) to the
0.1 N Na_2SO_4 solution increases the deformability of the specimen.
At more negative potentials sodium hexadecylbenzene sulfonate is
not adsorbed on the metal surface and the Δl vs φ curves determined in the pure solution and with the organic additive coincide.

The potential of zero charge of gold was determined by Beck
[50], from the extension of the electrode in the form of a thin ribbon.

It has been shown [38] that the dependence of the yield point
on potential is of the "electrocapillary" type, and the yield point

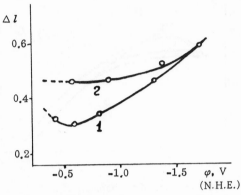

Fig. 46. Dependence of the relative elongation
of a lead single crystal on potential in aqueous
solutions: 1) 0.1 N Na$_2$SO$_4$; 2) 0.1 N Na$_2$SO$_4$ +
0.07 M C$_{16}$H$_{33}$C$_6$H$_4$SO$_3$Na [46].

maximum coincides with the potential of zero charge of the metal.
The influence of adsorption of surface-active substances on plastic
deformation of zinc, cadmium, tin, and gold was subsequently
studied by Masing and co-workers [51, 52].

The potential of zero charge can also be found from the minimum of the creep rate of the metal [53].

In a number of determinations of coefficients of friction [36,
37, 39, 54] it was shown that for platinum, nickel, cadmium, lead,
and copper the dependence of the coefficient of friction on potential
is also of the form of an electrocapillary curve. The decrease of
the coefficient of friction with increasing surface charge is attributed to increase of the disjoining action of the electric double
layer between the rubbing metal surfaces [39]. The potential corresponding to the maximum coefficient of friction is in satisfactory agreement with the electrocapillary maximum.

The methods discussed above make it possible to verify the
potentials of zero charge determined by other methods for many
metals. These methods may be used for measurements of potentials of zero charge in concentrated solutions or under conditions
involving reversible processes, when experimental difficulties
arise in determination of the potential of zero charge from the
minimum of the double layer capacity. However, the results

should be regarded with great caution, because products of electrochemical reactions may accumulate on the electrode surface
or in metal layers adjacent to the surface, changing its mechanical
properties. For example, atomic hydrogen, a possible reaction
product, increases brittleness sharply when it penetrates into the
metal.

Gokhshtein [55] has described a method for direct determination of variations of the surface tension of solid metals with the
potential. The potential of an electrode immersed in an electrolyte,
or of an L-shaped electrode touching the electrolyte surface, is
varied linearly with time. At the same time an alternating current
of small voltage amplitude $\Delta\varphi = \Delta\varphi_0 \cos \omega t$ is passed through the
electrode. Variation of the potential leads to variation of the sur-
face tension of the electrode, which is reflected by the Lippmann
equation. With sinusoidal variation of the potential, the variable
component of the surface tension may be represented as the sum
of harmonic vibrations with frequencies ω, 2ω, 3ω, etc. At
$\Delta\varphi < 0.1$ V, all the harmonics beyond the first may be neglected.
Variations of the surface tension cause vibration of the electrode,
which is transmitted to a piezoelectric element. A potential difference, which can be recorded, arises across the piezoelectric
element. At the point of zero charge a minimum $\left(\left[\frac{\Delta\sigma}{\Delta\varphi}\right]=0\right)$, is observed on the $\left[\frac{\Delta\sigma}{\Delta\varphi}\right]$ vs φ curve, as the amplitude of the first harmonic becomes zero. Change of sign of the surface charge is ac-
companied by a change of π in the phase of the first harmonic.
This is a very important point, because it makes it possible to
distinguish the potential of zero charge from potentials at which
arrests in changes of surface tension independent of the potential
of zero charge occur (this effect was noted by the author for the
system $Pt/I^- + I_2$). The results obtained by this method for lead
and thallium electrodes in various solutions [56, 58] are shown in
Fig. 47. The shift of the $\left[\frac{\Delta\sigma}{\Delta\varphi}\right]$ vs φ curves for different anions
is consistent with the results of electrocapillary measurements
at a mercury electrode, the behavior of which is very similar to
that of lead and thallium. Measurements were also performed
with silver; the potential of zero charge of silver was found to be
-0.75 V against the standard hydrogen electrode, which is close
to the value $\varphi_H = -0.7$ V found from the minimum of the differential capacity of the double layer [57]. Interpretation of the

results obtained by Gokhshtein's method with platinum metals is more complicated.

It has been shown recently [58] that this method can be used for studying adsorption of organic compounds on solid electrodes (Fig. 47, c).

Another group of methods for measurement of adsorption involves measurements of the contact angle between metals and solvents in presence of surface-active organic substances.

The shape of a drop of a liquid organic substance (or a gas bubble) on a horizontal flat electrode surface depends on the relation between the interfacial tensions: electrolyte−electrolyte (σ_{12}), electrode−organic liquid (gas) (σ_{13}), and electrolyte−organic liquid

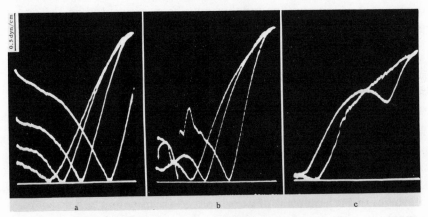

Fig. 47. Oscillograms representing the amplitude of the surface tension of a solid electrode vs the average potential ($\Delta\sigma$ vs φ_m) before and after introduction of various surface-active substances into the solution [56, 58]. a) Lead, shift of zero $\Delta\sigma$ on addition of halide anions to 1 N Na$_2$SO$_4$ (zero at 0.55 V): 0.15 N KCl (zero at 0.60 V); 0.15 N KBr (zero at 0.67 V); 0.15 N KI (zero at 0.78 V); electrode roughness factor 1.3; φ_m varied from −0.88 to −0.43 V vs S.H.E. (right to left) at the rate of 0.01 V/sec; frequency 4900 Hz, $\Delta i = 3.20 \cdot 10^{-2}$ A/cm^2; dimensions of working portion of the electrode 10 $\times$ 4 $\times$ 1.20 mm. b) Thallium; additions to 1 N Na$_2$SO$_4$ (zero at 0.74 V): 0.15 N KBr (zero at 0.82 V); 0.15 N KI (zero at 0.96 V); φ_m from −1.24 to −0.54 V vs N.H.E. (right to left); 0.02 V/sec; 4900 Hz, $\Delta i = 2.32 \cdot 10^{-2}$ A/cm^2; 10.4 $\times$ 5.4 $\times$ 1.0 mm. c) Lead in 1 N Na$_2$SO$_4$ before (1) and after (2) addition of 0.03 M C$_5$H$_{11}$OH: φ_m from −1.41 to −0.37 V vs N.H.E. (right to left); 0.03 V/sec; 4950 Hz; $\Delta i = 3.50 \cdot 10^{-2}$ A/cm^2; 10 $\times$ 4 $\times$ 1.20 mm (Δi is the assigned ac frequency).

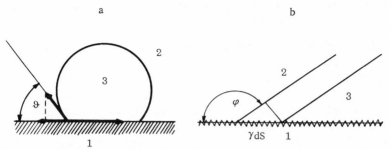

Fig. 48. Schematic diagram of interfacial forces at a three-phase boundary (a), and scheme demonstrating the influence of surface roughness on the contact angle (b): 1) Metal; 2) solution; 3) organic substance (gas).

(gas) (σ_{23}).* The equilibrium state of the drop is determined by the balance of these forces:

$$\cos \vartheta = \frac{\sigma_{13} - \sigma_{12}}{\sigma_{23}}, \qquad (V.2)$$

where ϑ is the contact angle (see Fig. 48a). An essential condition for validity of Eq. (V.2) is that the organic liquid (or gas) must be inert with respect to the metal and solution.

The magnitude of the contact angle depends on the electrode potential. This effect was first noted by Möller [61], and a correct theoretical interpretation was given by Frumkin and co-workers [62–64].

The value of σ_{23} is independent of the electrode potential. The quantity σ_{13} is the interfacial tension at the boundary between the metal with an adsorbed film of the electrolyte and the organic substance (gas). Therefore some variation of σ_{13} occurs when the potential varies, but this effect is much weaker than the variation of σ_{12}. The dependence of σ_{12} on the potential is represented by Lippmann's equation for electrocapillary effects. With this reasoning taken into account, in accordance with Eq. (V.2) the variation of the contact angle depends on variation of σ_{12}. The ϑ vs φ curve should have the form of an electrocapillary curve. The

---

*In certain studies of atmospheric corrosion [59, 60] the behavior of water drops on metal surfaces in a gaseous phase was investigated.

maximum value of the contact angle corresponds to the potential of zero charge of the electrode.

The shift of the potential of zero charge due to adsorption of surface-active organic substances can be estimated and their adsorbability on the metal surface qualitatively determined from changes in the position and shape of the ϑ vs φ curve. However, up to now this method has been used mainly for studying adsorption of organic substances at mercury surfaces [65, 66], as measurements on solid metals involve additional difficulties. They include the existence of hysteresis between the values of the contact angle during widening and narrowing of the base of the drop (or bubble).

Moreover, increase of the roughness factor (γ) (ratio of the true to the apparent metal area) leads to a change in the contact angle; this is evident from the following reasoning [67]. If the drop is in a state of equilibrium, the work done in extension of the drop base by an infinitesimal amount is zero. With the roughness factor taken into account the area of the drop base increases by γds (Fig. 48b), the metal−electrolyte interface decreases by the same amount, while the electrolyte−organic liquid interface increases by $\cos(180° - \vartheta)ds$. We then have

$$\sigma_{13}\gamma \, dS + \sigma_{23} \cos \vartheta' \, dS - \sigma_{12}\gamma \, dS = 0,$$
$$\cos \vartheta' = \frac{\gamma(\sigma_{13} - \sigma_{12})}{\sigma_{23}} = \gamma \cos \vartheta, \qquad (V.3)$$

where ϑ' is the experimental value of the contact angle and ϑ is the contact angle for a perfectly smooth surface.

In the case discussed the contact angle increases with increase of γ. The roughness factor has little or no influence on the contact angle only when ϑ is equal to or close to 90°. Therefore through polishing of the metal surface is necessary.

Bonnemay et al. [68] have described a method for determination of electrocapillary curves for solid electrodes by measurement of the rise of an electrolyte in a capillary made from the metal under investigation. The position of the meniscus was determined by transmission of x-rays through the capillary.

The relation between the capillary rise of the electrolyte (h) and the capillary radius (r) is given by the formula

$$h = \frac{2\sigma_{23} \cos \vartheta}{\rho g r} , \qquad (V.4)$$

where ϑ is the contact angle, ρ is the density of the electrolyte, σ_{23} is the liquid–gas interfacial tension, and g is the acceleration due to gravity. Putting (V.2) into (V.4), we obtain

$$h = \frac{2(\sigma_{13} - \sigma_{12})}{\rho g r} . \qquad (V.5)$$

Since σ_{13} is almost independent of the potential, after differentiating (V.5) with respect to potential we have:

$$dh = - \frac{2}{\rho g r} d\sigma_{12}. \qquad (V.6)$$

The Lippmann equation can be written in the following form:

$$\frac{d\sigma_{12}}{d\varphi} = - C\varphi, \qquad (V.7)$$

where φ is the electrode potential referred to the potential of zero charge. After substitution of (V.7) into (V.6) and second differentiation with respect to potential we can calculate the capacity of the double layer:

$$C = \frac{\rho g r}{2} \frac{d^2 h}{d\varphi^2} \quad \text{at} \quad C = \text{const.} \qquad (V.8)$$

It follows from Eq. (V.4) that if the capillarity laws are obeyed a minimum on the h vs φ curve corresponds to the potential of zero charge. The capillary surface must be carefully polished for the experiments. If the roughness factor of the surface is γ, the capacities calculated from Eq. (V.8) will be lower than the true capacity of the double layer by a factor γ, since in this case the perimeter of the three-phase boundary is $2\pi r \gamma$. This method has not so far been used for studying adsorption of organic substances.

2. Special Features of Methods for Measuring Double-Layer Capacity on Solid Metals

Methods for measurement of double-layer capacity with the aid of alternating current and of potential decay curves were discussed in Chapter I. These methods can also be used for investigation of adsorption phenomena at solid electrodes.

The technique of measurement of the capacity of the double layer on metals with a high hydrogen overpotential in the region of cathodic potentials does not differ in principle from measurements on mercury. In other cases capacity measurements with the aid of alternating current have to be carried out in presence of various electrochemical processes occurring at appreciable rates (evolution and ionization of hydrogen and oxygen, dissolution of metals, etc.). In the simplest case of an irreversible electrochemical process the capacity of the electric double layer can be found after conversion of the electrode impedance measured with a series circuit to a circuit with capacity and resistance in parallel. The following formulas are used for the conversion [9]:

$$C_2 = \frac{C_1}{1 + \omega^2 C_1^2 R_1^2}, \quad R_2 = R_1 \left(1 + \frac{1}{\omega^2 C_1^2 R_1^2} \right), \quad \text{(V.9)}$$

where C_1 is the capacity in the series circuit; R_1 is the resistance in the series circuit after subtraction of the solution resistance, $\omega = 2\pi\nu$ is the angular frequency, ν is the ac frequency, and C_2 and R_2 are the capacity and resistance of the parallel circuit.

The parallel connection of C_2 and R_2 is an equivalent circuit where C_2 is the capacity of the electric double layer and R_2 is the polarization resistance. This is valid if the electrode reactions are not accompanied by adsorption of intermediate reaction products (H, FeOH, etc.). If adsorption processes are taken into account, the equivalent circuits become more complicated [9, 69-80].

However, fairly often in measurements of double-layer capacity no correction is applied for the current expended in electrochemical reactions. Let us examine the conditions under which the capacity of the double layer can be measured without conversion to a parallel circuit. With a change of potential by $\Delta\varphi$ the alternating current is consumed for charging the capacity of the double layer and for electrode reactions:

$$\Delta I = C j\omega\Delta\varphi - \Delta i_c + \Delta i_a, \quad \text{(V.10)}$$

where C is the capacity of the double layer and $j = (-1)^{\frac{1}{2}}$. For

*The first term is obtained in this form after insertion into $C(d\Delta\varphi/d\Delta t)$ of the expression $\Delta\varphi = \Delta\varphi_0 \exp[i\omega t]$, where $\Delta\varphi_0$ is the ac voltage amplitude.

simplicity we assume that the anodic and cathodic reactions proceed irreversibly:

$$i_c = n_1 F k_1 \exp\left[-\frac{\alpha n_1 F \varphi}{RT}\right], \qquad (\text{V.11})$$

$$i_a = n_2 F k_2 \exp\left[\frac{\beta n_2 F \varphi}{RT}\right], \qquad (\text{V.12})$$

where α and β are the transfer coefficients of the cathodic and anodic reactions, and n_1 and n_2 are the number of electrons. Hence, assuming that $\Delta\varphi \ll RT/nF$, we have:

$$\Delta i_c = n_1 F k_1 \exp\left[-\frac{\alpha n_1 F\,(\varphi + \Delta\varphi)}{RT}\right] -$$
$$- n_1 F k_1 \exp\left[-\frac{\alpha n_1 F \varphi}{RT}\right] \simeq - i_c \frac{\alpha n_1 F \Delta\varphi}{RT}, \qquad (\text{V.13})$$

$$\Delta i_a \simeq i_a \frac{\beta n_2 F}{RT}\Delta\varphi. \qquad (\text{V.14})$$

Putting (V.13) and (V.14) into (V.10) and dividing the resultant equation by $\Delta\varphi$, we have:

$$\frac{1}{Z} = Cj\omega + i_c\frac{\alpha n_1 F}{RT} + i_a\frac{\beta n_2 F}{RT}$$

or, after certain transformations and separation of the imaginary and components:

$$Z = \frac{A}{A^2 + C^2\omega^2} + \frac{C\omega^2}{j\omega\,(A^2 + C^2\omega^2)}, \qquad (\text{V.15})$$

where

$$A = i_c\frac{\alpha n_1 F}{RT} + i_a\frac{\beta n_2 F}{RT} \ \text{ or }\ A = 2.3\left(\frac{i_c}{b_c} + \frac{i_a}{b_a}\right).$$

and b_c and b_a are the slopes of the φ vs log i curves. From a comparison of (V.15) with the impedance of the series circuit

$$Z = R_1 + \frac{1}{C_1 j\omega} \qquad (\text{V.16})$$

we find

$$C_1 = \frac{A^2}{C\omega^2} + C. \qquad (\text{V.17})$$

Hence $C_1 = C$ if $A^2/C\omega^2 \ll C$.

At sufficiently negative potentials, assuming C = 20 μF/cm^2,

$$A \simeq 2.3 \frac{i_c}{b_c} \; ; \; b_c = 0.12V,$$

$$\frac{i_c}{\gamma} \ll 0.6 \cdot 10^{-15} \, A/cm^2 \cdot Hz.$$

At sufficiently positive potentials

$$A \simeq 2.3 \frac{i_a}{b_a}, \quad b_a = 0.06V,$$

$$\frac{i_a}{v} \ll 0.3 \cdot 10^{-5} A/cm^2 \cdot Hz.$$

Figure 49 shows capacity curves calculated from (V.17) for various values of the spontaneous solution current (b_c = 0.12 V, b_a = 0.06 V, v = 800 Hz).* It is seen from the figure that at i = 10^{-5} A/cm^2 and high values of the spontaneous solution current distortion of the capacity curve begins even with small deviations of the potential from φ_{st}.

In a number of investigations impedance measurements have been used for determination of the resistance of the adsorption film [81, 82], the measurements being performed at low ac frequencies (1000 Hz and lower). The method essentially consists of measurement of the electrode impedance first in a pure electrolyte solution and then in presence of an inhibitor adsorbed on the electrode surface. The change of resistance after introduction of the inhibitor is taken as the resistance of the film. Here the authors assume that the electrode equivalent circuit consists of the double layer capacity and solution resistance connected in series. In reality the situation is more complicated because the equivalent circuit must be represented (in the simplest case) by the capacity of the double layer and the polarization resistance in parallel, with the solution resistance in series with them. Probably this partially explains the discrepancy between the values obtained by Machu by this method for the resistance of the adsorption film and the data reported by others [83]. Calculation of the film resistance by the method given in [81, 82] is possible only at high frequencies (of the order of 10-50 kHz), when the conductance

---

*The chosen values of b_c and b_a are ones most commonly met in studies of hydrogen evolution and metal dissolution.

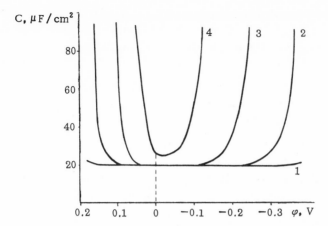

Fig. 49. Differential capacity curves calculated from Eq. (V.21) for capacity and resistance in series, at different values of the spontaneous solution current: 1) 10^{-6}; 2) 10^{-5}; 3) 10^{-4}; 4) 10^{-3} A/cm^2. The position of the stationary potential is indicated by the dash line.

of the polarization resistance branch may be neglected in comparison with that of the capacity of the double layer.

Adsorption of organic substances on silver, nickel, iron, and other metals has been studied by the ac method by various investigators [57, 84-88]. The following formula was used for calculation of θ from the lowering of the double-layer capacity on solid electrodes, as in the case of mercury:

$$\theta = \frac{C_0 - C}{C_0 - C'}, \tag{V.18}$$

where C_0 is the capacity of the double layer in the pure electrolyte solution; C is the capacity in the solution with addition of the organic substance; C' is the capacity of the double layer with limiting coverage of the electrode surface by the organic substance.

In a number of cases the kinetics of adsorption of an organic substance can be studied by investigation of the decrease of the capacity of the double layer with time. Adsorption of allylphenylthiourea on cobalt was studied by this method [88].

A serious disadvantage of measurements at solid electrodes is the dependence of capacity on frequency, which most workers attribute to surface roughness. The dependence of capacity on frequency is diminished sharply by polishing of the surface, surface fusion of the electrodes, or the use of single crystals for the experiments [15, 49, 89].

Some authors attribute variation of capacity with the ac frequency to penetration of the solution between the electrode and the insulating coating [89, 90].

The potential decay method is also used for measurement of capacity at solid electrodes [91-97]. A mathematical analysis of the method and assessments of its potentialities have been given in papers by Frumkin [98] and Grahame [99]. If the reaction

$$O + n\bar{e} \rightleftarrows R$$

occurs at the electrode and the electrode potential is shifted under the influence of external cathodic polarization by η_0 from the equilibrium value, when the external voltage source is disconnected the electric double layer undergoes discharge owing to the reaction occurring at the electrode surface. As was shown in Chapter I, the change of the electrode potential when the circuit is broken can be represented by the formula

$$\eta_0 - \eta = \frac{b}{2.3} \ln\left[\frac{ti_0}{bC} 2.3 + 1\right], \tag{V.19}$$

where i_0 is the current at η_0, η is the overpotential of the process at the instant t after breaking of the circuit, and b is the slope of the η vs log i polarization curve. It is assumed in integration that the capacity is independent of the electrode potential and that the rate of the reverse reaction indicated above may be ignored. If the electrode capacity depends on the potential, Eq. (V.19) can be applied only to small regions of the decay curve, on the assumption that the capacity is constant over a narrow range of potentials. The degree of surface coverage by the organic substance is calculated from Eq. (V.18). For calculation of the capacity of the double layer the potential is measured at the initial instant after breaking of the circuit, when

$$2.3 \frac{ti_0}{bC} \ll 1. \tag{V.20}$$

In this case Eq. (V.19) is simplified:

$$\eta_0 - \eta \simeq \frac{ti_0}{C}. \tag{V.21}$$

When the inequality (V.20) holds, simple calculation shows that
the potential must be recorded during a very short time interval,
which decreases with increase of the initial polarization current.
If $C = 20 \ \mu F /cm^2$, $b = 0.12$ V, and $i_0 = 10 \ mA/cm^2$, we have:

$$t \ll \frac{20 \cdot 10^{-6} \cdot 0.12}{2.3 \cdot 10^{-2}} \ sec = 10^{-4} \ sec.$$

Measurement of the potential during short intervals after dis-
connection of the current involves considerable experimental difficulties. Special electronic equipment is used for this purpose
[100-107].

If the reaction is partially reversible Eq. (V.20) is complicated
considerably and the potential decay curves become difficult to
analyze [99]. In this case the potential decays more slowly and
the calculated capacity is considerably higher than the capacity
of the double layer. High values of the capacity (200-350 $\mu F/cm^2$)
are obtained when ionization of atomic adsorbed hydrogen [93, 94]
or other intermediate products [96] occurs. High capacity values
may be obtained when electrochemical oxidation of adsorbed substances occurs [108, 109], and when the layer near the electrode
becomes alkaline if the electrode potential depends on the solution pH.

The applicability of the method is restricted by the influence
of the reverse process on the capacity. Usually the determinations
are performed under conditions such that the rate of the reverse
process is low and the measured capacity is the capacity of the
double layer.

Moreover, the rate of change of the potential with time increases with the increase of the preliminary polarization current.
At high current densities the adsorbed particles do not become
ionized immediately. Therefore the concentration of these particles during decay of potential does not correspond to the stationary concentration, and the adsorption isotherm calculated from
capacity data corresponds to the nonstationary state of the electrode. In certain cases differentiation of the φ vs t curves is help-
ful [110].

The decay method has been used for studying adsorption of
naphthoquinoline, acridine, and ethylquinoline on nickel [111],
of o-phenanthroline, phenylthiourea, and some other inhibitors on

iron [112], and of formic acid on platinum, palladium, and gold [108, 109], and also in studies of adsorption of certain organic substances on platinum at high anodic potentials [113]. Galvano-static pulse methods [114] are a variant of the potential decay method.

Measurement of adsorption of organic substances on metals of the platinum group is complicated by simultaneous adsorption of hydrogen or oxygen. Therefore analysis of the results obtained by impedance measurements is difficult, and the data are suitable mainly for qualitative characterization of adsorption [115-121]. Agreement between the results obtained by different workers for the same systems is poor owing to difficulties arising in imped-ance measurements and interpretation of the results in the case of metals of the platinum group.

Attempts at quantitative investigation of adsorption of amyl and methyl alcohols on smooth metals of the platinum group and gold with the aid of the impedance method have been made by Breiter [116, 119, 120]. For example, Eq. (V.18) was used for calculating the dependence of the degree of surface coverage on the potential from data on the lowering of capacity in presence of amyl alcohol at various concentrations. It was noted that the electrode capacity varied greatly with time; this was attributed both to the low rate of adsorption of the alcohol and to the de-crease in the activity of the smooth electrode with time. In an-other investigation [119], the adsorption of methanol on smooth platinum in $1\,N$ $HClO_4$ was studied as a function of the potential and concentration. The measurements were performed at an ac frequency of 300 Hz, with linear variation of the electrode poten-tial in the anodic and cathodic directions at a rate of 30 mV/sec between 0.1 and 1.6 V. For calculation of θ in methanol solu-tions, Breiter used Eq. (V.18) and (V.22):

$$\theta = \frac{C'}{C}. \tag{V.22}$$

Equation (V.18) can be used only if $C < C_0$, whereas in reality the reverse was true in a certain potential range. Therefore Breiter used Eq. (V.22); however, this is not based on any physically jus-tified picture of adsorption and is inapplicable at low θ. Com-parison of the results obtained by calculations from Eqs. (V.18) and (V.22) with the results of potentiodynamic measurements [122] showed approximate agreement (within 15%) between the values of θ obtained by the different methods only at $\theta > 0.7$. Breiter notes that quantitative data can be obtained by the impedance method at low ac frequencies only at the potentials of the double-layer region.

In another publication [120], Breiter presents data on adsorption of methanol (at concentrations from 10^{-3} to $1\ M$) in perchloric acid on Pt, Pd, Rh, Ir, and Au. These determinations were also carried out with linear variation of the electrode potential (30 mV/sec), in the potential range of 0.07-1.4 V for Pt, Rh, Ir, and Au, and 0.2-1.0 V for Pd, at 1 kHz frequency. Breiter notes decrease of the hydrogen pseudocapacity and of the double-layer capacity in presence of methanol, desorption of methanol at sufficiently high anodic potentials as the result of oxidation of the adsorbate, and considerable hysteresis of the curves obtained with variation of the potential in the anodic and cathodic directions. Adsorption was characterized by plots of the capacity decrease, $C_0 - C$, against the methanol concentration.

Adsorption of methanol on a polycrystalline gold electrode was also determined by Schmid and Hackerman [121] at 25 and 5° in $1\ N$ HClO$_4$. The measurements were performed with stepwise variation (Δi) of the polarizing current density, the resultant changes in the slope of the potential vs time curve [$\Delta\ (d\varphi/dt)$] being recorded. The ratio of these quantities was characterized as the differential electrode capacity.* The authors observed increase of the electrode capacity on addition of methanol, and the appearance of maxima on the curves; these maxima were attributed to desorption of methanol at sufficiently high positive and negative surface charges. The capacity was lower in presence of higher alcohols than in presence of methanol, but exceeded the double-layer capacity in the pure electrolyte solution in the region of adsorption of organic substances. The experimental data of Schmid and Hackerman [121] contradict Breiter's results. Their view that de-

---

*Pochekaeva [123] determined adsorption of allyl alcohol in 0.1 N H$_2$SO$_4$ from the increase in the slope of the charging curve of a Pt/Pt electrode in presence of the organic substance in the potential range of 0.3-0.6 V. As in this range adsorbed hydrogen and oxygen are present on the platinum surface in a solution of the pure electrolyte (see below), the adsorption of which alters in presence of the organic substance, this method is approximate. Moreover, oxidation of organic substances in the same potential range is possible.

sorption peaks may appear in the case of increased capacity during adsorption of organic substances contradicts the theoretical concepts discussed in Part I.

3. Study of Adsorption with the Aid of Electrochemical Oxidation or Reduction of Adsorbed Organic Substances

If an organic substance adsorbed on an electrode can be oxidized or reduced in a certain potential range, it is possible in principle to estimate the adsorption of the substance from the quantity of electricity consumed for its oxidation or reduction. The primary need in this method is to take into account possible errors in the results due to diffusion and electrochemical conversion of the organic substance from the solution volume during oxidation or reduction of the adsorbed layer. In order to avoid complications, either the test solution containing the organic substance is replaced by a solution of the pure electrolyte before oxidation or reduction of the adsorbed substance (the method of electrochemical oxidation or reduction in the adsorbed layer), or pulse methods are used for the measurements.

a. Method of Electrochemical Oxidation (or Reduction) in the Adsorbed Layer. This method was first used by Shlygin et al. [124, 125] and Pavela [126] in relation to the platinized platinum electrode. It was developed more fully by Frumkin, Podlovchenko, and Iofa [127, 128]. Niedrach [129] used it for investigating adsorption of hydrocarbons at a porous platinum – Teflon electrode.

The method can be summarized as follows. The electrode is immersed in a solution of the organic substance for some time for establishment of adsorption equilibrium. It is then washed with a fairly large amount of the pure electrolyte solution (dissolved hydrocarbons or CO and CO_2 may be removed by a current of inert gas) and subjected to anodic or cathodic polarization by direct current until the potentials of oxygen or hydrogen evolution are reached. The quantity of electricity consumed in polarization of the electrode after removal of the organic substance by washing

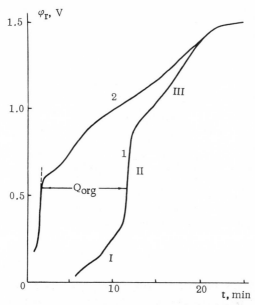

Fig. 50. Anodic charging curves of a Pt/Pt electrode in 1 N H_2SO_4 (1) and after contact of the electrode with butyl alcohol (2); Q_{org} represents the quantity of electricity consumed for oxidation of the adsorbed organic substance; current density in determinations of the charging curves $3 \cdot 10^{-4}$ A/cm^2.

is usually greater than the quantity of electricity required to reach the potentials of oxygen or hydrogen evolution in a solution of the pure electrolyte. The amount of substance adsorbed can be estimated from the difference in the quantity of electricity involved in oxidation or reduction processes.

Figure 50 shows the charging curves of a Pt/Pt electrode in a solution of the pure electrolyte* (1) and in presence of an adsorbed organic substance (2). The charging curve determined in the pure

---

*The technique and results of determinations of charging curves and potentiodynamic curves in solutions without added organic substances have been discussed in detail by Frumkin [21].

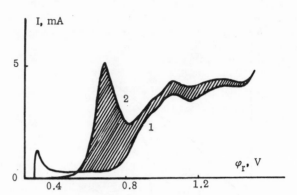

Fig. 51. Potentiodynamic I vs φ curves of a Pt/Pt electrode in 0.1 N H_2SO_4 (1) and after contact of the electrode with ethyl alcohol (2). Potential sweep rate v =
5 mV/sec.

electrolyte solution has three regions: hydrogen (I), double layer[*]
(II), and oxygen (III). In presence of an adsorbed substance the
course of the charging curve alters substantially. The charging
curves 1 and 2 coincide starting at a certain potential; this indicates complete oxidation of the adsorbed substances to products
which are not adsorbed on the electrode. Figure 50 also shows
graphical determination of the quantity of electricity Q_{org} consumed
for oxidation of the adsorbed substance.

In certain studies [131-133] of adsorption by the method described, the electrode potential was varied linearly with the aid
of a potentiostat, and the quantity of electricity consumed for
oxidation of the adsorption products was given by the hatched area
enclosed by the potentiodynamic curves 1 and 2 in Fig. 51.

The method of electrochemical oxidation (or reduction) in an
adsorbed layer is a striking illustration of the adsorption of or-

_____

[*]It followed even from the data of Shlygin et al. [3] that in sulfuric acid solutions
the presence of adsorbed gases is possible at potentials of the double-layer region.
Determinations with the aid of tracer atoms [130] show that the capacity of the
Pt/Pt electrode in the double-layer region is ~36 $\mu F/cm^2$, whereas a value of the
order of 70 $\mu F/cm^2$ is obtained from the slope of the charging curve in the 0.4-
0.7 V range. This confirms the conclusions in [3]. Marvet and Petrii [23] showed
that adsorbed oxygen is present on platinum at potentials of the double-layer region.

ganic substances itself. However, its application to quantitative investigations is limited by a number of conditions.

1. When the electrode is washed, the degree of desorption of the organic substance must be negligible and the adsorbate must not undergo any significant changes (molecular rearrangement, dimerization, etc.) [134, 135]. Therefore the method is applicable only in the case of irreversible adsorption of organic substances.

2. The adsorbed substance and the intermediate products of its conversion are removed during polarization of the electrode only by electrochemical oxidation or reduction.

3. The electrode surface must be fairly extensive. The presence even of small amounts of organic substances and oxygen in the electrolyte solution used for washing may distort completely the results obtained with smooth electrodes.

4. For estimation of surface coverage by this method it is generally necessary to know the composition of the adsorbed substance and the over-all reaction of its oxidation or reduction.

Pavela [126] determined for the first time the concentration dependence of adsorption of methanol on a platinized platinum electrode with the aid of the method of electrochemical oxidation in the adsorbed layer. Comparison of his data with the results obtained by Breiter and Gilman [122] in determinations of surface coverage by the potentiodynamic pulse method shows that the results of the two methods are in agreement at high surface coverages. This may be regarded as evidence that condition 1 is satisfied.* A similar conclusion follows from a comparison of the data of Breiter and Gilman with the results reported in [131], and from the data of Johnson and Kuhn [133].

Despite its substantial limitations, the method has been widely used for investigating adsorption of various organic substances on metals of the platinum group and their alloys (Pt, Rh, Pd, Pt + Ru, Pd + Ru, etc.), and has provided valuable information on the properties of the adsorbed particles. The great advantage of this method over the pulse methods described below is that it is pos-

---

*The differences at low coverages may be due to the fact that in Pavela's experiments possible oxidation of methanol by oxygen during washing of the electrode was not taken into consideration [136] and to removal of weakly bound methanol [137].

sible to study the behavior of adsorbed organic substances with electrode polarization by low current densities or at low rates of potential variation.

The possibilities of the method are extended considerably if it is used in conjunction with analytical methods. For example, Niedrach [129] and Podlovchenko and Stenin [138] analyzed the gaseous products formed during cathodic reduction of organic substances adsorbed on platinum. Johnson and Kuhn [139] correlated the quantity of electricity consumed for oxidation of an organic substance with the amount of CO_2 formed in the process. For determination of carbon dioxide, the adsorbed substance was oxidized in alkaline solution, which was then titrated with the use of phenolphthalein and methyl red indicators. Grubb and Lazarus [140] and Breiter [141] used gas chromatography for determination of CO_2 during galvanostatic oxidation of adsorbed organic substances. In an investigation of the adsorption of methanol by the method of electrochemical oxidation in the adsorption layer, Podlovchenko et al. [137] simultaneously analyzed the liquid phase for methanol, formaldehyde, and formic acid. Methyl alcohol with the C^{14} carbon isotope was used by Podlovchenko et al. [142] in analogous experiments. It was thus possible to use a radiometric method for determination of the CO_2 formed by oxidation of the products of methanol chemisorption. It will be shown later that with the aid of these methods it is possible to form conclusions regarding the nature of the products of chemisorption of organic substances on metals of the platinum group.

b. Potentiodynamic and Potentiostatic Pulse Methods. In 1962, Breiter and Gilman [122] proposed a method for studying adsorption of methanol on a smooth platinum electrode with the aid of rapid anodic potentiodynamic pulses.* By this method the electrode is held in the test solution at a given potential, and a rapid potentiodynamic sweep of potential is then applied. The dependence of the current on the electrode potential is recorded oscillographically. Breiter and Gilman [122] determined nonstationary anodic current−potential curves in $1\,N$ $HClO_4$ with various additions of methanol. These curves are shown in Fig. 52.

---

*The development of the potentiodynamic method is usually associated with the work of Will and Knorr [143]. However, Kolotyrkin and Chemodanov[144] used this method at the same time in an investigation of the platinum electrode.

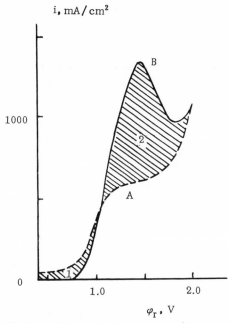

Fig. 52. Potentiodynamic i vs φ curves for a smooth
platinum electrode in 1 N HClO$_4$ (A) and in 1 N
HClO$_4$ + 1 M CH$_3$OH (B). The curves are measured
from the potential established in presence of methanol in open circuit. Potential sweep rate v = 800
V/sec (from data in [122]).

The quantity of electricity, determined from the difference between
the hatched areas 2 and 1 (Fig. 52), is

$$Q'_{org} = \int_0^t i_2 dt - \int_0^t i_1 dt. \qquad (V.23)$$

In the general case, Q'_{org} is given by the following equation:

$$Q'_{org} = Q_{org} + Q''_{org} + \Delta Q_c + \Delta Q_o, \qquad (V.24)$$

where Q_{org} is the charge corresponding to oxidation of the substance adsorbed on the electrode during the anodic pulse; Q''_{org}
is the charge corresponding to oxidation of substance reaching the
electrode by diffusion during the pulse; ΔQ_c is the difference between the quantities of electricity for charging the double-layer

capacity in absence and in presence of the adsorbed substance; ΔQ_0 is the difference between the charges for oxidation of the surface in absence and in presence of the organic substance.

The applicability of the potentiodynamic pulse technique to studies of adsorption of organic substances is based on the following conditions.

1. The potential sweep rate is so high that diffusion of the organic substance from the solution volume during the pulse can be neglected ($Q_{org}^{''} \sim 0$).

2. All the adsorbed substance is oxidized during the pulse and at the end of the sweep the electrode surface is free from organic particles.

3. The double-layer capacity, and the adsorption pseudo-capacity due to formation of an oxide layer on the electrode surface, do not change substantially in presence of organic molecules on the surface ($\Delta Q_C \ll Q_{org}$; $\Delta Q_0 \sim 0$).

4. The number of electrons required for oxidation of an adsorbed particle is independent of the surface coverage.

5. The adsorbed substance and the intermediate reaction products are removed from the surface only by electrochemical oxidation.

Fulfillment of the above conditions means that $Q_{org}^{'} = Q_{org}$. In this case the surface coverage θ by the organic substance is calculated from the formula

$$\theta = \frac{Q_{org}}{Q_{org}^{max}}, \qquad (V.25)$$

where Q_{org}^{max} is the maximum charge required for oxidation of the adsorbed substance.

The anodic galvanostatic pulse method can be used instead of the rapid potentiodynamic method. The galvanostatic and potentiodynamic techniques do not differ in principle and are based on the same assumptions.

To verify that the first condition is satisfied, the potentiodynamic curves for the test solution are recorded at different sweep rates and $Q_{org}^{'}$ is plotted against the sweep rate. It is thus possible to determine the sweep rate at which the limiting value of Q'_{org} is reached. It was shown [122] that sweep rates $v > 200$ V/sec must be used for determination of adsorption in presence of 1 M CH_3OH. Substantially lower sweep rates may be used at lower concentrations of the organic substance in solution and with electrodes having an extensive surface. For example, sweep rates from 0.1 to 0.4 V/sec are sufficient for investigation of adsorption of ethane and methane on porous platinum – Teflon electrodes [145]. In these cases high sweep rates (> 0.92 V/sec) could not be used because the shape of the potentiodynamic curves was greatly distorted as the result of ohmic potential drops in the electrode itself.

To verify that condition 2 is satisfied, I vs φ curves determined during the first and subsequent potentiodynamic pulses in the solution are compared. If condition 2 is fulfilled, the curve obtained during the second pulse should coincide with the curve for the pure electrolyte solution. Whereas, as was shown by Breiter and Gilman [122, 146], condition 2 is satisfied even at sweep rates up to 2000 V/sec in adsorption of methanol and CO on smooth platinum, in the case of oxidation of adsorbed ethylene and acetylene [147] independence of the charge, corresponding to oxidation of the adsorbed layer, was attained at $v = 10$ V/sec in solutions saturated with a mixture of argon and hydrocarbon, the concentration of the latter being 1%. At higher sweep rates oxidation of the substance during the first anodic pulse was incomplete and Q'_{org} diminished. Thus, the measured charge Q'_{org} was only a certain fraction (a) of the true charge corresponding to the amount of substance adsorbed:

$$Q'_{org} = a Q_{org}. \qquad (V.26)$$

The value of a ($a \leq 1$) could be calculated for each sweep rate in solutions of low adsorbate concentrations, and then used for calculation of Q_{org} from the results of determinations in more concentrated solutions.

The greatest errors arise if condition 3 is not fully satisfied. Thus, Khazova et al. [148] showed that methanol at high concentrations has an appreciable influence on adsorption of oxygen at a platinum surface. For more accurate determination of the quantity of electricity consumed for oxidation of the surface during the anodic pulse, Brummer [149, 150] used a combination of anodic and cathodic pulses (current reversal method). Immediately after oxidation of the adsorbed substance by the anodic pulse, a cathodic pulse is applied to the electrode. The curve obtained during the latter pulse has plateaus corresponding to reduction of the oxide layer formed during the anodic pulse and to adsorption of hydrogen. The amount of adsorbed oxygen at the potential of current reversal is found from the magnitude of the first plateau while the extent of oxidation of the adsorbed substance is estimated from the second (see below). Objections were raised [151] to this method, based on the fact that the oxygen adsorbed on platinum is not reduced completely during a rapid cathodic pulse, and overestimated values are therefore obtained for Q_{org}. Subsequently Brummer [152] examined the potentialities of the current reversal method critically and found conditions for obtaining data by this method in agreement with the results given by the anodic galvanostatic method described above. He also proposed a method for introducing a correction for charging of the double-layer capacity [153].* Podlovchenko et al. [142] based this correction on the results of determinations with labeled atoms.

It must be emphasized that when the anodic potentiodynamic pulse method is used for studying adsorption of organic substances only the total quantity of electricity consumed in oxidation of the adsorbed substance can be determined. As the determinations are performed at high potential sweep rates, the resultant I vs φ curves are of nonequilibrium character and cannot be used directly for drawing inferences regarding the oxidation mechanism. This also applies to cases where rapid potentiodynamic pulses are used for studying the oxidation mechanism of organic substances present in solution. Conway, Gileadi, et al. [77, 156, 157] have discussed the possibilities of the potentiodynamic and galvanostatic methods for studying electrochemical oxidation and adsorption of organic substances.

4. Methods for Studying Adsorption

of Organic Substances Based on

Adsorptional Displacement

Oikawa and Mukaibo [158] and Franklin and Sothern [159], in studying adsorption of acetic acid and nitriles on platinum, de-

---

*The question of correction for charging of the double-layer capacity in chronopoten-tiometric and chronoamperometric measurements is the subject of a number of recent publications (e.g., [154, 155]).

termined the decrease of the amount of hydrogen adsorbed on the surface when the electrode was brought into contact with a solution of the organic substance. This method is known in the literature as the cathodic pulse method. — $cp M$

By this method, the quantity of electricity $_sQ_H$ required for formation of a monolayer of adsorbed hydrogen atoms on the electrode in a solution of the pure electrolyte is first determined with the aid of a cathodic galvanostatic or potentiodynamic pulse. The amount of hydrogen adsorbed, $_0Q_H$, after exposure of the electrode in a solution containing the organic substance at a given potential is then found similarly. The coverage of the electrode surface by the organic substance can be found from Eq. (V.27):

$$\theta = \frac{_sQ_H - _0Q_H}{_sQ_H}. \tag{V.27}$$

It was pointed out by Breiter [116] that the use of Eq. (V.27) implies that all the sites on which hydrogen is adsorbed are also accessible to adsorption of the organic substance. In reality, however, owing to steric hindrance the electrode remains able to adsorb a certain amount Q_H of hydrogen when the limiting surface coverage by the organic substance has been reached. Accordingly, Eq. (V.28) was proposed [160] for calculation of the surface coverage:

$$\theta = \frac{_sQ_H - _0Q_H}{_sQ_H - Q_H}. \tag{V.28}$$

The cathodic pulse method can be used if the following conditions are satisfied.

1. Desorption or additional adsorption of the organic substance on the electrode surface does not occur during the pulse.

2. The organic substance is not reduced in the course of cathodic polarization of the electrode during the pulse.

3. No appreciable amounts of hydrogen diffuse into the metal during the pulse.

4. Complete coverage of the surface by adsorbed atomic hydrogen is attained during the cathodic pulse in the pure electrolyte solution. If a monolayer of adsorbed hydrogen is not formed on the surface, the method is applicable only if the organic substance is adsorbed only on the sites where hydrogen is adsorbed.

5. The amount of adsorbed hydrogen decreases only as the result of occupation of some of the sites by the adsorbed organic substance, and no decrease of hydrogen adsorption occurs owing to decrease of its bonding energy with the surface in presence of the organic substance, etc.; i.e., if no long-range effects occur [8, 21].

Bagotskii and Vasil'ev [160] used a potential sweep rate v ~ 40 V/sec in their study of methanol adsorption by the potentiodynamic cathodic pulse method. The current density used in galvanostatic measurements under analogous conditions was 0.06 A/cm^2 [122], which was sufficient for fulfillment of the first condition. According to Gilman [147], sweep rates of the order of 300 V/sec must be used in the case of solutions saturated with ethylene and acetylene. The results obtained at v < 60 V/sec are clearly erroneous owing to reduction and desorption of adsorbed ethylene and acetylene. Adsorption of propane was investigated by the cathodic galvanostatic pulse method at current densities of the order of 100-200 mA/cm^2 by Brummer et al. [161].

The accuracy in determination of surface coverage by the cathodic pulse method depends to a considerable extent on the accuracy in determination of the quantities of electricity expended in side processes: ionization of oxygen (if the potential at which adsorption is determined lies in the oxygen region), charging of the double layer, and evolution of molecular hydrogen. Various methods have been used [146, 147, 160] for determining the quantity of electricity expended in hydrogen evolution (one of these methods is evident from Fig. 53).

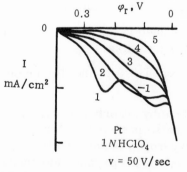

Fig. 53. Cathodic potentiodynamic curves for a smooth platinum electrode in 1 N HClO$_4$ saturated with ethane, determined after different time intervals during adsorption of ethane at φ = 0.4 V: 1) 10 sec (curve coincides with the curve for the pure electrolyte solution); 2) 20 sec; 3) 100 sec; 4) 500 sec; 5) 1000 sec. Potential sweep rate v = 50 V/sec. ΔQ_H is found as the difference between the areas enclosed by curve 1 and by the other curves (from data in [165]).

The cathodic pulse method and the anodic pulse method described above measure different properties of adsorbed organic molecules and are mutually complementary [162]. Thus, the cathodic pulse method is sensitive to the manner of adsorption of the organic substance, regardless of the nature of the organic particles adsorbed. It can therefore be used for distinguishing between one- and two-site adsorption of molecules [146] and for establishing the nature of the distribution of the organic substance on the surface between sites having different bond energies H_{ads}. The anodic pulse method, on the other hand, measures the charge required for oxidation of the adsorbed substance and, as will be discussed later, makes it possible to reveal partial oxidation or dehydrogenation of organic molecules during adsorption.

If measurements by the cathodic pulse method are carried out in the region of hydrogen adsorption potentials, they can be used for determining the decrease of the amount of hydrogen adsorbed only at potentials in the range between the potential under investigation and the potential of evolution of molecular hydrogen. As a result of this, the values obtained may be too low [163]. Therefore a combination of the anodic and cathodic pulse methods must be used for determination of the adsorption of an organic substance in the region of hydrogen adsorption. Brummer and Turner [164], in determination of the adsorption of an organic substance by the cathodic pulse method, after contact of the electrode with the organic substance at potentials corresponding to the hydrogen region applied a potential of 0.5 V (vs a reversible hydrogen electrode) to the platinum electrode for a short interval (of the order of a few msec), when the previously adsorbed hydrogen was ionized; the cathodic curve was then determined from $\varphi = 0.5$ V. It can be shown by special experiments that brief application of 0.5 V to the electrode does not affect the results.

For investigation of adsorption effects at a platinum electrode, Gilman [146] devised the multipulse potentiodynamic method (the MPP method) which comprises a whole series of operations whereby a reproducible state of the electrode surface is produced and changes of the electrode properties during adsorption can be measured. This method was initially used for investigating adsorption of CO on platinum. To explain the principle of the method, Fig. 54 shows the sequence of the pulses used by Gilman [165] in a study of adsorption of ethane on a smooth platinum microelectrode.

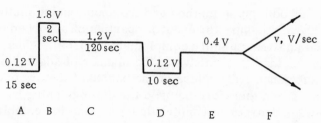

Fig. 54. Sequence of rectangular potentiostatic pulses used by
Gilman [165] in investigating adsorption of ethane on smooth
platinum.

During stage A a low anodic potential is applied to the electrode
and the solution is saturated with ethane. A potential of 1.8 V is
then applied to the electrode; at this potential the surface becomes
covered with a layer of adsorbed oxygen, which prevents adsorption of the hydrocarbon, while the previously adsorbed substance
is oxidized. The passive oxide layer previously formed on the
surface is retained during stage C, so that the hydrocarbon is not
adsorbed. At the same time, gaseous oxygen is not evolved at
1.2 V, so that the solution can be brought into equilibrium with
ethane. Stage D is necessary for reduction of the oxide layer.
Stage E, the duration of which is determined by the particular problem, is followed by anodic or cathodic potentiodynamic potential
sweep.

The MPP method has also been applied to a conducting
Teflon−platinum black semimicroelectrode [145] and to a rhodium
electrode [166]. The applicability of the MPP method to quantitative studies of adsorption of organic substances has been demonstrated under conditions where adsorption is limited by diffusion [147].

Other variants of the method based on adsorptional displacement
have also been used. For example, Girina, Fioshin, and Kazarinov
[167] determined adsorption of acetate ions on a platinum electrode
at high anodic potentials from the decrease of adsorption of SO_4^{2-}
ions (labeled with the S^{35} isotope) as the result of displacement by
CH_3COO^- ions. Kazarinov and Mansurov [168] determined adsorption of naphthalene and of methyl and hexyl alcohols on platinum
from changes in the adsorption of labeled SO_4^{2-}, PO_4^{3-}, and Na^+
in presence of these organic compounds.

5. Study of Adsorption of Organic

Substances with the Aid of

Radioactive Tracers

Methods involving the use of radioactive tracers have been widely used in recent years for studying adsorption of solid electrodes [169]. Critical reviews of radioactivity methods for studying adsorption have been published [170, 171].

The simplest but the least satisfactory method consists of measurement of the activity of the surface of an electrode taken out of a solution after adsorption of a radioactive substance. The electrode is put into an electrolyte containing a radioactive organic substance, and is then withdrawn and washed with a solvent to remove traces of radioactive solution from the surface. The activity of the metal surface is compared with a standard and the amount of substance adsorbed is found. A defect of the method is that adsorption is measured under conditions when adsorption equilibrium is disturbed. Moreover, the substance is partially removed when the specimen is washed with solvent; this lowers the accuracy of the measurements. This method can be used for determining adsorption of stably chemisorbed substances. It was applied in studies of adsorption of stearic acid, perfluorooctanoic acid, and thiourea on iron, nickel, copper, and other metals [18, 20, 172-174].

Bockris and co-workers developed and used two methods for determining the adsorption of organic substances on solid electrodes with the aid of labeled atoms.

The first [175] was based on a principle first put foward by Joliot [176] (Fig. 55). By this technique the amount of substance adsorbed on the electrode is estimated from the activity of the side of the electrode away from the solution. A thin foil of the metal is placed on the counter window, and the counter is put above the solution. When the metal comes into contact with the solution the activity increases sharply owing to adsorption of the organic substance. The amount of substance adsorbed is found from the difference between the activities after and before contact of the electrode with the solution. The activity before contact can be found by extrapolation of the dependence of activity on the distance between the counter and the solution to zero distance. The background activity of the solution can also be determined directly if

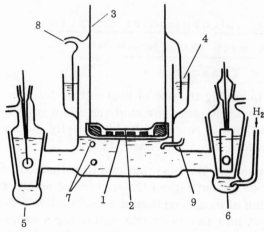

Fig. 55. Schematic diagram of cell for determinations with
the aid of radioactive tracers by the foil method: 1) Working
electrode; 2) counter window; 3) sides of counter; 4) water
seal; 5) auxiliary electrode; 6) reference electrode; 7) gas
inlet; 8) gas outlet; 9) inlet for solution sample (from data
in [179]).

a labeled compound which is not adsorbed specifically on the test
electrode is chosen. In a number of cases a solution of sodium
carbonate containing C^{14} can be used for this purpose. This method is suitable for metals which can be obtained in the form of very
thin foil (about 20 μ) or can be applied in a thin layer on foil made
from another metal. It is therefore sometimes described as the
foil method. Very thin electrodes must be used in order to reduce
adsorption of the radiation in the metal layer to a minimum. Flannery and Walker [177] increased the sensitivity of the method by
depositing a layer of the metal on mica, Teflon or other material
which absorbs radiations only slightly, using this layer as the
electrode. High sensitivity is achieved only if the thickness of
such an electrode does not exceed a few 1000 Å. The preparation
of such thin layers, which must at the same time have adequate
strength and behave as equipotential electrodes, constitutes the
main experimental difficulty in this technique.

The foil method is used with radioactive tracers emitting
soft β-radiation (C^{14}, S^{35}) and solutions in which the adsorbate
concentration does not exceed 10^{-3} M. At higher concentrations

of organic substances the increase of radioactivity near the electrode as the result of adsorption is small in comparison with the strong background activity. This technique has been used for studying adsorption of thiourea [175, 178, 179] and of naphthalene, benzene, and phenanthrene [180, 181] on gold, of ethylene [182, 183] and of benzene [184] on platinum black deposited on gold foil, of butane [177] on a sputtered platinum film on mica, and of methyl alcohol [185] on platinum.

For determination of adsorption on metals which cannot be obtained in the form of very thin films, Green et al. [171] devised a technique whereby the amount of substance adsorbed is determined after withdrawal of the electrode from the solution (the tape method). The electrode, in the form of a continuous metal tape (about 50 μ thick), is drawn through the electrolytic cell. The tape is held in the solution for a time sufficient for establishment of adsorption equilibrium; it is then withdrawn from the cell and passed between two proportional counters, which determine the total amount of radioactive material present on the tape. The thickness of the liquid layer (and therefore the amount of radioactive substance in it) is calculated from the change in the capacity of a condenser between the plates of which the tape passes. The amount of substance adsorbed by the metal surface can be found by subtraction of the activity due to the substance in the solution film wetting the tape from the total activity of the tape. The errors in determination by the tape method depend on the radioactivity of the solution carried out by the tape and on inaccurate potentiostatic control due to ohmic losses when the tape leaves the cell (the potential of the tape is not controlled after it leaves the cell). This method has been used for determination of adsorption of n-decylamine on Cu, Ni, Fe, Pb, and Pt [186] and of naphthalene on Cu, Ni, Fe, and Pt [187]. The tape method was not used in later investigations with radioactive tracers, probably because it is less accurate than the foil method.

Kazarinov [188] devised a method for measuring adsorption in which, for measurement of the radioactivity, the electrode is lowered to a membrane of thin glass, polyethylene, mica, or other material, which acts as the floor of the cell (Fig. 56). The measured radioactivity is the sum of the activities of the adsorbed substance, the layer of solution between the membrane and the electrode, the substance adsorbed on the membrane, and the back-

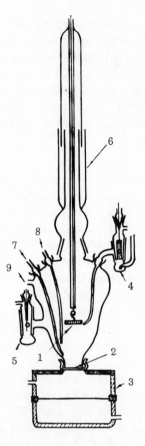

Fig. 56. Apparatus for measurement with the aid of radioactive tracers: 1) Working electrode; 2) cell bottom covered
with polythene film; 3) counter;
4) reference electrode; 5) auxiliary electrode; 6) water seal;
7) inlet for solution sample;
8 and 9) gas inlet and outlet
(from data in [188]).

ground activity of the solution. Methods for determination of the
various quantities needed for finding the amount of radioactive substance from the total activity measured are described in the paper. This method was used for determination of the adsorption of
hexyl alcohol on a Pt/Pt electrode. Podlovchenko, Stenin, and
Kazarinov [142] used this technique for studying the kinetics of
adsorption and desorption of methyl alcohol on platinum.

In contrast with the methods discussed earlier, adsorption
measurements with the aid of radioactive tracers are direct methods. However, in the studies which have been reported the measurements with the aid of labeled atoms were in general less accurate than pulse methods. Moreover, radioactivity methods give
only the total amount of carbon adsorbed on the electrode (in the

case of adsorption of organic substances containing C^{14}). As will be shown later, adsorption of organic compounds on metals of the platinum group is sometimes accompanied by changes in the nature of the compounds, and radiometric methods are therefore insensitive to the structure of the adsorbed layer. Difficulties also arise in radioactivity methods owing to evolution of hydrogen, oxygen, or carbon dioxide during the adsorption measurement, as the evolving gas may prevent access of the organic substance to the electrode. If phase oxides are formed during the determinations, capture of the organic substance by the growing oxide layer on the electrode may lead to anomalously high adsorption values.

The most reliable results in measurements of adsorption on solid electrodes may be obtained with the aid of a combination of electrochemical and radiometric methods. This is illustrated by the work of Gileadi et al. [188a].

6. Determination of Adsorption from the Weight Increase of the Adsorbent or from the Decrease of the Adsorbate Content in Solution

This group of methods is based on determination of the change in the weight of the adsorbent or in the concentration of the adsorbate in the solution. Hackerman and co-workers [189, 190] used the adsorbent weight method for determination of adsorption. This method is suitable for studying adsorption of organic substances of low volatility from organic solvents having high vapor pressures. Hackerman et al. [189-191] studied adsorption of higher alcohols, acids, esters, amines, nitriles, thiourea derivatives, and some other substances on steel powder. Benzene was used as the solvent. The apparatus is shown in Fig. 57.

The tube (2) containing the powder was put into the vessel (1) containing a solution of the surface-active substance in benzene (5). The tube had a porous bottom (4) which ensured unimpeded contact between the solution and the powder. After adsorption equilibrium had been established the tube with the powder was taken out of the vessel and the powder was dried and weighed. The difference between the weights before and after adsorption was equal to the weight of substance adsorbed. In dilute solutions the solute con-

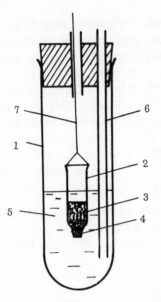

Fig. 57. Apparatus for determination of adsorption from the increase in the weight of the adsorbent. 1) Glass vessel; 2) glass tube; 3) adsorbent; 4) porous glass partition; 5) solution; 6) tube for changing the solution (from data in [189]).

centration may decrease as the result of adsorption. In order to maintain the solute concentration in the solution constant, the depleted portion of solution was replaced through the tube (6) by a fresh portion. According to the authors, the method is accurate within 5%. It should be pointed out that drying of the powder in air before the adsorption measurements is undesirable, as it leads to partial passivation of the iron surface and to changes in its adsorptional and electrochemical properties [192].

If the specific surface of the adsorbent is large relative to the solution volume, adsorption of surface-active substances is accompanied by appreciable changes of the solution concentration. This effect was used by Frumkin et al. [3, 4] in studies of adsorption of inorganic ions on platinized platinum. If the total volume of the solution is known, the amount of substance adsorbed can be easily calculated from the change of concentration.

Adsorption of organic substances has been determined from changes of their concentration in solution after adsorption on metal powders [17, 193-195] or on metal gauzes [196, 197]. Various methods, dependent on the nature of the substance and the solution concentration, are used for determining changes of the concentration.

For example, the concentration of the organic substance has been determined spectrophotometrically from the degree of lowering of the maximum in the absorption spectrum [196-198]. Conway, Barradas, and Zawidzki [196] used this method for studying adsorption of acridine, quinoline, and other substances on copper, nickel, and silver over a wide concentration range $(5 \cdot 10^{-7}\text{-}10^{-4} M)$. Cavallaro, Felloni, and Trabanelli [198] determined adsorption of tolylthiourea and diphenylthiourea on iron in hydrochloric acid solutions.

In more concentrated solutions of surface-active substances, the concentration can be determined directly from the weight of the adsorbate after evaporation of the solvent. The difference between the weights of the solute before and after adsorption corresponds to the amount adsorbed. This method was used for studying adsorption of certain fatty acids from benzene on nickel and platinum [191]. The concentration of the dissolved substance has been determined by titration [199, 200], and found from the lowering of the electrocapillary maximum on mercury [201, 202]. Balezin and co-workers [203, 204] used a chromatographic method for determination of adsorption of sodium benzoate, dicyclohexylamine, and other substances on iron and magnetite. Other methods for determination of solute concentration in the course of adsorption processes have also been proposed [193, 194, 195].

The above method for measurement of adsorption by determination of changes in solute concentration is of great interest to investigators, because it permits study of adsorption effects without disturbance of adsorption equilibrium, whereas in a number of other methods (certain radiochemical methods, determination of adsorption from the weight of the adsorbent) the measurements are made after removal of the adsorbent from the solution. This may introduce appreciable errors in investigations of reversible adsorption processes.

However, this method can be used for adsorption measurements only in relatively dilute solutions of the organic substance.

7. Study of Adsorption with the Aid of the Electrode Photoelectric Effect

The production of current in electrochemical systems when one of the electrodes is exposed to light, first discovered in 1839 by Becquerel [206], has recently come into use for adsorption studies. This effect may be due to a number of essentially different causes [207-214]. It was pointed out by Barker [211] that transfer of electrons from the metal after absorption of light quanta (photoemission) into the solution must play a significant part in the electrode photoelectric effect. This viewpoint has been developed in a number of publications [212, 214, 215]. Photoemission of electrons from metals into solutions has been named the electrode photoelectric effect [216].

A quantitative theory of the electrode photoelectric effect has been put forward by Gurevich, Brodskii, and Levich [216-218]. In this theory the "5/2" law has been derived for the photocurrent I:

$$I \sim (\omega - \omega_0)^{5/2}, \qquad (V.29)$$

where ω_0 is the red boundary of the photoelectric effect and ω is the frequency of the radiation. The shift of the red boundary of the photoelectric effect due to polarization of the electrode should conform to the relation

$$h\omega_0(\varphi) = h\omega_0(0) + e\varphi. \qquad (V.30)$$

where $\omega_0(0)$ is the red boundary in absence of polarization. A number of theoretical aspects of the electrode photoelectric effects have been examined by Gurevich and Rotenberg [219-221).

The theory is in quantitative agreement with experiment. In illustration, data correlated with Eqs. (V.29) and (V.30) are shown in Fig. 58. The work function of a high-energy electron emitted from the metal into the solution at the potential of zero charge can be found from the potential corresponding to intersection of the linear plot with the abscissa axis and the wavelength with the aid of Eq. (V.30). If the work function is known, the point of zero charge can be found by the electrode photoelectric effect method.

In dilute solutions $I^{0.4}$ is approximately proportional to $(\varphi - \psi_0)$ where ψ_0 is the potential of the Helmholtz outer plane. As the derivative $d\psi_0/d\varphi$ has a maximum at the potential of zero charge, $dI^{0.4}/d\varphi$ is then a minimum. Therefore the potential of zero charge can be determined directly by study of the photoelectric effect in dilute solutions [222].

Experimental data on the dependence of the electrode photoelectric effect on the composition of the solution and the nature of

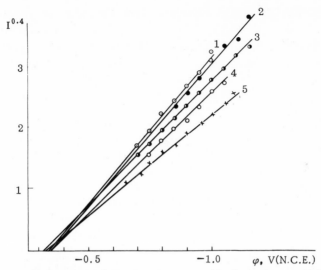

Fig. 58. Dependence of the photocurrent on the electrode potential in 1 N Na$_2$SO$_4$ + 0.01 N H$_2$SO$_4$: 1) mercury; 2) 3% thallium amalgam; 3) 18% indium amalgam; 4) 50% indium amalgam; 5) lead (from data in [222]. The photocurrent is expressed in arbitrary units.

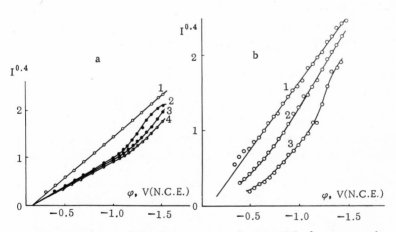

Fig. 59. Dependence of the photocurrent on the potential of a mercury electrode: a) in 0.1 N KF saturated with N$_2$O (1), and in presence of [(C$_4$H$_9$)$_4$N]Br at concentrations: 10^{-5} M (2), 10^{-4} M (3), and 10^{-3} M (4) (from data in [226a]); b) in 0.1 N KCl (1), and in presence of 0.04 M iso-C$_5$H$_{11}$OH (2), and 0.02 M C$_6$H$_{13}$COOH (3) (from data in [226]). The photocurrent is expressed in arbitrary units.

the electrode have been reported [223-226a]. It has been found [226, 226a] that when substances which are adsorbed at the electrode are added to the solutions the course of the $I^{0.4}$ vs φ curves alters substantially (Fig. 59). Extrapolation to I = 0 of the regions of the curves corresponding to adsorption of organic substances shows that the red boundary is independent of the presence of organic substances in the solution, and of the nature of the metal (Fig. 58). Although the nature of this influence of the organic substance on the electrode photoelectric effect is not fully clear theoretically, the above data indicate that this method may be used for studying adsorption.

The mercury electrode was used in studies of adsorption with the aid of the electrode photoelectric effects in order to verify the theory and to elucidate the specific features of the method in relation to the simplest systems. As adsorption of organic substances on mercury can be studied with high accuracy by other methods, it is apparently to be expected that the electrode photoelectric effect method will prove most useful for studying adsorption on solid electrodes.

8.   Certain Other Methods for Studying Adsorption on Solid Electrodes

The methods discussed above do not exhaust all the techniques used in studying adsorption of organic substances.

Electron diffraction has been used in a number of investigations [227-229]. For example, Bowden and Moore [229] observed that a close-packed adsorption layer of ethyl stearate on platinum melts when the temperature is raised to 34°C (mp of ethyl stearate), while on cadmium fusion ("disorientation") of the ethyl stearate monolayer occurred at 110°C. The latter temperature is close to the melting point of cadmium stearate. It was concluded from these observations that adsorption of ethyl stearate on platinum is due to physical forces, while in the case of cadmium chemisorption occurs.

The ellipsometric method [230-232] has also been used for determination of packing densities of molecules in adsorption layers and for measuring the lengths of adsorbed molecules. It was used for detection of thin layers of calomel on mercury [233] and for studying the state of adsorbed oxygen on platinum [234, 235].

These methods provide deeper insight into adsorption phenomena but have not been widely used as yet.

The use of infrared spectroscopy in studies of adsorption from solutions has been reported in the literature [12, 236, 237].

Bonnemay and Lamy [238] have shown that electron paramagnetic resonance can be used for determining the isotherms of hydrogen adsorption on platinum in electrolyte solutions.

Adsorption of organic substances on electrodes leads to decrease in the rates of various electrochemical reactions [239].

In numerous investigations (e.g., see [240-243]) attempts have been made to use data on reaction kinetics in presence of surface-active substances for calculation of the degree of surface coverage by organic substances. These calculations were based (in most cases) on the assumption that the adsorbed molecules shield a part of the electrode surface mechanically, lowering the reaction rate in these regions virtually to zero in comparison with the rate on a clean surface. The formula

$$\theta = 1 - \frac{i}{i_0}, \qquad (V.31)$$

where i and i_0 is the current in presence and in absence of the organic substance at constant potential, is used for determination of θ. The current at limiting coverage of the electrode surface is neglected. Constancy of the slope of the polarization curve with increasing concentration of the organic substance is an essential condition for the calculation. In calculations with the aid of Eq. (V.31) it is not taken into account that adsorbed molecules alter the charge of the electrode surface and may have a retarding or accelerating effect on reactions at open regions of the surface [244] owing to changes of the ψ_1-potential. However, even if this is taken into account the adsorption isotherms of organic substances, derived from kinetic data, differ appreciably from the results of direct adsorption measurements [242, 243].

Bagotskii, Vasil'ev, et al. [160, 245] investigated the influence of adsorption of various organic and inorganic substances on the oxidation rates of methanol, formic acid, and molecular hydrogen at a smooth platinum electrode in $1 N$ HClO$_4$ and $1 N$ LiOH. Equation (V.31) was used for calculating the surface coverage. In this way data were obtained on adsorption of higher aliphatic alcohols (from n-C_3H_7OH to $C_7H_{15}OH$) and of isomeric alcohols (iso-$C_5H_{11}OH$ and tert-$C_5H_{11}OH$). The dependences of the surface coverage by organic and inorganic substances on the electrode potential are in qualitative agreement with results obtained by other methods, e.g., with the aid of radioactive tracers [188].

Vagramyan and Solov'eva [246] studied the kinetics of adsorption of aliphatic alcohols on zinc by determination of the increase of the overpotential for zinc deposition with time. It was assumed that the increase of the overpotential is due only to shielding of the surface by alcohol molecules. Similar investigations with silver, copper, and a number of other metals in presence of organic additives have been carried out by Matulis and Bodnevas [247] and many others.

It is possible to make a comparative assessement of the adsorption rates of organic substances by determination of polarization curves with the use of an electrode the surface of which is cleaned at a definite rate. With rapid cleaning the organic substance is not adsorbed on the surface and has no influence on the rates of electrode processes. This method was used by Tomashov and Vershinina [248] for studying adsorption of tetrabutylammonium cations and Br⁻ anions from sulfuric acid solutions on nickel.

References

1. A. N. Frumkin and A. I. Shlygin, Dokl. Akad. Nauk SSSR, 2:176 (1934).
2. A. I. Shlygin and A. N. Frumkin, Acta Physicochim. URSS, 3:791 (1935).
3. A. I. Shlygin, A. N. Frumkin, and V. I. Medvedovskii, Acta Physicochim. URSS, 4:911 (1936).
4. A. N. Frumkin and A. I. Shlygin, Acta Physicochim. URSS, 5:819 (1936); Izv. Akad. Nauk SSSR, No. 5, 773 (1936).
5. B. V. Érshler and M. A. Proskurnin, Acta Physicochim. URSS, 6:195 (1937).
6. B. V. Érshler, Acta Physicochim. URSS, 7:327 (1937).
7. B. V. Érshler, G. A. Deborin, and A. N. Frumkin, Acta Physicochim. URSS, 8:565 (1938).
8. A. I. Shlygin and B. V. Érshler, Acta Physicochim. URSS, 11:45 (1939); A. I. Shlygin, É. Razumovskaya, and K. Rozental', Zh. Fiz. Khim., 13:1079 (1939); B. V. Érshler, Zh. Fiz. Khim., 13:1092 (1939).
9. P. Dolin and B. V. Érshler, Acta Physicochim. URSS, 13:747 (1940); Zh. Fiz. Khim., 14:886 (1940); P. Dolin, B. V. Érshler, and A. N. Frumkin, Acta Physicochim. URSS, 13:779 (1940); Zh. Fiz. Khim., 14:907 (1940).
10. G. A. Deborin and B. V. Érshler, Zh. Fiz. Khim., 14:708 (1940).

11.   J. A. V. Butler and G. Armstrong, Proc. Roy. Soc. (London), A137:604 (1932);
       G. Armstrong, F. Himsworth, and J. A. V. Butler, Proc. Roy. Soc. (London),
       A143:89 (1933); J. Pearson and J. A. V. Butler, Trans. Faraday Soc., 34:1163
       (1938).
12.   B. J. Piersma, Electrosorption (edited by E. Gileadi), Plenum Press, New York
       (1967), p. 19.
13.   S. Brunauer, P. H. Emmett, and E. Teller, J. Am. Chem. Soc., 60:309 (1938).
14.   P. Chinebault and A. Schürenkämper, J. Phys. Chem., 69:2300 (1965).
15.   Tsa Ch'uan-Hsin and Z. A. Iofa, Dokl. Akad. Nauk SSSR, 131:137 (1960);
       A. N. Frumkin, N. S. Polyanovskaya, and N. B. Grigor'ev, Dokl. Akad. Nauk
       SSSR, 157:1445 (1964); U. V. Pal'm, V. É. Past, and R. Ya. Pullerits, Élek-
       trokhimiya, 2:604 (1966).
16.   H. A. Smith and J. F. Fuzek, J. Am. Chem. Soc., 68:229 (1946).
17.   H. A. Smith and K. A. Allen, J. Phys. Chem., 58:449 (1954).
18.   J. J. Bordeaux and N. Hackerman, J. Phys. Chem., 61:1323 (1957).
19.   J. Kivel, F. C. Albers, D. A. Olsen, and R. E. Johnson, J. Phys. Chem., 67:1235
       (1963).
20.   R. J. Adams, H. L. Weisbecker, and W. J. McDonald, J. Electrochem. Soc.,
       111:774 (1964).
21.   A. N. Frumkin, Advances in Electrochemistry and Electrochemical Engineering
       (edited by P. Delahay), Vol. 3, Interscience, New York (1963), p. 287.
22.   R. Fisher, H. Chon, and J. Aston, J. Phys. Chem., 68:3240 (1964).
23.   R. V. Marvet and O. A. Petrii, Élektrokhimiya, 3:901 (1967).
24.   G. P. Khomchenko, A. I. Pletyushkina, and G. D. Vovchenko, Catalytic
       Liquid-Phase Reactions [in Russian], Izd. Akad. Nauk Kaz. SSR, Alma-Ata
       (1963), p. 295.
25.   M. R. Tarasevich, K. A. Radyushkina, and R. Kh. Burshtein, Élektrokhimiya,
       3:455 (1967).
26.   A. A. Balandin, Vikt. I. Spitsyn, L. I. Barsova, A. E. Agronomov, and N. P.
       Dobrosel'skaya, Zh. Fiz. Khim., 41:2623 (1967).
27.   A. I. Pletyushkina, G. P. Khomchenko, and G. D. Vovchenko, Zh. Fiz. Khim.,
       40:711 (1966).
28.   A. A. Balandin, Vikt. I. Spitsyn, L. I. Barsova, and M. D. Navalikhina, Izv.
       Akad. Nauk SSSR, Ser. Khim., 1514 (1966).
29.   A. D. Semenova, A. I. Pletyushkina, and G. D. Vovchenko, Vestn. Mosk.
       Gos. Univ., Ser. Khim., No. 4, 45 (1968).
30.   R. Kh. Burshtein, M. R. Tarasevich, and V. S. Vilinskaya, Élektrokhimiya,
       3:349 (1967).
31.   M. Breiter, K. Hoffmann, and C. Knorr, Z. Elektrochem., 61:1168 (1957);
       T. B. Warner and S. Schuldiner, J. Electrochem. Soc., 111:992 (1964).
32.   S. Gilman, Electroanalytical Chemistry (edited by A. J. Bard), Vol. 2, Marcel
       Dekker, Inc., New York (1967), p. 111.
33.   A. N. Frumkin, Dokl. Akad. Nauk SSSR, 154:1432 (1964).
34.   P. A. Rebinder and N. A. Kalinovskaya, Zh. Fiz. Khim., 5:332 (1934).
35.   P. A. Rebinder and E. K. Venstrem, Zh. Fiz. Khim., 19:1 (1945); Acta
       Physicochim. URSS, 19:36 (1944); Dokl. Akad. Nauk SSSR, 68:329 (1949).
36.   J. O'M. Bockris and R. Parry-Jones, Nature, 17:930 (1953).

37.   F. P. Bowden and D. Tabor, in: Properties of Metallic Surfaces, Inst. Metals (1953).

38.   V. I. Likhtman, E. D. Shchukin, and P. A. Rebinder, Physicochemical Mechanics of Metals [in Russian], Izd. Akad. Nauk SSSR, Moscow (1962).

39.   E. K. Venstrem, V. I. Likhtman, and P. A. Rebinder, Dokl. Akad. Nauk SSSR, 107:105 (1956).

40.   D. I. Leikis and E. K. Venstrem, Dokl. Akad. Nauk SSSR, 112:97 (1957).

41.   V. D. Kuznetsov, Zh. Prikl. Fiz., 6:33 (1929).

42.   V. D. Kuznetsov and V. V. Lavrent'eva, Z. Krist., 80:54 (1931).

43.   P. A. Rebinder and N. A. Kalinovskaya, Zh. Tekh. Fiz., 2:726 (1932).

44.   A. N. Frumkin and A. V. Gorodetskaya (Gorodetskaja), Z. Phys. Chem., 136:451 (1928).

45.   F. L. Kukoz and S. A. Semenchenko, Élektrokhimiya, 1:1454 (1965); 2:74 (1966).

46.   E. K. Venstrem and P. A. Rebinder, Zh. Fiz. Khim., 26:1847 (1952).

47.   V. I. Likhtman, P. A. Rebinder, and G. V. Karpenko, Influence of Surface-Active Media on Metal Deformation Processes [in Russian], Izd. Akad. Nauk SSSR, Moscow (1954).

48.   T. I. Borisova, B. V. Érshler, and A. N. Frumkin, Zh. Fiz. Khim., 22:925 (1948).

49.   T. I. Borisova and B. V. Érshler, Zh. Fiz. Khim., 24:337 (1950).

50.   T. R. Beck, Extended Abstracts of the 19th Meeting of CITCE, Detroit, USA (1968), p. 214.

51.   A. Pfützenreuter and G. Masing, Z. Metallkunde, 42:361 (1951).

52.   W. Klinkenberg, K. Lücke, and G. Masing, Z. Metallkunde, 44:362 (1955).

53.   V. I. Likhtman, L. A. Kochanova, D. I. Leikis, and E. D. Shchukin, Élektrokhimiya (in press).

54.   D. N. Staicopolous, J. Electrochem. Soc., 108:900 (1961).

55.   A. Ya. Gokhshtein, Zavod. Lab., 32:815 (1966); USSR Patents Nos. 178,161 and 179,043; Élektrokhimiya, 2:1061, 1318 (1966); Dokl. Akad. Nauk SSSR, 174:394 (1967); Fiz. Tekhn. Poluprov., 1:1486, 1787 (1967); Élektrokhimiya, 4:248, 619, 665, 886 (1968); Dokl. Akad. Nauk SSSR, 185:385 (1968).

56.   A. Ya. Gokhshtein, Élektrokhimiya (in press).

57.   D. I. Leikis, Dokl. Akad. Nauk SSSR, 135:1429 (1960).

58.   A. Ya. Gokhshtein, Élektrokhimiya (in press); Electrochim. Acta (in press).

59.   J. E. Berger, J. Phys. Chem., 69:2598 (1965).

60.   R. A. Erb, J. Phys. Chem., 69:1306 (1965).

61.   H. G. Möller, Ann. Phys., 27:665 (1908); Z. Phys. Chem., 65:226 (1908).

62.   A. N. Frumkin, A. V. Gorodetskaya, B. N. Kabanov, and N. I. Nekrasov, Zh. Fiz. Khim., 3:351 (1932).

63.   A. V. Gorodetskaya and B. N. Kabanov, Zh. Fiz. Khim., 4:529 (1933).

64.   A. N. Frumkin, V. S. Bagotskii, Z. A. Iofa, and B. N. Kabanov, Kinetics of Electrode Processes [in Russian], Izd. Mosk. Gos. Univ. (1952).

65.   B. Kabanov and M. Ivanishchenko, Izv. Akad. Nauk SSSR, Otd. Mat. Estestv. Nauk, 755 (1936).

66.   E. Cherneva and A. V. Gorodetskaya, Zh. Fiz. Khim., 13:111 (1939).

67.   R. N. Wenzel, Ind. Eng. Chem., 28:988 (1936).
68.   M. Bonnemay, G. Brönoel, P. Jonville, and E. Levart, Compt. Rend., 260:5262 (1965).
69.   B. V. Érshler, Zh. Fiz. Khim., 22:683 (1948).
70.   D. C. Grahame, J. Electrochem. Soc., 99:370C (1952).
71.   H. Gerischer, Z. Phys. Chem., 198:286 (1951); Z. Elektrochem., 55:98 (1951).
72.   M. Breiter, H. Kammermaier, and C. Knorr, Z. Elektrochem., 60:37, 119 (1956).
73.   J. H. Sluyters, Rec. Trav. Chim., 79:1092 (1960).
74.   K. J. Vetter, Elektrochemische Kinetik, Springer-Verlag, Berlin-Göttingen-Heidelberg (1961).
75.   V. V. Batrakov and Z. A. Iofa, Élektrokhimiya, 1:123 (1965).
76.   D. I. Leikis, Élektrokhimiya, 1:472 (1965).
77.   B. E. Conway, E. Gileadi, and H. Angerstein-Kozlowska, J. Electrochem. Soc., 112:341 (1965); B. E. Conway, N. Marincic, D..Gilroy, and E. Rudd, J. Electrochem. Soc., 113:1144 (1966); E. Gileadi and B. E. Conway, in: Modern Aspects of Electrochemistry, Vol. 3, Plenum Press, New York (1964), p. 347; B. E. Conway, Theory and Principles of Electrode Processes, Ronald Press, New York (1965).
78.   D. I. Leikis and D. P. Aleksandrova, Élektrokhimiya, 3:865 (1967).
79.   I. M. Novosel'skii, Élektrokhimiya, 4:1077 (1968); The Double Layer and Adsorption on Solid Electrodes [in Russian], Izd. Tartusk. Univ., Tartu (1968), p. 103.
80.   E. A. Ukshe, Élektrokhimiya, 4:1116 (1968).
81.   W. Machu and V. K. Gouda, Werkstoffe Korrosion, 13:745 (1962).
82.   H. Grubitsch and F. Hilbert, Werkstoffe Korrosion, 17:289 (1966).
83.   J. O'M. Bockris and B. E. Conway, J. Phys. Colloid Chem., 53:527 (1949).
84.   P. J. Hilson, Trans. Faraday Soc., 48:462 (1952).
85.   V. V. Losev, Dokl. Akad. Nauk SSSR, 88:499 (1953).
86.   T. Murakawa and N. Hackerman, Corrosion Sci., 4:387 (1964).
87.   Z. A. Iofa (Jofa), V. V. Batrakov, and Cho Ngok Ba, Electrochim. Acta, 9:1645 (1964).
88.   V. V. Batrakov, Gamil Khanas Avad, and Z. A. Iofa, Zashchita Metallov (in press).
89.   R. de Levie, Electrochim. Acta, 9:1231 (1964); 10:113 (1965).
90.   D. I. Leikis, É. S. Sevast'yanov, and L. L. Knots, Zh. Fiz. Khim., 38:1833 (1964); R. de Levie, J. Electroanal. Chem., 9:117 (1965).
91.   A. N. Frumkin and N. A. Aladzhalova, Zh. Fiz. Khim., 18:493 (1944).
92.   Ya. M. Kolotyrkin, Zh. Fiz. Khim., 20:667 (1946); 31:659 (1957).
93.   V. É. Past and Z. A. Iofa, Zh. Fiz. Khim., 33:1230 (1959).
94.   V. V. Sobol' and Z. A. Iofa, Dokl. Akad. Nauk SSSR, 138:1151 (1961).
95.   Ya. V. Durdin, L. Kish, and V. I. Kravtsov, Proceedings of the Fourth Conference on Electrochemistry [in Russian], Izd. Akad. Nauk SSSR (1959), p. 102.
96.   J. O'M. Bockris and H. Kita, J. Electrochem. Soc., 108:676 (1961).
97.   U. V. Pal'm and V. É. Past, Dokl. Akad. Nauk SSSR, 146:1374 (1962); Elektrokhimiya, 1:602 (1965).

98.   A. N. Frumkin, Acta Physicochim. URSS, 18:23 (1943); Discussions Faraday Soc., 1:57 (1947).

99.   D. C. Grahame, J. Phys. Chem., 57:257 (1953).

100. A. Hickling, Trans. Faraday Soc., 33:1540 (1937).

101. A. Hickling and F. Salt, Trans. Faraday Soc., 36:1226 (1940); 37:450 (1941).

102. N. A. Fedotov, Zh. Fiz. Khim., 25:3 (1951).

103. V. É. Past and Z. A. Iofa, Dokl. Akad. Nauk SSSR, 106:1050 (1956); Zh. Fiz. Khim., 33:913 (1959).

104. H. Fischer, M. Seipt, and G. Morlock, Z. Elektrochem., 59:440 (1955).

105. M. Seipt, Proceedings of the 7th Meeting of CITCE, Lindau (1955), p. 57.

106. U. V. Pal'm, V. É. Past, and V. R. Reében, Zh. Fiz. Khim., 35:1136 (1961).

107. V. I. Kravtsov, Zh. Fiz. Khim., 35:1144 (1961).

108. B. E. Conway and M. Dzieciuch, Can. J. Chem., 41:21, 38, 55 (1963).

109. B. E. Conway, E. Gileadi, and M. Dzieciuch, Electrochim. Acta, 8:143 (1963).

110. H. Angerstein-Kozlowska and B. E. Conway, J. Electroanal. Chem., 7:109 (1964).

111. O. Völk and H. Fischer, Electrochim. Acta, 5:112 (1961).

112. H. Jamaoka, W. Lorenz, and H. Fischer, Fundamental Problems of Modern Theoretical Electrochemistry, Proceedings of the 14th Meeting of CITCE [Russian translation], "Mir," Moscow (1965), p. 464.

113. L. A. Mirkind and M. Ya. Fioshin, Dokl. Akad. Nauk SSSR, 154:1163 (1964).

114. W. R. Busing and W. W. Kauzmann, J. Chem. Phys., 20:1129 (1952).

115. V. L. Kheifets and B. S. Krasikov, Dokl. Akad. Nauk SSSR, 94:101 (1954).

116. M. W. Breiter, J. Electrochem. Soc., 109:42 (1962).

117. É. S. Sevast'yanov and D. I. Leikis, Izv. Akad. Nauk SSSR, Ser. Khim., 450 (1964).

118. R. A. Rightmire, R. L. Rowland, D. L. Boos, and D. L. Beals, J. Electrochem. Soc., 111:242 (1964).

119. M. W. Breiter, Electrochim. Acta, 7:533 (1962).

120. M. W. Breiter, Electrochim. Acta, 8:973 (1963).

121. G. M. Schmid and N. Hackerman, J. Electrochem. Soc., 110:440 (1963).

122. M. W. Breiter and S. Gilman, J. Electrochem. Soc., 109:622 (1962).

123. T. I. Pochekaeva, Zh. Fiz. Khim., 35:1606 (1961).

124. A. I. Shlygin and M. E. Manzhelei, Uch. Zap. Kishinevsk. Univ., 8:13 (1953).

125. A. I. Shlygin, Proceedings of the Third Conference on Electrochemistry [in Russian], Izd. Akad. Nauk SSSR, Moscow (1953), p. 322.

126. T. O. Pavela, Ann. Acad. Sci. Fennicae, Ser. A, 59 (1954).

127. A. N. Frumkin and B. I. Podlovchenko, Dokl. Akad. Nauk SSSR, 150:349 (1963).

128. B. I. Podlovchenko and Z. A. Iofa, Zh. Fiz. Khim., 38:211 (1964).

129. L. W. Niedrach, J. Electrochem. Soc., 111:1309 (1964).

130. V. E. Kazarinov and N. A. Balashova, Dokl. Akad. Nauk SSSR, 157:1174 (1964).

131. Hira Lal, O. A. Petrii, and B. I. Podlovchenko, Élektrokhimiya, 1:316 (1965).

132. J. Giner, Electrochim. Acta, 9:63 (1964).

133. P. R. Johnson and A. T. Kuhn, J. Electrochem. Soc., 112:599 (1965).

134. D. V. Sokol'skii, Hydrogenation in Solutions [in Russian], Izd. Akad. Nauk
      Kaz. SSR, Alma-Ata (1962).
135. A. V. Shashkina and I. I. Kulakova, Zh. Fiz. Khim., 35:1846 (1961).
136. B. I. Podlovchenko, O. A. Petrii, and E. P. Gorgonova, Élektrokhimiya, 1:182
      (1965).
137. B. I. Podlovchenko, A. N. Frumkin, and V. F. Stenin, Élektrokhimiya, 4:339
      (1968).
138. B. I. Podlovchenko and V. F. Stenin, Élektrokhimiya, 3:649 (1967).
139. P. R. Johnson and A. T. Kuhn, J. Electrochem. Soc., 112:599 (1965).
140. W. T. Grubb and M. E. Lazarus, J. Electrochem. Soc., 114:360 (1967).
141. M. W. Breiter, Electrochim. Acta, 12:1213 (1967); J. Electroanal. Chem., 14:407
      (1967); 15:221 (1967).
142. B. I. Podlovchenko, V. F. Stenin, and V. E. Kazarinov, the Double Layer and
      Adsorption on Solid Electrodes [in Russian], Izd. Tartusk. Univ., Tartu (1968),
      p. 121; Élektrokhimiya (in press).
143. F. Will and C. Knorr, Z. Elektrochem., 64:258, 270 (1960).
144. Ya. M. Kolotyrkin and A. N. Chemodanov, Dokl. Akad. Nauk SSSR, 134:128
      (1960).
145. L. W. Niedrach, S. Gilman, and I. Weinstock, J. Electrochem. Soc., 112:1161
      (1965); L. W. Niedrach, J. Electrochem. Soc., 113:645 (1966).
146. S. Gilman, J. Phys. Chem., 67:78 (1963).
147. S. Gilman, Trans. Faraday Soc., 62:466 (1966).
148. O. Khazova, Yu. B. Vasil'ev, and V. S. Bagotskii, Élektrokhimiya, 1:82 (1965).
149. S. B. Brummer, J. Phys. Chem., 69:562 (1965).
150. S. B. Brummer and J. I. Ford, J. Phys. Chem., 69:1355 (1965).
151. S. Gilman, J. Phys. Chem., 70:2880 (1966).
152. S. B. Brummer and K. Cahill, J. Electroanal. Chem., 16:207 (1968).
153. S. B. Brummer, J. Phys. Chem., 71:2838 (1967).
154. P. J. Lingane, Anal. Chem., 39:485 (1967); W. T. de Vries, J. Electroanal.
      Chem., 17:31 (1968); 18:469 (1968); M. L. Olmstead and R. S. Nicholson,
      J. Phys. Chem., 72:1650 (1968).
155. M. Casselli, G. Ottombrini, and P. Papoff, Electrochim. Acta, 13:241 (1968);
      R. S. Rodgers and L. Meites, J. Electroanal. Chem., 16:1 (1968).
156. S. Srinivasan and E. Gileadi, Electrochim. Acta, 11:321 (1966).
157. E. Gileadi, G. Stoner, and J. O'M. Bockris, J. Electrochem. Soc., 113:585
      (1966).
158. M. Oikawa and T. Mukaibo, J. Electrochem. Soc. Japan, 20:568 (1962).
159. T. C. Franklin and R. D. Sothern, J. Phys. Chem., 58:951 (1954).
160. V. S. Bagotskii and Yu. B. Vasil'ev, Electrochim. Acta, 11:1439 (1966);
      Advances in the Electrochemistry of Organic Compounds [in Russian], Izd.
      "Nauka," Moscow (1966), p. 38.
161. S. B. Brummer, J. I. Ford, and M. J. Turner, J. Phys. Chem., 69:3424 (1965).
162. E. Gileadi, J. Electroanal. Chem., 11:137 (1966).
163. M. W. Breiter, J. Electrochem. Soc., 112:1244 (1965).
164. S. B. Brummer and M. J. Turner, Hydrocarbon Fuel Cell Technology (edited
      by S. Baker), Academic Press, New York—London (1965), p. 409.

CHAPTER 5

165. S. Gilman, Trans. Faraday Soc., 61:2546 (1965).
166. S. Gilman, J. Phys. Chem., 71:4330, 4339 (1967).
167. G. P. Girina, M. Ya. Fioshin, and V. E. Kazarinov, Élektrokhimiya, 1:478 (1965).
168. V. E. Kazarinov and G. N. Mansurov, Élektrokhimiya, 2:1338 (1966).
169. N. A. Balashova and V. E. Kazarinov, Usp. Khim., 34:1721 (1965).
170. V. E. Kazarinov and N. A. Balashova, Peredovoi Nauchno-Tekhn. Opyt. No. 17-63-883/7, GOSINTI, Moscow (1963), p. 18.
171. M. Green, D. A. J. Swinkels, and J. O'M. Bockris, Rev. Sci. Instr., 33:No.1, 20 (1962).
172. J. W. Shepard and J. P. Ryan, J. Phys. Chem., 60:127 (1956).
173. A. I. Fil'ko, Uch. Zap. Mosk. Gos. Ped. Inst. im. V. I. Lenina, 146:62 (1960).
174. T. K. Ross and D. H. Jones, Symposium European sur les Inhibiteurs de Corrosion, Ferrara, 1960, Compt. Rend. (1961), p. 163.
175. E. A. Blomgren and J. O'M. Bockris, Nature, 186:No. 4721, 305 (1960).
176. F. Joliot, J. Chim. Phys., 27:119 (1930).
177. R. J. Flannery and D. C. Walker, Hydrocarbon Fuel Cell Technology (edited by S. Baker), Academic Press, New York–London (1965), p. 335.
178. J. O'M. Bockris, Z. Phys. Chem., 214:1 (1960).
179. H. Wroblowa and M. Green, Electrochim. Acta, 8:679 (1963).
180. M. Green and H. Dahms, J. Electrochem. Soc., 110:1075 (1963).
181. M. Green and H. Dahms, J. Electrochem. Soc., 110:466 (1963).
182. H. Dahms, M. Green, and I. Weber, Nature, 196:1310 (1962).
183. E. Gileadi, B. T. Rubin, and J. O'M. Bockris, J. Phys. Chem., 69:3335 (1965).
184. W. Heiland, E. Gileadi, and J. O'M. Bockris, J. Phys. Chem., 70:1207 (1966).
185. R. E. Smith, H. B. Urbach, J. H. Harrison, and N. L. Hatfield, J. Phys. Chem., 71:1250 (1967).
186. J. O'M. Bockris and D. A. J. Swinkels, J. Electrochem. Soc., 111:736 (1964).
187. J. O'M. Bockris, M. Green, and D. A. J. Swinkels, J. Electrochem. Soc., 111:743 (1964).
188. V. E. Kazarinov, Élektrokhimiya, 2:1170 (1966).
188a. E. Gileadi, L. Duic, and J. O'M. Bockris, Electrochim. Acta, 13:1915 (1968).
189. E. L. Cook and N. Hackerman, J. Phys. Colloid Chem., 55:549 (1951).
190. N. Hackerman and A. H. Roebuck, Ind. Eng. Chem., 46:1481 (1954).
191. A. C. Makrides and N. Hackerman, Ind. Eng. Chem., 47:1773 (1955).
192. N. A. Shurmovskaya and R. Kh. Burshtein, Zh. Prikl. Khim., 30:1176 (1957).
193. E. B. Greenhill, Trans. Faraday Soc., 45:625 (1949).
194. G. A. Emel'yanenko, S. P. Tat'yanina, and E. A. Vlasova, Zh. Prikl. Khim., 37:688 (1964).
195. M. S. Platonov, Zh. Russ. Fiz.-Khim. Obshch., 61:1055 (1929).
196. B. E. Conway, R. G. Barradas, and T. Zawidzki, J. Phys. Chem., 62:676 (1958).
197. B. E. Conway and R. G. Barradas, Transactions of Symposium on Electrode Processes, J. Wiley and Sons, Philadelphia–New York (1959), p. 229.
198. L. Cavallaro, L. Felloni, and G. Trabanelli, Symposium European sur les Inhibiteurs de Corrosion, Ferrara, 1960, Compt. Rend. (1961), p. 111.
199. E. Heymann and E. Bye, Kolloid-Z., 59:153 (1932).

200. C. C. Nathan, Corrosion, 12:161t (1956).
201. Z. A. Iofa, É. M. Lyakhovetskaya, and K. Sharifov, Dokl. Akad. Nauk SSSR, 84:543 (1952).
202. G. S. Parfenov and P. I. Chervyakov, Proceedings of Scientific Conference [in Russian], Omsk (1963), p. 210.
203. S. A. Balezin, P. G. Kuznetsov, and I. A. Podol'nyi, in: Corrosion Inhibitors [in Russian], "Sudostroenie," Moscow—Leningrad (1965), p. 5.
204. S. A. Balezin, I. M. Zhuravlev, and I. A. Podol'nyi, in: Corrosion Inhibitors [in Russian], "Sudostroenie," Moscow—Leningrad (1965), p. 18.
205. E. Hutchinson, Trans. Faraday Soc., 43:439 (1947).
206. E. Becquerel, Compt. Rend., 9:145, 561 (1839).
207. M. Volmer and W. Moll, Z. Phys. Chem., 161:401 (1932).
208. P. Hillson and E. K. Rideal, Proc. Roy. Soc., A199:295 (1949).
209. H. Mauser and U. Sproesser, Chem. Ber., 97:2260 (1964).
210. M. Hevrovsky, Z. Phys. Chem., 52:1 (1967); Proc. Roy. Soc., A301:411 (1967).
211. G. Barker and A. Gardner, Fundamental Problems of Modern Theoretical Electrochemistry [Russian translation], Izd. "Mir," Moscow (1965), p. 118.
212. G. C. Barker, A. W. Gardner, and D. S. Sammon, J. Electrochem. Soc., 113:1182 (1966).
213. P. Berg, Extended Abstracts of the 18th CITCE Meeting (1967), p. 105.
214. P. Delahay and V. S. Srinivasan, J. Phys. Chem., 70:420 (1966).
215. P. Delahay, J. Electrochem. Soc., 113:1198 (1966).
216. Yu. Ya. Gurevich, A. M. Brodskii, and V. G. Levich, Élektrokhimiya, 3:1302 (1967).
217. A. M. Brodskii and Yu. Ya. Gurevich, Dokl. Akad. Nauk SSSR, 178:868 (1968).
218. A. M. Brodskii and Yu. Ya. Gurevich, Zh. Éksperim. i Teor. Fiz., 54:213 (1968).
219. Yu. Ya. Gurevich and Z. A. Rotenberg, Élektrokhimiya, 4:529 (1968).
220. Z. A. Rotenberg and Yu. Ya. Gurevich, Élektrokhimiya, 4:984 (1968).
221. Z. A. Rotenberg, Yu. Ya. Gurevich, and Yu. V. Pleskov, Élektrokhimiya, 4:1086 (1968).
222. Yu. V. Pleskov and Z. A. Rotenberg, J. Electroanal. Chem., 20:1 (1969).
223. L. I. Korshunov, Ya. M. Zolotovitskii, and V. A. Benderskii, Élektrokhimiya, 4:499 (1968).
224. Z. A. Rotenberg and Yu. V. Pleskov, Élektrokhimiya, 4:826 (1968).
225. V. P. Sharma, P. Delahay, G. G. Susbielles, and G. Tessari, J. Electroanal. Chem., 16:285 (1968).
226. L. I. Korshunov, Ya. M. Zolotovitskii, and V. A. Benderskii, Élektrokhimiya (in press).
226a. Z. A. Rotenberg and Yu. V. Pleskov, Élektrokhimiya (in press).
227. N. Hackerman and E. E. Glenn, J. Phys. Chem., 54:497 (1950).
228. J. V. Sanders, Research (London), 2:586 (1949).
229. F. P. Bowden and A. C. Moore, Trans. Faraday Soc., 47:900 (1951).
230. A. Rothen, Rev. Sci. Instr., 28:283 (1957).
231. J. Krüger, Corrosion, 22:88 (1966).
232. A. C. Hall, J. Phys. Chem., 70:1702 (1966).

233.   J. O'M. Bockris, M. A. V. Devanathan, and A. K. N. Reddy, Proc. Roy. Soc.,
       A279:327 (1964).
234.   A. K. N. Reddy, M. A. Genshaw, and J. O'M. Bockris, J. Electroanal. Chem.,
       8:406 (1964).
235.   A. K. N. Reddy, M. A. Genshaw, and J. O'M. Bockris, J. Chem. Phys., 48:671
       (1968).
236.   E. Gileadi and B. Piersma, Modern Aspects of Electrochemistry, Vol. 4
       (edited by J. O'M. Bockris), Plenum Press, New York (1966), p. 47.
237.   R. V. Scowen and J. Leja, Can. J. Chem., 45:2821 (1967).
238.   M. Bonnemay and C. Lamy, Compt. Rend., 265:C, 695 (1967); Extended
       Abstracts of 19th CITCE Meeting, Detroit, Michigan, USA (1968), p. 220.
239.   A. N. Frumkin, Electrochim. Acta, 9:465 (1964); Dokl. Akad. Nauk SSSR,
       85:373 (1952).
240.   W. Müller and W. Lorenz, Z. Phys. Chem., 27:23 (1961).
241.   T. P. Hoar and R. D. Holliday, J. Appl. Chem., 3:502 (1953).
242.   A. Aramata and P. Delahay, J. Phys. Chem., 68:880 (1964).
243.   K. Venkatesan, N. V. Nikolaeva-Fedorovich, and B. B. Damaskin, Zh. Fiz.
       Khim., 34:129 (1965).
244.   R. Parsons, Advances in Electrochemistry and Electrochemical Engineering
       (edited by P. Delahay), Vol. 1, Interscience, New York (1961), p. 2.
245.   S. S. Beskorovainaya, Yu. B. Vasil'ev, and V. S. Bagotskii, Élektrokhimiya,
       2:1029 (1965).
246.   A. T. Vagramyan and Z. S. Solov'eva, Zh. Fiz. Khim., 24:1252 (1950).
247.   Yu. Yu. Matulis and A. I. Bodnevas, Izv. Akad. Nauk SSSR, Otd. Khim. Nauk,
       No. 4, 577 (1954).
248.   N. D. Tomashov and L. P. Vershinina, Electrochim. Acta (in press).

Chapter 6

Adsorption of Organic Substances on Metals of High Hydrogen Overpotential

1. Experimental Data on Adsorption of Organic Substances on Metals of High Hydrogen Overpotential

Solid metals of high hydrogen overpotential, such as lead, thallium, tin, and bismuth, approximate to the mercury electrode in double-layer structure and may be described as "mercury-like" [1].

Differential capacity vs electrode potential curves recorded with single crystals and with sintered lead, thallium, bismuth, and cadmium electrodes [2-7] are of the same form as in the case of mercury. With lead and bismuth [4, 6] even quantitative agreement was observed at potentials close to φ_n; this is illustrated in Fig. 60, which gives C vs φ curves for a drop-like polycrystalline bismuth electrode and a liquid mercury electrode.

The behavior of such metals is also analogous to that of mercury in solutions containing organic substances.

The maximum adsorption of organic substances corresponds to the region close to the potential of zero charge. Table 9 gives what are, in our opinion, the most reliable values of φ_n for a number of metals.

At potentials close to φ_n the C vs φ curve has a minimum, while at some distance from φ_n desorption of the organic sub-

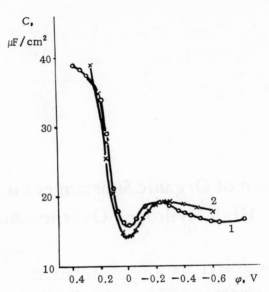

Fig. 60. Dependence of the capacity of a mercury (1) and bismuth (2) electrode in 0.01 N HCl on the potential. The potentials are taken from φ_n of the metals (from data in [5]).

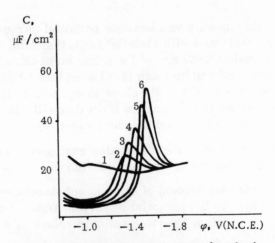

Fig. 61. Differential capacity curves for a lead electrode in 0.1 N H_2SO_4. 1) Pure solution; with n-amyl alcohol [M]: 2) 0.025; 3) 0.035; 4) 0.05; 5) 0.075; 6) 0.1 (from data in [7]).

TABLE 9

Metal	$\varphi_n(v)$ (N.H.E.)	Composition of solution	Method
Cd	−0.72	0.01-0.001 N NaF	Capacity minimum [7]
	−0.70	0.005 N KCl	Hardness minimum [8]
Tl	−0.67 ± 0.05	0.01-0.001 N NaF	Capacity minimum [9]
Ag	−0.70	0.002-0.01 N Na$_2$SO$_4$	Capacity minimum [10]
	−0.75 to − 0.80	2-0.2 N Na$_2$SO$_4$	Interfacial tension minimum [11]
	−0.80	0.1 N HCl	Charging current [12]
Ga	−0.69	0.01 N HClO$_4$	Capacity minimum [13]
	−0.70	0.006 N HCl	Capacity minimum [13]
	−0.70 to −0.69	0.01 N KCl 0.01 N NaClO$_4$	Electrocapillary maximum[14]
Pb	−0.60	0.01 N NaF	Capacity minimum [15]
	−0.64	0.01 N Na$_2$SO$_4$	Capacity minimum [4]
Sn	−0.46	0.001 N KClO$_4$	Capacity minimum [16]
	−0.38	0.01 N KCl	Immersion [17]
Bi	−0.40	0.01 N KCl	Capacity minimum [6]
Hg	−0.19	0.01 N NaF	Capacity minimum [18]
	−0.19	0.01 N NaF	Electrocapillary maximum and streaming electrode [19]
Sb	−0.17	0.001 N K$_2$SO$_4$	Capacity minimum [20]
	−0.17	0.001 N H$_2$SO$_4$	Capacity minimum [20]
Cu	−0.05	1 N Na$_2$SO$_4$	Wetting minimum [21]
	−0.03	0.02 N Na$_2$SO$_4$	Capacity minimum [22]
	+ 0.07	0.01 N KCl	Immersion

stance occurs. In the case of aliphatic alcohols desorption peaks correspond to this process.

As an example, Fig. 61 shows differential capacity curves obtained by Grigor'ev and Machavariani [23] for a polished poly-crystalline lead electrode in presence of n-amyl alcohol. Desorption peaks in the cathodic potential region were also obtained for thallium, lead, antimony, bismuth, and cadmium in presence of a number of aliphatic alcohols [2, 6, 23-26]. Investigation of the adsorption of alcohols on these metals in the region of positive potentials for determination of the second desorption peak was difficult owing to oxidation of the surface.

Measurements in acid solutions make it possible in a number of cases to observe a second desorption peak, corresponding to potentials which are positive in relation to φ_n. In this way Pullerits, Pal'm, and Past [27, 28] were able to obtain a second desorption peak on bismuth in presence of certain aliphatic alcohols (n-propyl, n-butyl, n-amyl, etc.). The results of determinations in amyl alcohol solutions are shown in Fig. 62. Increase of the alcohol concentration extends the adsorption region, while the potential corresponding to the desorption peak for lead and bismuth electrodes is shifted as in the case of mercury (see Fig. 63).

Desorption peaks of organic substances have also been observed for tin, bismuth, and lead in 0.5 N Na_2SO_4 solutions containing a mixture of β-naphthol, thymol, and diphenylamine [29].

At the same time, studies of adsorption effects on metals of high overpotential reveal certain differences from mercury, which

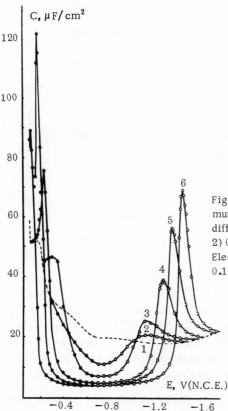

Fig. 62. Differential capacity curves for a bismuth electrode in n-amyl alcohol solutions of different concentrations. 1) Pure electrolyte; 2) 0.01; 3) 0.02; 4) 0.05; 5) 0.1; 6) 0.2 M. Electrolytes: 0.1 N K_2SO_4 (white circles) and 0.1 N H_2SO_4 (black circles) (from data in [25]).

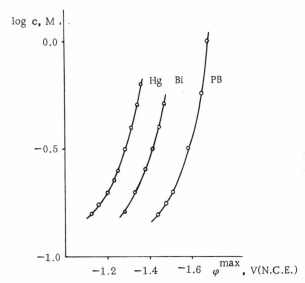

Fig. 63. Dependence of the desorption peak potential on the n-butyl alcohol concentration in 0.1 N Na_2SO_4 (from data in [7, 26]).

depend on the nature of the metals. For example, comparison of C vs φ curves determined for lead, thallium, and antimony (discussed more fully in Section 3) shows that desorption peaks of aliphatic alcohols are observed at less negative potentials than in the case of mercury. The differences become greater in presence of other organic compounds (e.g., tetraalkylammonium salts); in particular, there are no desorption peaks on the C vs φ curves.

This effect was observed in studies of adsorption of tetraalkylammonium ions on zinc and bismuth [5, 30]. The capacity of the double layer decreases in the region of maximum adsorption of tetrabutylammonium (TBA) cations on zinc and bismuth. However, when the cations are desorbed there are no desorption peaks on the C vs φ curves (see Figs. 64 and 65); this can probably be attributed to the nonequilibrium nature of the C vs φ curves.

In other respects the behavior of tetraalkylammonium cations on these metals is similar to their behavior on mercury. Adsorption of the cations increases with the length of the carbon chain in the series tetramethylammonium (TMA) < tetraethylammonium

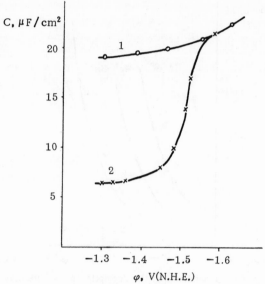

Fig. 64. Differential capacity curves for a zinc single
crystal in 1 N HCl with addition of TBA: 1) 0; 2)
$1 \cdot 10^{-3}$ M (from data in [4]).

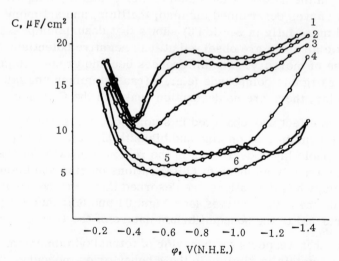

Fig. 65. Capacity curves for bismuth in 0.005 N K$_2$SO$_4$ with
additions of TBA: 1) 0; 2) 10^{-7}; 3) 10^{-6}; 4) 10^{-5}; 5) 10^{-4}; 6)
10^{-2} M (from data in [28]).

(TEA) < tetrabutylammonium (TBA). In dilute potassium sulfate solutions adsorption of TMA and TEA makes the electric double layer less diffuse, and φ_n shifts into the region of positive potentials. Similar results were obtained earlier for a mercury electrode by Damaskin and Nikolaeva-Fedorovich [31].

Similarities and differences are found in studies of the adsorption of thiourea on lead and mercury [15, 32, 33]. In the presence of thiourea in 0.1 N NaF the capacity of the double layer on lead increases, as it does on mercury. However, the increase on lead is less pronounced than on mercury. As in the case of mercury the C vs φ curves determined in dilute solutions have characteristic minima, which shift into the region of more negative potentials with increase of the thiourea concentration. Repulsive forces predominate between the adsorbed thiourea molecules on mercury and lead, and in both cases the adsorption of thiourea is satisfactorily represented by a logarithmic isotherm [15, 34]. It has been established experimentally that at a sufficiently high thiourea concentration the adsorption potential difference on lead, as on mercury, increases linearly with the logarithm of the adsorbate concentration. Grigor'ev and Krylov [35] explained this relationship on the basis of a statistical-mechanical treatment of repulsive dipole—dipole interaction.

Adsorption of dibutylsulfonaphthalene on cadmium [36, 37] and of diphenylamine and α- and β-naphthol on lead [38] has also been studied by measurement of double layer capacities. It was shown that the adsorbabilities of the organic compounds depend significantly on the electrode potential, but desorption peaks were not observed.

Well-defined dependence of decylamine and naphthalene on the potential of a lead electrode was demonstrated by the tracer atom method [39, 40].

Among the data discussed, recent investigations with bismuth [6, 27, 28, 30] and lead [15, 23] should be specially noted; their results can be compared quantitatively with data for the mercury electrode.

For example, the results of determinations with lead and bismuth [23, 27] indicate that the theoretical concepts discussed in Chapters III and IV can be applied to quantitative assessment of the adsorption behavior of alcohols on solid metals of high hydrogen overpotential. According to these concepts, the value of $(\varphi - \varphi_m)^2$ and the maximum height of the desorption peak on the C vs φ curve should, in the first approximation, vary linearly with the logarithm of the concentration of the organic substance; this is confirmed experimentally for lead and bismuth [23, 27, 28].

The methods of calculation described in Part I of this book were used for determining the principal adsorption parameters characterizing the isotherms of alcohol adsorption on bismuth and lead. Values of the attraction constant (a) were found from the shape of the experimental adsorption isotherm determined at the potential of maximum adsorption, and from the width of the desorption peaks and the slope of the φ^{max} vs log c curve at the desorption potential (φ^{max}). The values of φ_N (the adsorption potential difference), A, and B_0 were determined from the dependence of φ^{max} on log c. In the case of ethyl alcohol φ_N was determined from the ε vs φ curve. Values of $-\Delta \bar{G}_A^0$ corresponding to $\varphi = 0$ and $\theta = 1$ were calculated from the B_0 values. The calculated parameters, given in Table 10, have the same dimensions as the values in Table 8.

The table shows that attractive interaction and the free energy of adsorption of organic molecules increase with the length of the hydrocarbon chain.

Comparison of the parameters in Tables 8 and 10 shows that mercury, bismuth, and lead have almost the same values of a_0, A, C', Γ_m, and φ_N. The increase of $-\Delta \bar{G}_A^0$ (~0.8 kcal/mole) due to

TABLE 10. Basic Adsorption Parameters for Normal Aliphatic Alcohols at 25 °C (According to [23] and [28])

Metal	Formula	a_0	C'	φ_N	A	$\Gamma_m \cdot 10^{10}$	S	B_0	$\Delta \bar{G}_A^0$
	n-C_2H_5OH	0.92	6.6	0.36	(1.73)	(7.0)	(23.7)	0.46	1.87
	n-C_3H_7OH	1.10	5.2	0.30	1.39	5.61	29.7	1.72	2.65
Bi	n-C_4H_9OH	1.18	4.8	0.27	1.30	5.25	31.7	7.16	3.47
	n-$C_5H_{11}OH$	1.32	4.2	0.23	1.22	4.92	33.8	28.0	4.26
	n-$C_6H_{13}OH$	1.40	3.8	0.20	1.10	4.43	37.6	118	5.09
	n-$C_7H_{15}OH$	1.50	3.3	0.18	1.08	4.35	38.4	495	5.92
Pb	n-$C_5H_{11}OH$	1.40	4.2	0.22	–	–	–	25	4.3

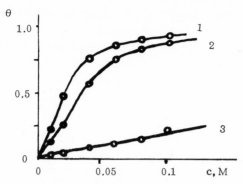

Fig. 66. Isotherms of n-amyl alcohol adsorption on
a bismuth electrode at different values of φ: 1)
−1.4; 2) −1.2; 3) −1.1 V (saturated calomel electrode). The points represent experimental data; the
continuous lines are calculated (from data in [26]).

increase of the length of the alcohol molecule by one CH_2 group is
also virtually the same in each case. Certain differences revealed
by the comparison will be discussed later in relation to the influence of the nature of the metal on adsorption.

Figure 66, which shows the adsorption isotherms of n-amyl
alcohol, is a good illustration of the applicability of the theoretical
concepts developed for the mercury electrode to solid metals of
high hydrogen overpotential. It is seen that the experimental and
calculated isotherms almost coincide. The parameters given in
Table 10 were used for calculation of the isotherms. The experimental values of θ were determined from the lowering of the
double-layer capacity with the aid of Eq. (V.18) in the region of
maximum adsorption, and from the equation

$$\varepsilon = \varepsilon_0(1 - \theta) + \varepsilon'\theta \qquad \text{(VI.1)}$$

at potentials differing substantially from the potential of maximum
adsorption. This method of determining θ is discussed in Chapter III.

2.   Data on Adsorption of Organic

Substances on Silver and Copper

From their adsorption properties, silver and copper may be
placed between metals of high hydrogen overpotential and metals

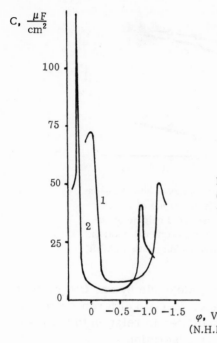

Fig. 67. Differential capacity curves for sil
(1) and mercury (2) in 1 N Na_2SO_4 saturated
with hexyl alcohol (from data in [13]).

of the iron group, although on the whole they are closer to metals
of high hydrogen overpotential.

Comparison of differential capacity curves for polycrystalline
silver and single crystals in pure electrolyte solutions at potentials close to φ_n with corresponding curves for mercury shows
that the electric double layers on these metals are similar in
structure [41-45].

Differential capacity curves for silver and copper in presence
of aliphatic alcohols have desorption peaks. In the case of silver,
desorption peaks were found for hexyl alcohol [10, 25], and in the
case of copper, for propyl and amyl alcohols [46] and a mixture
of β-naphthol, thymol, and diphenylamine [29]. Leikis and Sevast'-
yanov [10, 25] showed that the desorption peak for hexyl alcohol
on silver (as on most solid metals of high hydrogen overpotential)
is observed at less negative potentials relative to φ_n than in the
case of mercury and is less pronounced (see Fig. 67).

The C vs φ curves for copper and for metals of high hydrogen
overpotential in solutions of organic amines do not have desorption

peaks [47], although capacity measurements and the radioactive
tracer method indicate considerable dependence of the adsorbability of amines on the electrode potential [39, 40, 47].

According to published data [48-51] the adsorbability of organic compounds is lower on copper than on metals of high hydrogen overpotential. Adsorption of organic compounds on the
cationic type on copper increases in presence of halide ions [51].
The adsorption of these substances on silver also increases in
presence of Cl⁻ ions [52]. At the same time, Fujii et al. [48-50]
noted that the adsorbability of cetyl mercaptan on copper powder
in 5% HCl is low.

Determinations of adsorption of quinoline in $0.1 N$ HCl on
silver, carried out by Barradas and Conway [52], showed that the
shape of the adsorption isotherm varies in accordance with the
electrode potential. At fairly negative potentials the isotherm
was L-shaped, while at more positive potentials it became S-
shaped; this indicated increased attractive interaction between the
adsorbed molecules with increase of potential. Isotherms for adsorption of indole, certain acridine derivatives, and a number of
other compounds on silver could not be obtained over a wide concentration range because of the low solubility of these substances
in water.

Sigmoid isotherms were also obtained in studies of adsorption
of caprylic, capric, and lauric acids on copper from $0.1 N$ Na_2SO_4
[52a]. Attractive interaction between the adsorbate molecules increased with the hydrocarbon chain length.

It has been found [53, 54] that the dependence of surface coverage on aniline concentration in the case of copper powder and on
the phenol concentration in the case of copper powder at the stationary potential is in satisfactory agreement with the Langmuir
isotherm.

Hansen and Clampit [55] used the method of capacity measurement by the decay of potential for investigating adsorption of caproic, enanthic, and caprylic acids and of heptyl alcohol on copper
and silver in $0.1 N$ $HClO_4$ + $0.001 N$ HCl. They found that the adsorbability of organic acids increases with the length of the carbon
chain in the following sequence: caproic < enanthic < caprylic.
Comparison of the adsorbabilities of heptanol, n-heptanal, and
enanthic acid revealed an unusual sequence: the adsorption of

heptanol on copper was higher than that of enanthic acid, but lower than that of the corresponding aldehyde.

In the investigation cited θ was calculated from the decrease of the double-layer capacity in presence of the organic substance with the aid of the formula

$$\theta = 1 - C/C_0. \tag{VI.2}$$

Formula (VI.2) is obtained from Eq. (V.18) if it is assumed that the capacity of the double layer is zero at limiting coverage. This assumption permits calculations only at low surface coverage, and therefore the results obtained at high and moderate coverages by Hansen and Clampit [55] are inaccurate.

Studies of adsorption of thiourea on copper have been fairly numerous [56-61], as thiourea is used as a surface-active additive for improving the properties of electrolytic copper deposits. Adsorption of thiourea on copper is accompanied by decomposition of its molecules with formation of sulfide, and at low thiourea concentrations (10^{-5} M) mainly its decomposition products are adsorbed [61]. At high concentrations (10^{-3} M) predominant adsorption of undecomposed thiourea occurs. The copper electrode is similar in this respect to metals of the iron group.

Machu and Gouda [62] studied adsorption of di(o-tolyl)thiourea on a number of metals, including copper and silver. Smialowski and Szklarska-Smialowska [63] investigated adsorption of dibenzyl sulfoxide on copper.

3. Influence of the Nature of the Metal on Adsorption of Organic Substances

As was noted above, in passing from mercury to solid electrodes certain peculiarities are observed, which stress the influence of the nature of the metal on adsorption, in addition to certain common relationships in the adsorption of organic substances. These distinctions may be due to a number of factors, the influence of which is examined more fully below.

a. Role of the State of Aggregation. The first question to arise is whether the observed differences are due to the different states of aggregation of the metal.

The first experimental investigations of this point were carried out by Gorodetskaya and Proskurnin [64] with solid and liquid mercury in 50% K_2CO_3 solution containing ethyl alcohol. Experiments showed that the shape of the differential capacity curves and the capacity of the double layer in presence of the organic compound are virtually independent of the state of aggregation of mercury. Admittedly, the C vs φ curves determined with the liquid electrode in the pure electrolyte solution showed a capacity increase in the cathodic potential region at a potential lower by about 0.2 V than with the solid electrode. This increase was attributed to formation of an amalgam. Some differences between the C vs φ curves were also observed in presence of ethyl alcohol at very negative and positive potentials.

Since elucidation of the role of the state of aggregation in adsorption processes was of fundamental significance, this problem was studied repeatedly. For example, Kheifets and Krasikov [65] studied adsorption on nonyl alcohol from a solution of $2 N$ H_2SO_4 in 80% C_2H_5OH on liquid and solid mercury. They found a significant difference of behavior between the liquid and solid metals: in the case of solid mercury, desorption of nonyl alcohol was observed at more negative potentials and was incomplete.

Subsequently Leikis and Sevast'yanov [66] studied adsorption of hexyl alcohol on liquid and solid gallium and on Wood's alloy at negative surface charges in $1 N$ Na_2SO_4. They concluded that at negative surface charges adsorption of the alcohol is independent of the state of aggregation of the electrode. Solidification of Wood's alloy led only to some increase of capacity, which was attributed to increase of the electrode area as the result of a change in the specific volume of the metal [67]. However, Bagotskaya and Morozov showed that there is a difference in the course of the C vs φ curves at more positive charges of the gallium surface. It was found that the C vs φ curves measured with a solid electrode do not have the capacity minimum at low concentrations, due to the diffuse nature of the double layer. According to Frumkin [1], this difference can be explained as follows. When gallium solidifies, crystallites of different orientations, which apparently have different φ_N, appear on the metal surface. This leads to disappearance of the minimum on the C vs φ curves.

It follows from the results of the investigations discussed above that a change in the state of aggregation has some influence

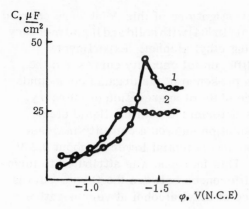

Fig. 68. Differential capacity curves for 1 N KCl saturated with hexyl alcohol: 1) Silver electrode; 2) silver electrode after deposition of mercury (2 mg/cm^2) (from data in [9]).

on the adsorptional and electrochemical properties of electrodes, although possibly this influence is not apparent over the entire range of potentials studied.

b. Surface Heterogeneity. The surface of a solid metal is energetically heterogeneous. Individual regions of a solid metal (crystal faces of low reticular density, dislocations, vacancies, and other defects) have higher free surface energy. The concentration of such regions and their distribution over the surface must influence the adsorption properties of the metal. This question will be discussed more fully later in relation to metals of the iron group. The influence of surface heterogeneity is also observed in the case of metals of high hydrogen overpotential and silver and copper.

It is noted in a number of publications [2, 10, 25, 26, 46] that the desorption peaks for solid metals are more extended over the potential and are lower than for mercury. The change of peak shape is probably due to the energetic heterogeneity of the surface, which is formally equivalent to decrease of attractive interaction between adsorbed organic molecules. Dagaeva et al. [9] produced surface heterogeneity artificially by electrodeposition of a certain amount of mercury on a silver surface. After this treatment the metal surface became heterogeneous and the hexyl alcohol desorption peak on the C vs φ curve became substantially lower and broader (Fig. 68).

The influence of the state of the surface on the position of the potential of the desorption peak was noted by Dobren'kov and

Golovin [26]. The desorption peak of isoamyl alcohol on cadmium subjected to chemical polishing with subsequent cathodic polarization as observed at a more positive potential (by about 0.1 V) than in the case of an unpolished electrode. However, in another investigation [23] no appreciable difference in behavior was found between polished and etched lead electrodes.

The heterogeneity factor of a solid surface can be estimated on the assumption that attractive interaction between adsorbate molecules is the same on mercury as on a solid metal [26].

For representation of adsorption processes on a uniformly heterogeneous surface, Temkin [68] proposed an isotherm which, in the general case, is written as

$$\theta = \frac{1}{f} \left[\frac{1 + B_{max}C}{1 + B_{min}C} \right] , \qquad (VI.3)$$

where B_{max} and B_{min} are adsorption equilibrium constants corresponding to the highest and lowest values of the energy of adsorption, which are connected with the energy of adsorption by the relation $B = \exp\left[-\Delta G_A/RT\right]$; f is the heterogeneity factor, characterizing the difference between the minimum and maximum values of the energy of adsorption.

In the region of medium surface coverages, where $B_{max}c \gg 1$ and $B_{min}c \ll 1$, Eq. (VI.3) is simplified:

$$\theta = \frac{1}{f} \ln \left[B_{max}C \right] . \qquad (VI.4)$$

It would seem that experimental adsorption isotherms determined for solid electrodes must conform to Eq. (VI.4). However, in a number of studies of the adsorptional properties of silver and copper [53, 54] it was found that the surface coverage obeys the Langmuir isotherm.

Conformity of the coverage of a heterogeneous surface by an organic substance to the Langmuir isotherm may be partially explained by the action of two mutually compensating factors. On the one hand, the free energy of adsorption decreases with increasing coverage; on the other, it increases owing to attractive forces between the adsorbed molecules; i.e., mutual compensation of the effects of the exponential terms in the Temkin isotherm (VI.4)

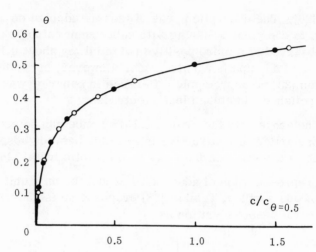

Fig. 69. Calculated adsorption isotherms: white circles represent
the Frumkin isotherm ($a = -2.6$); black points represent the Temkin isotherm ($f = 9.2$).

and the Frumkin isotherm [69] occurs. The Frumkin isotherm
may be written in the form:

$$Bc = \frac{\theta}{1-\theta} \exp(-2a\theta), \tag{VI.5}$$

where

$$B = B_0 \exp\left[-\frac{\displaystyle\iint C_0 d\varphi^2 + C'\varphi\left(\varphi_N - \frac{\varphi}{2}\right)}{RT\,\Gamma_m} \right],$$

and φ is the electrode potential measured relative to φ_n at $\theta = 0$.

If the surface heterogeneity is less pronounced, as appears
to be the case for metals of high hydrogen overpotential [23, 26, 28]
and in certain instances for silver [52], attractive interaction between the adsorbed molecules is not fully compensated and the adsorption conforms to the Frumkin isotherm.

It should be noted that the logarithmic adsorption isotherm
can also be represented by the Frumkin isotherm if it is assumed
that $a < 0$ (Fig. 69).

c. Competitive Adsorption of Organic Molecules and Water. As adsorption of an organic substance is
accompanied by displacement of water molecules, the difference
between the adsorbabilities of organic compounds on mercury and
on other metals may be attributed to changes in the energy of interaction of the organic substance and of water molecules with the
electrode surface. The concepts of competitive adsorption have
been developed fully by Bockris, Swinkels, and Green [39, 40].*

The change in the ratio of the adsorption energies of organic
substances and water in passing from mercury to metals of high
hydrogen overpotentials can be seen from a comparison of the
values of B_0 and $-\Delta \bar{G}_A^0$ for mercury, bismuth, and lead, given in
Tables 8 and 10. The values of B_0 and $-\Delta \bar{G}_A^0$ are lower on bismuth
than on mercury for all the alcohols studied with the exception of
ethanol; this indicates that the hydrophilic properties of bismuth
and lead are more pronounced than those of mercury. The other
differences revealed by comparison of Tables 8 and 10 cannot be
explained at present, owing primarily to the insufficient experimental data.

The energy of adsorption of water is apparently higher on
copper than on metals of high hydrogen overpotential. The low
adsorbability of cetyl mercaptan on mercury has been attributed
to this [48-50]. The authors consider that when dibutyl ether is
added to the solution it forms hydrogen bonds with water molecules
and weakens the bonding between the solvent molecules and the
metal surface. As a result, the mercaptan molecules displace
water molecules from a considerable part of the copper surface
and are adsorbed on the metal.

Finally, as was pointed out by Frumkin [70], metals of the
iron group are even more hydrophilic. This subject will be considered more fully in the next chapter. Here we confine ourselves
to a description of the adsorption activity of hexyl alcohol. Melik-
Gaikazyan [71] found that hexyl alcohol molecules are adsorbed
on mercury from $1 N$ KCl saturated with hexyl alcohol over a wide
range of potentials, from 0 to -0.9 V (standard hydrogen electrode).
A monomolecular layer and, in a narrower region of potentials,

---

*It should be pointed out that the authors did not take into account the change in the
capacity of the double layer on replacement of organic molecules by water molecules.

even multimolecular layers of hexyl alcohol are formed. The
situation is entirely different in the case of iron. According to
capacity measurements, iron does not adsorb hexyl alcohol to any
appreciable extent [72]. Evidently, the energy of adsorption of
water molecules on iron is higher than that of the organic compound; therefore, hexyl alcohol molecules are unable to displace
water molecules from the electric double layer.

Differences in the adsorption behavior of metals are also
due to different values of φ_N, C_0, C', a, and A for different metals.
For example, $\varphi_N = \varphi_{dip}^{org} - \varphi_{dip}^{H_2O}$ will in general be different
for different metals. However, the similar values of φ_N for
aliphatic alcohols on mercury, lead, and bismuth (see Tables 8
and 10) merely show that if the values of φ_{dip}^{org} and $\varphi_{dip}^{H_2O}$ change in
passing from one metal to another, they change to roughly the
same extent. The values of C_0 determined on lead and bismuth
at fairly negative potentials are appreciably higher than on mercury [4, 20]. As the result of these differences in the adsorption
parameters of the Frumkin isotherm, the potentials of the desorption peaks of aliphatic alcohols, taken relative to φ_n of the
electrode, are not the same on solid metals as on mercury (Table
11).

The desorption peaks of the alcohols are observed at $\theta \simeq 0.5$.
For this state, assuming that a and c in the Frumkin isotherm
are equal, we can write:

$$B = B_0 \exp\left[-\frac{\iint C_0 d\varphi^2 + C'\varphi\left(\varphi_N - \frac{\varphi}{2}\right)}{RT\,\Gamma_m} \right].$$

Even if Γ_m and B_0 for different metals are equal, the desorption potential of an organic substance is determined by the
relation

$$\iint C_0 d\varphi^2 + C'_\varphi(\varphi_N - \varphi/2) = \text{const.}$$

However, the main differences between the positions of the
desorption potentials for different metals depend on the values
of B_0.

Thus, desorption of organic substances at the same potentials can be observed only if all the parameters of the Frumkin
equation are equal.

TABLE 11. Potentials of the Desorption Peaks of Aliphatic
Alcohols Relative to the Potentials of Zero Charge of the Metals

Metal	Composition of solution	φ^{max} (V)
Pb	0.1 N NaF + ethyl alcohol (4 M)	−0.62 [23]
Hg	0.1 N NaF + ethyl alcohol (4 M)	−0.92 [73]
Pb	0.1 N NaF + n-butyl alcohol (0.5 M)	−0.53 [23]
Hg	0.1 N NaF + n-butyl alcohol (0.5 M)	−0.83 [73]
Tl	1 N KCl + n-amyl alcohol (0.1 M)	−0.40 [2]
Pb	0.1 N NaF + n-amyl alcohol (0.1 M)	−0.44 [24]
Bi	0.1 N K$_2$SO$_4$ + n-amyl alcohol (0.1 M)	−0.70 [27]
Hg	0.1 N NaF + n-amyl alcohol (0.1 M)	−0.74 [73]
Ag	1 N Na$_2$SO$_4$ + n-hexyl alcohol (sat.)	−0.5 [25]
Pb	1 N Na$_2$SO$_4$ + n-hexyl alcohol (sat.)	−0.5 [25]
Sb	1 N Na$_2$SO$_4$ + n-hexyl alcohol (sat.)	−0.50 [25]
Hg	1 N KCl + n-hexyl alcohol (sat.)	−0.73 [71]
Bi	0.1 N KCl + n-octyl alcohol (sat.)	−0.67 [6]
Hg	1 N KNO$_3$ + n-octyl alcohol (sat.)	−0.60 [74]

d. Role of Adsorbed Gases and Ions. In addition to the factors already considered, the specific nature of given metals is manifested in the fact that as we pass from mercury to other metals the energy of adsorption of ions present in solution changes; moreover, adsorbed hydrogen and oxygen may appear on the metal surface.

The influence of adsorption of halide ions and oxides on adsorption of organic substances has been noted in the literature [51, 52, 75].

If hydrogen and oxygen are adsorbed reversibly, the relationships of their simultaneous adsorption with organic substances can be derived thermodynamically (see Chapter IX).

It should be noted in conclusion that attempts have been made in the literature to compare the adsorptional properties of metals on the basis of some of their physical properties [76] or electrochemical characteristics.

For example, in a number of publications [77–90] the concepts originally developed by Antropov [77–80] were used for assessment of the adsorbability of organic compounds on various metals.

250

CHAPTER 6

According to these concepts, the conditions of adsorption of organic compounds on all metals are the same at the same rational potential [91] or the potential on the Φ-scale (in Antropov's notation). Therefore data obtained for adsorption of organic substances on mercury can then be extended to other metals.

These concepts made it possible to explain a number of experimental results and once again emphasized the important role of the potential of zero charge in adsorption phenomena.

However, Frumkin [1] showed that the structure of the electric double layer in electrolyte solutions without addition of organic substances at potentials equally distant from φ_N may in general be different on different metals. The data presented above show that in presence of organic substances factors appear which produce additional changes in the structure of the double layer at equal potentials on Antropov's Φ-scale.

Thus, caution must be exercised in applying Antropov's concepts to adsorption of organic compounds on different metals.

References

1. A. N. Frumkin, J. Res. Inst. Catalysis, Hokkaido Univ., 15:61 (1967).
2. T. I. Borisova and B. V. Érshler, Zh. Fiz. Khim., 24:337 (1950).
3. D. I. Leikis and B. N. Kabanov, Tr. Inst. Fiz. Khim., Novye Metody Fiz. Khim. Issled., No. 2, 5 (1957).
4. K. V. Rybalka and D. I. Leikis, Élektrokhimiya, 3:383 (1967).
5. Tsa Ch'uan-Hsin and Z. A. Iofa, Dokl. Akad. Nauk SSSR, 131:137 (1960).
6. U. V. Pal'm, V. É. Past, and R. Yu. Pullerits, Élektrokhimiya, 2:604 (1966).
7. V. Ya. Bartenev, É. S. Sevast'yanov, and D. I. Leikis, Élektrokhimiya, 4:745 (1968).
8. E. K. Venstrem, V. I. Likhtman, and P. A. Rebinder, Dokl. Akad. Nauk SSSR, 107:105 (1956).
9. I. G. Dagaeva, D. I. Leikis, and É. S. Sevast'yanov, Élektrokhimiya, 3:891 (1967).
10. D. I. Leikis, Dokl. Akad. Nauk SSSR, 135:1429 (1960).
11. A. Ya. Gokhshtein, Élektrokhimiya, 2:1061, 1318 (1966).
12. T. N. Andersen, R. S. Perkins, and H. Eyring, J. Am. Chem. Soc., 86:4496 (1964).
13. A. N. Frumkin, N. B. Grigor'ev, and I. A. Bagotskaya, Élektrokhimiya, 2:329 (1966).
14. A. M. Morozov, N. B. Grigor'ev, and I. A. Bagotskaya, Élektrokhimiya, 2:1235 (1966).
15. N. B. Grigor'ev and D. N. Machavariani, Élektrokhimiya (in press).
16. A. N. Frumkin, Svensk Kem. Tidskr., 77:300 (1965).
17. B. Jakuszewski and Z. Kozlowski, Roczniki Chem., 36:1573 (1962); 38:96 (1963).

18. B. B. Damaskin, N. V. Nikolaeva-Fedorovich, and A. N. Frumkin, Dokl. Akad. Nauk SSSR, 121:129 (1958).

19. D. C. Grahame, E. Coffin, J. Cummings, and U. Roth, J. Am. Chem. Soc., 74:1207 (1952); D. C. Grahame, J. Am. Chem. Soc., 76:4819 (1954); J. Electrochem. Soc., 98:343 (1951).

20. V. Past, U. Pal'm, K. Pal'ts, R. Pullerits, and M. Khaga, in: The Double Layer and Adsorption on Solid Electrodes [in Russian], Tartu, Est.SSR (1968), p. 114.

21. M. Bonnemay, G. Bronoël, P. Jonville, and E. Levart, Compt. Rend., 260:5262 (1965).

22. V. L. Kheifets and B. S. Krasikov, Dokl. Akad. Nauk SSSR, 109:586 (1956).

23. N. B. Grigor'ev and D. N. Machavariani, in: The Double Layer and Adsorption on Solid Electrodes [in Russian], Tartu,Est.SSR (1968), p. 61; Elektrokhimiya in press).

24. T. I. Borisova, B. V. Érshler, and A. N. Frumkin, Zh. Fiz. Khim., 22:925 (1948).

25. É. S. Sevast'yanov and D. I. Leikis, Izv. Akad. Nauk SSSR, Ser. Khim., No. 3, 450 (1964).

26. G. A. Dobren'kov and V. A. Golovin, in: The Double Layer and Adsorption on sorption on Solid Electrodes [in Russian], Tartu, Est.SSR (1968), p. 125.

27. R. Ya. Pullerits, U. V. Pal'm, and V. É. Past, Élektrokhimiya, 4:728 (1968).

28. R. Ya. Pullerits, U. V. Pal'm, and V. É. Past, in: The Double Layer and Adsorption on Solid Electrodes [in Russian], Tartu,Est.SSR (191968), p. 125.

29. M. Loshkarev, A. Krivtsov, and A. Kryukova, Zh. Fiz. Khim., 23:221 (1949).

30. U. V. Pal'm, V. É. Past, and R. Ya. Pullerits, Élektrokhimiya, 3:376 (1967).

31. B. B. Damaskin and N. V. Nikolaeva-Fedorovich, Zh. Fiz. Khim., 35:1279 (1961).

32. F. M. Schapink, M. Oudeman, K. W. Leu, and J. N. Helle, Trans. Faraday Soc., 56:415 (1960).

33. A. M. Morozov, N. B. Grigor'ev, and I. A. Bagotskaya, Élektrokhimiya, 3:585 (1967).

34. R. Parsons, Proc. Roy. Soc., A261:79 (1961).

35. N. B. Grigor'ev and V. S. Krylov, Élektrokhimiya, 4:763 (1968).

36. G. A. Dobren'kov, Tr. Kazansk. Khim.-Tekhnol. Inst., 34:186 (1965).

37. V. A. Golovin, Tr. Kazansk. Khim.-Tekhnol. Inst., 34:207 (1965).

38. M. Loshkarev and V. Chernenko, Zh. Fiz. Khim., 34:1060 (1960).

39. J. O'M. Bockris and D. A. J. Swinkels, J. Electrochem. Soc., 111:736 (1964).

40. J. O'M. Bockris, M. Green, and D. A. J. Swinkels, J. Electrochem. Soc., 111:743 (1964).

41. I. G. Dagaeva, D. I. Leikis, and É. S. Sevast'yanov, Élektrokhimiya, 2:820 (1966); J. Electrochem. Soc., 113:1341 (1966).

42. É. S. Sevast'yanov and T. Vitanov, Élektrokhimiya, 3:408 (1967).

43. L. Ramaley and C. G. Enke, J. Electrochem. Soc., 112:947 (1965).

44. M. A. Vorsina and A. N. Frumkin, Dokl. Akad. Nauk SSSR, 24:918 (1939); Zh. Fiz. Khim., 19:171 (1945).

45. R. J. Watts-Tobin, Phil. Mag., 61:133 (1961).

46. V. L. Kheifets, B. S. Krasikov, V. V. Sysoeva, and N. V. Guseva, Vestn. Leningr. Gos. Univ., No. 4, 127 (1957).

47.   B. S. Krasikov and M. V. Pevnitskaya, Vestn. Leningr. Gos. Univ., No. 10, 133 (1958).

48.   S. Fujii and K. Aramaki, Third International Congress on Metal Corrosion (abstracts of papers) [Russian translation], Moscow (1966), p. 175.

49.   S. Fujii and K. Kobayashi, Corrosion Eng., 14:449 (1965); Extended Abstracts, 17th CITCE Meeting, Tokyo (1966), p. 2.

50.   K. Aramaki and S. Fujii, Corrosion Eng., 14:542 (1965).

51.   S. Biallozor, Roczniki Chem., 40:1091 (1966).

52.   R. G. Barradas and B. E. Conway, J. Electroanal. Chem., 6:314 (1963).

52a. G. Kretzschmar, Monatsber. Deut. Akad. Wiss. Berlin, 9:609 (1967).

53.   R. J. Newmiller and R. B. Pontius, J. Phys. Chem., 64:584 (1960).

54.   F. I. Khaibullin, in: Physicochemical and Technological Investigations of Mineral Ores [in Russian], Izd. "Nauka," Tashkent (1965), p. 158.

55.   R. S. Hansen and B. H. Clampit, J. Phys. Chem., 58:908 (1954).

56.   B. Ke, J. J. Hoekstra, B. C. Sison, Jr., and D. Trivich, J. Electrochem. Soc., 106:382 (1959).

57.   J. Llopis, J. M. Gamboa, I. Arizmendi, and F. Alonso, J. Electrochem. Soc., 109:368 (1962).

58.   J. Llopis and J. M. Gamboa, Radioisotopes in Scientific Research, Vol. 2, Pergamon Press, p. 478.

59.   A. A. Sutyagina, Zavod. Lab., 24:43 (1958).

60.   S. M. Kochergin and L. L. Khonina, Zh. Prikl. Khim., 36:673 (1963).

61.   S. S. Kruglikov, Yu. I. Sinyakov, and N. T. Kudryavtsev, Élektrokhimiya, 2:100 (1966).

62.   W. Machu and V. K. Gouda, Werkstoffe Korrosion, 13:745 (1962).

63.   M. Smialowski and Z. Szklarska-Smialowska, Bull. Acad. Polon. Sci., 1:155 (1953)

64.   A. V. Gorodetskaya and M. A. Proskurnin, Zh. Fiz. Khim., 12:411 (1938).

65.   V. L. Kheifets and B. S. Krasikov, Dokl. Akad. Nauk SSSR, 94:101 (1954).

66.   D. I. Leikis and É. S. Sevast'yanov, Dokl. Akad. Nauk SSSR, 144:1320 (1962).

67.   S. Karpachev, N. Ladygin, and V. Zykov, Zh. Fiz. Khim., 17:75 (1943).

68.   M. I. Temkin, Zh. Fiz. Khim., 25:296 (1941).

69.   A. N. Frumkin, Tr. Inst. im. L. Ya. Karpova, 4:56 (1925); 5:3 (1926); Z. Phys., 35:792 (1926).

70.   A. N. Frumkin, "Fundamental problems of modern theoretical electrochemistry," Proceedings of 14th CITCE Conference [Russian translation], Izd. "Mir," Moscow (1965), p. 302.

71.   V. I. Melik-Gaikazyan, Zh. Fiz. Khim., 26:560 (1952).

72.   Z. A. Iofa (Jofa), V. V. Batrakov, and Cho Ngok Ba, Eletrokhimiya, 9:1645 (1964).

73.   B. B. Damaskin, A. A. Survila, and L. E. Rybalka, Élektrokhimiya, 3:146 (1967).

74.   D. C. Grahame, J. Am. Chem. Soc., 68:301 (1946).

75.   R. A. Erb, J. Phys. Chem., 69:1306 (1965).

76.   H. Fischer, Z. Elektrochem., 54:459 (1950).

77.   L. I. Antropov, Zh. Fiz. Khim., 25:1495 (1951); Proceedings of Conference on Electrochemistry [in Russian], Izd. Akad. Nauk SSSR, Moscow (1953), p. 380; Tr. Novocherk. Politekhn. Inst., 25/39:5 (1954).

78. L. I. Antropov and S. N. Banerjee, J. Indian Chem. Soc., 35:531 (1958); 36:451 (1959).
79. L. I. Antropov, J. Sci. Ind. Res. (New Delhi), 18B:314 (1959); First International Congress on Metal Corrosion, Butterworths, London (1961), p. 147.
80. L. I. Antropov, Kinetics of Electrode Processes and Null Points of Metals, Council of Scientific and Industrial Research, New Delhi (1960).
81. C. P. De, Nature, 180:803 (1957).
82. H. C. Gatos, Nature, 181:1060 (1958).
83. D. M. Brasher, First International Congress on Metal Corrosion, London (1961), p. 56.
84. E. Blomgren and J. O'M. Bockris, J. Phys. Chem., 63:1475 (1959).
85. E. Blomgren, J. O'M. Bockris, and C. Jesch, J. Phys. Chem., 65:2000 (1961).
86. G. G. Lysenko and O. S. Afanas'ev, Dokl. Akad. Nauk Ukr. SSR, 1091 (1961).
87. M. A. Loshkarev and M. P. Sevryugina, Nauchn. Tr. Dnepropetr. Khim.-Tekhn. Inst., 2(12):97 (1961).
88. Z. Ostrowski, Werkstoffe Corrosion, 9:522 (1960).
89. G. Montel, Corrosion et Anticorrosion, 11:420 (1963).
90. H. Fischer and W. Seiler, Corrosion Sci., 6:159 (1966).
91. D. C. Grahame, Chem. Rev., 41:441 (1943).

Chapter 7

Adsorption of Organic Compounds
on Metals of the Iron Group

Because of certain peculiarities in the behavior of metals of the iron group, a separate chapter is devoted to their adsorption properties.

First, the energy of adsorption of water molecules is considerably higher on metals of the iron group than on metals of high hydrogen overpotential, and on silver and copper. This conclusion follows from measurements of differential capacity, which indicate that hexyl alcohol is not adsorbed to any appreciable extent on iron and cobalt [1] and only slightly on nickel [2, 3]. However, even in the case of nickel the capacity of the electric double layer decreases from 19 $\mu F/cm^2$ only to 15-16 $\mu F/cm^2$ in a solution of hydrochloric acid saturated with hexyl alcohol.

Second, the partially filled d-orbitals of these metals have a significant influence on the mechanism of adsorption of organic compounds.

Third, the surfaces of iron, cobalt, and nickel are oxidized fairly easily in air or in aqueous solutions, while reduction of the oxides formed proceeds very slowly under cathodic polarization. As will be shown below, the presence of an oxide film alters the adsorption activity of the electrode surface substantially.

Fourth, in contrast to metals of high hydrogen overpotential, in the case of metals of the iron group a potential region in which they behave virtually as ideally polarized electrodes cannot be defined.

The foregoing characteristics undoubtedly complicate studies of adsorption effects on metals of the iron group and lower the experimental accuracy. The available experimental data relating to metals of the iron group are largely qualitative and sometimes contradictory.

1. Dependence of Adsorption on the Potential and Position of the Point of Zero Charge

Some data on the dependence of adsorption of organic substances on the potential are available in the literature for metals of the iron group. The capacity curves obtained by Hillson [2] for nickel indicate that maximum adsorption of hexylammonium atoms in hydrochloric acid solutions occurs at $\varphi \simeq -0.42$ V (N.H.E.). At more positive potentials adsorption diminishes, and when a potential between -0.2 to -0.15 V is reached the organic cations are desorbed.

Data on the adsorption of decylamine and naphthalene on iron and nickel [4, 5] also indicate that adsorption depends on the electrode potential. In illustration, Fig. 70 shows data for adsorption

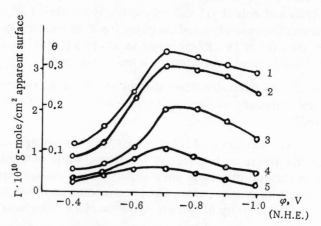

Fig. 70. Adsorption of n-decylamine on nickel from 0.9 N $NaClO_4$, pH = 12; n-decylamine concentration (M): 1) 7.5 · 10^{-5}; 2) 5 · 10^{-5}; 3) 2.5 · 10^{-5}; 4) 1 · 10^{-5}; 5) 0.5 · 10^{-5} (from data in [4]).

of decylamine in nickel. Curves of similar form were also obtained for an iron electrode. According to Bockris et al. [4, 5], the shift of the potential of maximum adsorption (φ_m) of iron and nickel in the negative region relatively to $\varphi_n = -0.47$ V (also according to their data) is due to differences in the adsorption energy of the water molecule dipoles at positive and negative charges of the electrode surface. Details of this explanation may be found in the original paper [4]. In our view, however, the shift of φ_m may be attributed to the influence of adsorbed hydrogen and metal oxidation products on adsorption of organic substances, as some of the adsorption measurements were performed on the positive side of the stationary potential. The stationary potentials of nickel and iron in the $NaClO_4$ solutions studied (pH = 12) are approximately -0.7 V [6, 7], and are close to the values of φ_m, which are -0.7 to -0.9 V and -0.8 V respectively for decylamine on these metals.

Comparison of the adsorption behavior of organic compounds on metals of the iron group with their behavior on mercury and on the solid metals discussed in Chapter VI reveals additional peculiarities. For example, the C vs φ curves for metals of the

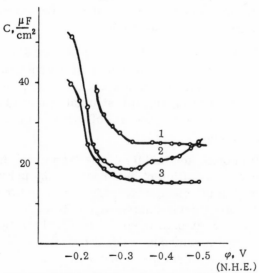

Fig. 71. Differential capacity curves for iron in $1\ N\ H_2SO_4$: 1) Pure solution; 2) $5 \cdot 10^{-3}\ M$ nonanesulfonic acid, introduced at $\varphi = -0.5$ V; 3) the same, introduced at $\varphi = -0.25$ V (from data in [8]).

iron group in presence of aliphatic alcohols do not have desorption peaks; this is primarily due to low adsorbability of the alcohols.

Moreover, the dependence of organic compounds on the potential is less pronounced with metals of the iron group than with metals of high hydrogen overpotential. A possible explanation is that some of the adsorbate molecules form stronger chemical bond with the surface metal atoms. The nature of these bonds depends on the metal, the organic substance, and the potential at which adsorption is studied. The adsorption energy of molecules or ions which form chemical bonds with the metal is probably much higher than the adsorption energy of water molecules, and therefore complete desorption of the organic substance does not occur even at a considerable distance from φ_n of the metal.

Differential capacity curves determined for a polished zone-refined iron electrode in $1 N$ H_2SO_4 are shown in Fig. 71 [8]. It is seen that nonanesulfonic acid is adsorbed only at potentials on the positive side of -0.5 V, and maximum adsorption is reached at -0.3 to -0.32 V (S.H.E.). In the potential range from -0.25 to -0.3 V the adsorption is virtually irreversible. For example, if the substance is introduced at the stationary potential $(-0.25$ V), desorption of the acid does not occur at a sufficient distance from the potential of maximum adsorption.

In the light of the views put forward in Part I on the structure of the electric double layer in presence of organic substances, it is to be expected that maximum adsorption of organic substances on metals of the iron group should occur at potentials close to φ_n. Examination of data on φ_n of metals of the iron group is of interest in this connection.

The position of φ_n for metals of the iron group has not yet been determined with sufficient accuracy, owing primarily to difficulties in preparation of the electrode surfaces for the determinations. The large deviations between the φ_n values given by different authors are evident from Table 12. Investigations of adsorption of organic substances make it possible to define the region of presumed φ_n values more precisely.

---

*These considerations refer to the simplest case, when no other adsorbed particles (e.g., gases) are present on the electrode surface. It was shown by Frumkin [9] that in presence of adsorbed hydrogen or oxygen the potentials of maximum adsorption of a substance may differ significantly from φ_n.

TABLE 12. Values of the Potential of Zero Charge for Metals
of the Iron Group

Metal	φ_n, V (N.H.E.)	Solution	Method
Fe	−0.40	0.003 N HCl	Crossed filaments [10]
	−0.37	0.001 N H_2SO_4	Capacity minimum [11]
	0.00	−	Work function [12]
Co	−0.58 to −0.68	−	Work function [13]
	−0.34	H_2SO_4 (pH = 1.17)	Capacity minimum [14]
	−0.40	H_2SO_4 (pH = 2.06)	Capacity minimum [14]
Ni	−0.23	0.0012 N HCl + $1 \cdot 10^{-3}$ N $LaCl_3$	Polarization curves [15]
	−0.22	H_2SO_4 + Na_2SO_4 (pH = 0.94)	Capacity minimum [16]
	−0.37	H_2SO_4 + Na_2SO_4 (pH = 6)	Capacity minimum [16]
	+ 0.08	−	Work function [13]

Since iron is the metal which has been studied most fully in this group and the position of φ_n of iron is often discussed in the literature, it is appropriate here to consider the φ_n of iron in greater detail.

According to Antropov [12], φ_n of iron should be approximately 0 V, which differs noticeably from the other values in Table 12. His conclusion is based on calculation of φ_n from the electronic work function (W_e):

$$\varphi_n = W_e - 4.31. \qquad \text{(VII.1)}$$

The expression (VII.1) can be derived with the assumption that the shift of potential due to adsorption of water from aqueous solutions on the metal surface and the shift of potential due to displacement of the electron cloud are independent of the nature of the metal at φ_n [17].

Calculation of φ_n from formula (VII.1) is very rough. This is due, first, to the inadequate reproducibility of the values of the work function found by different methods. Thus, according to Fomenko [18], the reproducibility of W_e is 0.1-0.3 eV, which is equivalent to an error of 0.1-0.3 V in calculation of φ_n by Eq. (VII.1). It should be pointed out that W_e is determined under conditions differing greatly from those used in determination of φ_n in aqueous solutions (the metal is degassed and oxides are removed).

Second, the assumption that the potential shifts due to adsorption of water on different metals are equal is a crude approximation. For example, in solutions free from surface-active ions the magnitudes of this shift for mercury and gallium differ by about 0.3 V [19, 20]. Therefore the values for φ_n of iron found by capacity measurements [11] or by the crossed filament method [10] are probably more trustworthy.

The position of φ_n of iron between -0.3 and -0.4 V is confirm by data on adsorption of cationic and anionic organic compounds from solutions in $1\,N\,H_2SO_4$. Thus, at the stationary potential (φ_{st}) of iron, -0.25 to -0.27 V, tribenzylamine present in cationic form in $1\,N\,H_2SO_4$ is adsorbed very weakly on iron (Fig. 72). Anions (e.g., nonanesulfonic acid) are adsorbed much better under these conditions (see Fig. 71). If we assume, with Antropov, that $\varphi_n = 0$ V, then the metal surface is negatively charged at φ_{st}, and the adsorption properties of these cations should conform to the opposite relationship. Data on joint adsorption of organic and inorganic ions indicate that the φ_n values are on the negative side of φ_{st}.

Lorenz and Fischer [21] recently concluded on the basis of capacity determinations that φ_n of iron is 0 V, although its exact position cannot be determined owing to the presence of surface-active OH^- anions in solution. According to these authors, the minimum at $\varphi = -0.4$ V on the C vs φ curve determined in 0.01 and 0.001 $N\,H_2SO_4$ corresponds to the potential of the "uncharged surface" (in Antropov's terminology) in presence of adsorbed OH^- ions, which shift φ_n by approximately 0.4 V into the negative region. In presence of Cl^- ions ($1\,N$ HCl) the C vs φ curves have a minimum corresponding to $\varphi = -0.12$ V, which is attributed by the author to maximum diffuseness of the ionic layer of the double layer. Lorenz and Fischer consider that halide and OH^- ions are adsorbed reversibly both on iron and on mercury, and therefore adsorption of these anions leads to displacement of φ_n determined from the position of the differential capacity minimum into the negative potential region.

In our opinion, the minimum observed by Lorenz and Fischer in sulfuric acid solutions is not related to the diffuse nature of the double layer, as the minimum was not intensified by dilution of the electrolyte. The well-defined minimum observed by them at more

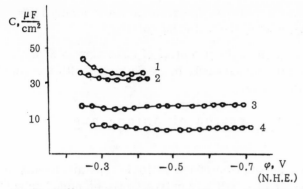

Fig. 72. Differential capacity curves of iron in 1 N H_2SO_4:
1) Pure solution; 2) $1 \cdot 10^{-3}$ M tribenzylamine; 3) $1 \cdot 10^{-2}$ N
KI;4) $1 \cdot 10^{-3}$ M tribenzylamine + $1 \cdot 10^{-2}$ N KI.

positive potentials in $1 N$ HCl is probably due to chemisorption of
Cl⁻ ions. This process begins to an appreciable extent at more
positive potentials [22]. As was shown by Frumkin [23] and Iofa
[24], with irreversible chemisorption of anions the φ_n of iron
determined from the differential capacity minimum should shift
into the positive region of potentials.

Differences in the character of adsorption of halide ions in
iron and mercury can also be established from capacity and
polarization measurements: adsorption of anions on mercury
leads to increase of the double-layer capacity, while in the case
of iron the capacity decreases [25]; halide ions accelerate evolution of hydrogen on mercury and retard it on iron.

In another publication Lorenz, Sarropulos, and Fischer [26]
support the view that the true position of φ_n corresponds to more
positive potentials than −0.4 V by capacity data determined in sulfuric acid solutions with high contents of ethyl alcohol, dioxane, or
acetone. However, they disregarded the appearance of a diffusion
potential at the boundary between the test solution and the solution
of the reference electrode, and therefore their values of the potential are indefinite.

It must be specially noted that the concepts of Lorenz and
Fischer cannot explain the increased adsorption of organic compounds of the cationic type on iron with increasing concentration

of halide ions and in passing from Cl^- to I^- ions. These questions will be examined in the following section.

The question of the potential of zero charge of metals of the iron group must undoubtedly be reexamined with the aid of the tracer atom method.

2. Joint Adsorption of Anions and Cations

Organic substances of the cationic type are hardly adsorbed at all on iron and cobalt and only slightly on nickel in sulfuric and perchloric acid solutions at the stationary potential [27-33]. Adsorption of organic cations and of onium compounds, which exist in cation form in acid solutions, increases sharply on introduction of halide ions into sulfuric or perchloric acid solutions, and in solutions of hydrohalic acids [1, 3, 22, 24, 25, 27-29, 34-37].

In accordance with the views of Frumkin [23] and Iofa [24] the increased adsorption of organic cations can be attributed to changes of the surface charge of the metal in presence of halide ions. The surface of iron is positively charged in $1N$ H_2SO_4 at a stationary potential of -0.25 to -0.27 V (N.H.E.), and ions of like charge are not adsorbed effectively on the surface. Chemisorption of halide ions on iron shifts φ_n toward more positive potentials. The metal surface at the stationary potential becomes negatively charged, and the cations are attracted by electrostatic forces to the metal surface. These concepts may be used for explaining joint adsorption of anions and cations on other metals of the iron group.

Let us examine more fully the influence of chemisorbed ions on the position of φ_n. When irreversible chemisorption of anions occurs and stable metal—halogen covalent bonds are formed, the adsorbed ions enter the metallic part of the double layer, and the charge of the ions forms a part of the charge of the metal surface. The resultant dipoles shift φ_n toward more positive values. As a result, φ_n in presence of irreversibly chemisorbed anions corresponds to more positive potentials than φ_n in a solution not containing surface-active anions. The surface of a metal with adsorbed halide ions can also be regarded as the surface of an electrode of different composition (iron with a definite content of iron

halide on the surface), having a more positive potential of zero
charge than pure iron.

Shifts of φ_n in the positive direction are also observed as the
result of chemisorption of oxygen (a fraction of a monolayer) on
platinum [38, 39] and other metals of the platinum group [40, 41].
When oxygen is adsorbed on iron an increase of the work function
is reported if the oxygen atoms do not penetrate into the metal
lattice [42].

In the case of mercury, chloride and other halide anions enter
the ionic part of the double layer, and therefore adsorption of these
anions shifts φ_n into the negative region [43].

It might seem at first sight that the whole problem reduces to
a normal difference between the definitions of the potential of zero
charge in the two cases. However, this difference has a physical
basis. With sufficiently stable chemisorptive bonds, the adsorbed
anions are attached almost inseparably to the metal surface, and
this must influence the results of experimental determinations
of φ_n.

The adsorption and inhibiting action of organic cations and of
organic compounds of the onium (ammonium, oxonium, etc.) type
on metals of the iron group increase in the sequence from Cl⁻ to
Br⁻ and especially I⁻ ions; this is due to the more stable chemi-

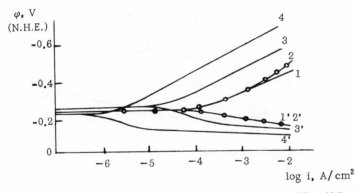

Fig. 73. Polarization curves of zone-refined iron in 1 N H_2SO_4: 1) Pure
solution; 2) $1 \cdot 10^{-3}$ M TBAS; 3) $1 \cdot 10^{-4}$ N KI; 4) $1 \cdot 10^{-4}$ N KI + 1 ·
10^{-3} M TBAS (from data in [1]).

TABLE 13. Increase of the Hydrogen Overpotential
$(\Delta\eta)$ at $i_c = 1 \cdot 10^{-3}$ A/cm^2

Metal	Solution	$\Delta\eta$, mV
Fe	1 N H$_2$SO$_4$ + 1 $\cdot$ 10^{-3} M TBAS + 1 $\cdot$ 10^{-4} N KI	240 [28]
α-Co	1 N H$_2$SO$_4$ + 1 $\cdot$ 10^{-3} M TBAS + 1 $\cdot$ 10^{-4} N KI	100
Ni	1 N H$_2$SO$_4$ + 1.5 $\cdot$ 10^{-3} M TBAS + 1 $\cdot$ 10^{-4} N KI	65 -70 [3]

sorption of I$^-$ ions because of the easy deformability of their electron shells. Differential capacity curves demonstrate the high
adsorbability of tribenzylamine in presence of I$^-$ ions on zone-refined iron (Fig. 72). The polarization curves in Fig. 73 also indicate strong retardation of hydrogen evolution and dissolution of
iron in the simultaneous presence of tetrabutylammonium sulfate
(TBAS) and I$^-$ ions.

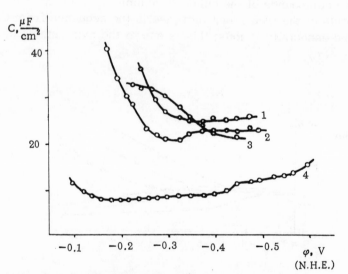

Fig. 74. Differential capacity curves of iron in 1 N H$_2$SO$_4$: 1) Pure
solution; 2) 3 $\cdot$ 10^{-4} M tribenzylamine; 3) 2 $\cdot$ 10^{-4} N H$_2$S; 4) 2 $\cdot$
10^{-4} N H$_2$S + 3 $\cdot$ 10^{-4} M tribenzylamine (from data in [28]).

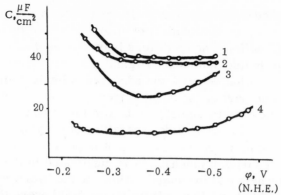

Fig. 75. Differential capacity curves of iron in 1 N H_2SO_4;
1) Pure solution; 2) $1 \cdot 10^{-3}$ M tribenzylamine; 3) $1 \cdot 10^{-2}$ M
benzenesulfonic acid; 4) $1 \cdot 10^{-3}$ M tribenzylamine + $1 \cdot$
10^{-2} M benzenesulfonic acid (from data in [1]).

Adsorption of organic cations and their influence on the hydrogen overpotential in presence of halide ions decrease in the series iron > cobalt > nickel (Table 13).

This relationship is apparently due to decrease of the adsorption of halide ions and of the shift of φ_n into the positive region in the sequence from iron to nickel. A possible cause of this sequence may be progressive filling of the 3d sublevel with electrons in the same series: iron < cobalt < nickel.

Adsorption of organic compounds of the cationic type also increases in presence of HS⁻ and NCS⁻ ions [28, 44-46] (Fig. 74). The mechanism of joint adsorption of HS⁻ or NCS⁻ ions with organic cations is the same as in the case of halide ions and organic substances. These results account for the more effective action of cationic inhibitors on acid corrosion of steels containing sulfur in comparison with pure iron [24, 27]. When the steels dissolve, sulfide ions appear in the solution and are adsorbed together with the organic cations.

Iofa and co-workers [1] observed increased adsorption of organic cations in presence of organic anions. Figure 75 shows the results of capacity measurements in $1N$ H_2SO_4 containing tribenzylamine and benzenesulfonic acid. It is seen that increased adsorption of tribenzylamine in presence of benzenesulfonic acid is accompanied by broadening of the region of adsorption potentials. However, the mechanism of adsorption of organic cations and aromatic sulfonic acids differs from the mechanism of coadsorption of halide anions and organic cations. Apparently, in distinction from halide ions, the adsorbed sulfonic acid anions are in the ionic part of the double layer. Therefore joint adsorption of organic cations and anions is similar in character to coadsorption of halide or sulfide ions with tetraalkylammonium cations, observed on mercury [48, 49]. It is possible that the difference in the adsorption mechanism is manifested in the different effects of organic and inorganic cations on the hydrogen overpotential. Figure 76 shows that sulfosalicylic acid lowers the hydrogen overpotential. Adsorption of I^- ions, on the other hand, raises the hydrogen overpotential (see Fig. 73). In simultaneous presence of cations and anions the hydrogen overpotential increases in both cases.

Halide ions also influence adsorption of neutral organic substances (hexyl alcohol, camphor) and of organic compounds of the anionic type. Apparently, in presence of halide ions the surface of iron is rendered hydrophobic in addition to undergoing a

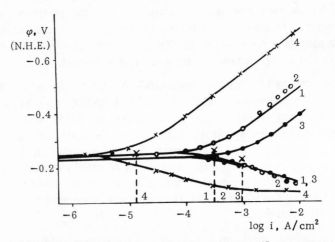

Fig. 76. Polarization curves of zone-refined iron in 1 N H_2SO_4:
1) Pure solution; 2) $1 \cdot 10^{-3}$ M tributylamine; 3) $1 \cdot 10^{-2}$ M sulfosalicylic acid; 4) $1 \cdot 10^{-3}$ M tributylamine + $1 \cdot 10^{-2}$ M sulfosalicylic acid. The dash lines indicate the spontaneous solution currents.

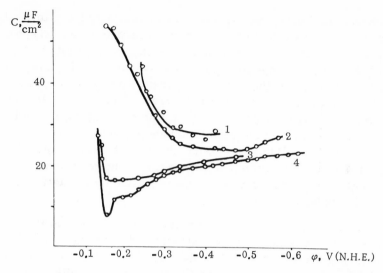

Fig. 77. Differential capacity curves of iron in 1 N H_2SO_4: 1) Pure solution; 2) $2 \cdot 10^{-4}$ M camphor; 3) $2 \cdot 10^{-3}$ N KI; 4) $2 \cdot 10^{-4}$ M camphor + $2 \cdot 10^{-3}$ N KI.

change of charge. This leads to some increase in the adsorption of neutral organic substance. Thus, in the case of camphor the adsorption increases in a narrow range of potentials before the desorption potential is reached (Fig. 77). Both factors influence adsorption of organic anions, but the influence of the surface charge predominates and organic cations (sulfonic acids, propionic acid) are not adsorbed on iron to any appreciable extent in presence of halide ions [34, 36]. These results are in agreement with the concepts put forward by Frumkin and Iofa [23, 24].

Joint adsorption of halide ions and organic cations was later studied by Hackerman et al. [50, 51]; their results in the main confirmed the work of Frumkin and Iofa [23, 24].

The high adsorbability of tetraalkylammonium halides and of ethylquinolinium iodide [37, 52], and of thiourea and its derivatives [53] is due to a considerable extent to joint adsorption of the organic and inorganic ions formed by dissociation or decomposition of the organic compounds in aqueous solutions.

Adsorption of thiourea on metals of the iron group has been studied most fully [54-67].

Several different mechanisms have been postulated in the literature for decomposition of thiourea in presence of iron [37, 54, 68-73]. However, in all the proposed schemes an invariable product of thiourea decomposition is hydrogen sulfide, which is evolved at the metal surface and can be detected with the aid of radiometric measurements if labeled thiourea (S^{35}) is used. Adsorption of thiourea apparently proceeds in two stages. First, its molecules are adsorbed on the metal surface by virtue of the free electron pair of the sulfur atom. The adsorbed thiourea molecules than undergo slower chemical changes. Thiourea derivatives also decompose with formation of hydrogen sulfide by the action of the hydrogen evolved at the metals.

Apparently, decomposition products of thiourea and its derivatives (S^{2-} ions and organic cations) and chemically unchanged molecules are both adsorbed on the surface of iron and other metals. Molecules are adsorbed on the metal surface with formation of chemical bonds between sulfur and metal atoms. In a number of cases adsorption of thiourea and its derivatives may also be caused by interaction of NH_2 groups with the metal, since the nitrogen atom, like the sulfur atom, has a free electron pair. Infrared spectra confirm that metal−nitrogen and metal−sulfur bonds may exist in complex compounds formed by thiourea with metals [74, 75].

Differential capacity curves determined for a cobalt electrode in presence of allylphenylthiourea (APTU) [76] are shown in Fig. 78a. In the cathodic region the amount of substance adsorbed is virtually independent of the potential; this indicates formation of stable chemisorption bonds between APTU molecules and cobalt surface atoms. On the other hand, adsorption of tetrabutylammonium iodide (TBAI) is determined mainly by electrostatic forces and in dilute solutions of TBAI it depends on the electrode potential (Fig. 78b). The appreciable variation of the amount of TBAI adsorbed with the potential is due to variation of the adsorbability of I^- ions (Fig. 78b, curve 5). The bonding between I^- ions and metal surface atoms gradually becomes stronger. Therefore when the potential is changed in the reverse direction only partial desorption of I^- ions occurs, and hysteresis is observed between the forward and reverse C vs φ curves. The sharp increase of capacity in the anodic region is due to reversible formation of an adsorption compound of the CoOH type during anodic dissolution of the metal [77].

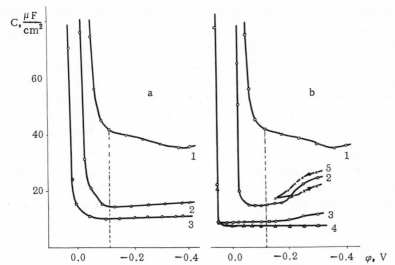

Fig. 78. Dependence of the differential capacity on potential, determined for a cobalt electrode in 1 N H_2SO_4. a: 1) Pure solution; 2) $2 \cdot 10^{-5}$ M APTU. b: 1) Pure solution; 2) $1 \cdot 10^{-6}$; 3) $1 \cdot 10^{-5}$; 4) $5 \cdot 10^{-3}$ M TBAI; 5) $3.5 \cdot 10^{-6}$ M KI. The stationary potential is indicated by dash lines (from data in [76]).

In addition to the example discussed above, adsorption of tolyl-thiourea and sym-diphenylthiourea on iron in hydrochloric acid solutions has been studied by Cavallaro, Felloni, and Trabanelli [53], of phenylthiourea by Fischer et al. [29, 30] and by Heidemeyer and Kaesche [78], and of di(o-tolyl)thiourea by Machu and Gouda [79].

3. Dependence of the Adsorption

of Organic Substances on Concentration

According to the literature, the Langmuir, Freundlich, and Temkin isotherms have been used for representing the experimental dependence of the adsorption of organic substances on concentration on metals of the iron group.

Langmuir isotherms were obtained in studies of adsorption of stearic acid and some aliphatic amines from organic solvents on steel and iron, and in studies of adsorption of o-phenanthroline and phenylthiourea in aqueous hydrochloric acid solutions [29, 30]. In the latter studies [29, 30] surface coverage was determined from the decrease of double-layer capacity measured by the potential decay method. Data obtained by the tracer atom method on adsorption of decylamine [4], naphthalene [5], and phenylthiourea [78] also confirm that the Langmuir isotherm may be appreciable despite the heterogeneity of the surface. Admittedly, Bockris et al. [4, 5] used a more complicated Langmuir isotherm, derived for the case when the organic molecules displace n water molecules from the surface.

As already discussed in Chapter VI, conformity of experimenta data to the Langmuir isotherm may be explained by mutual compensation of the exponential terms in the Frumkin and Temkin isotherms.

The relation (VII.2), which is a modified Langmuir isotherm,

$$\frac{c}{\Gamma} = \frac{1}{B\Gamma_m} + \frac{c}{\Gamma_m}$$ (VII.2)

is often used as the criterion of applicability of the Langmuir isotherm to adsorption effects. When the dependence of c/Γ on c is linear, the Langmuir isotherm is obtained. Dependence of this type was observed in determinations of adsorption of phenol from aqueous solutions on iron and nickel [80].

Gatos [81] investigated adsorption of sodium benzoate on steel in NaCl solutions (pH = 7.5) and found that at low surface coverages, corresponding to between 0.07 and 0.16 of a monolayer, the adsorption conforms to the Freundlich isotherm.

However, adsorption of most of the organic substances studied on metals of the iron group obeys the Temkin isotherm. Logarithmic adsorption isotherms were obtained by Volk and Fischer [82] for butyne-2-diol-1,4 (see Fig. 86), propargyl alcohol, n-ethylquinoline, and β-naphthoquinoline on nickel. Hackerman et al. [83] note that adsorption of poly(4-vinylpyridine), poly(4-vinylpiperidine), and poly(ethyleneamine) on steel powder is represented by the Temkin isotherm.

Iofa and Batrakov [37, 84] showed from capacity measurements by the ac method that adsorption of allylphenylthiourea in $1 N H_2SO_4$ and of tribenzylmethylammonium sulfate in $1 N$ HCl on iron obeys the Temkin isotherm. The experimental results are

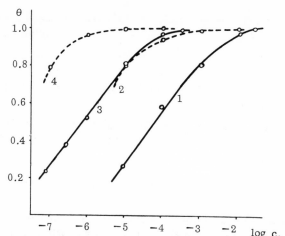

Fig. 79. Isotherms of adsorption of tribenzylmethylammonium sulfate (1, 2) from 1 N HCl ($\varphi = -0.4$ V) and of allylphenylthiourea (3, 4) from 1 N H$_2$SO$_4$ ($\varphi = -0.25$ V vs S.H.E.) on an iron electrode, calculated from decrease of the double-layer capacity (1, 3) and from polarization curves (2, 4).

shown in Fig. 79. Adsorption isotherms calculated from polarization curves are shown in the same figure. Curves 2 and 4 were calculated from the decrease in the rate of hydrogen evolution during cathodic polarization in presence of the organic substance, on the assumption that the adsorbed molecules shield the surface mechanically and the rate of hydrogen evolution on the covered regions may be neglected. It follows from Fig. 79 that calculation from the polarization curves gives overestimated values for θ.

The kinetics of adsorption of organic substances on metals of the iron group has been studied little. Direct studies of the kinetics of adsorption of acetic and stearic acid from organic solvents on iron and certain other metals, carried out with the aid of radioactive tracers, have been reported [85, 86].

Batrakov et al. [76] studied the variation of surface coverage of a cobalt electrode with time during adsorption of allylphenylthiourea from 1 N H$_2$SO$_4$. Variations of the amount of substance adsorbed were determined from the decrease of the double-layer capacity after addition of the organic substance to the electrolyte. The coverage was calculated from Eq. (V.18). The results are in good agreement with the formula of Roginskii and Zel'dovich [87] for the kinetics of slow chemical adsorption of a substance on a heterogeneous surface:

$$\frac{d\theta}{dt} = k \exp[-b\theta], \qquad \text{(VII.3)}$$

where k and b are constants. Written in integral form for a long adsorption time, when $kt/b \gg 1$, Eq. (VII.3) becomes

$$\theta = \frac{1}{b} \ln\left[\frac{k}{b}t\right]. \qquad \text{(VII.4)}$$

Figure 80 shows that an experimental plot of θ vs log t is virtually linear. It was also shown in the same investigation [76] that the rate of adsorption of allylphenylthiourea depends on the electrode potential, reaching the maximum value near the stationary potential (−0.1 V vs N.H.E.).

Fil'ko [54] used the radioactive tracer method for investigating exchange between adsorbed thiourea molecules on iron and thiourea molecules in $2\,N\,H_2SO_4$. The concentrations of radioactive thiourea previously adsorbed on the iron and of inactive

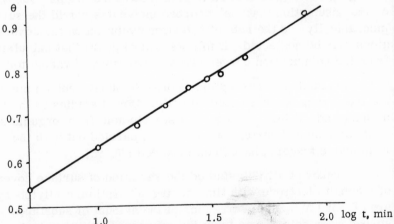

Fig. 80. Dependence of the surface coverage on the logarithm of the adsorption time, determined for electrolytic cobalt in 1 N H_2SO_4 with addition of 5 · 10^{-6} M of allylphenylthiourea at $\varphi = -0.3$ V vs S.H.E. (from data in [76]).

thiourea in 2 N H_2SO_4 were varied for this purpose. In solutions containing 10 moles of thiourea per liter, the activity of the metal fell sharply during the first hour after immersion in the solution. The decrease of the radioactivity of the metal was attributed mainly to exchange of adsorbate molecules between the electrode surface and the solution, rather than to decrease of the adsorbate concentration as the result of surface dissolution of the iron (see Section 8 of this chapter), because control experiments in pure 2 N H_2SO_4 solutions showed that the radioactivity of the metal alters little during the same time.

4. Influence of Mechanical and Thermal Treatment on the Adsorption Properties of Electrodes

Preliminary treatment has a substantial influence on the substructure of a metal. For example, rolling leads to definite orientation of the crystal faces relatively to the metal surface (surface grain orientation and increased concentration of dislocations). Annealing is accompanied by recovery and recrystallization processes, leading to considerable decrease in the concentration of lattice defects and to reorientation of the crystal faces. These changes in the state of the surface cannot fail to affect the adsorption and electrochemical properties of the metal, because the free surface energy changes with changes in texture and defect concentration. The influence of surface texture on electrochemical processes has been reviewed by Kochergin [88].

Increase of the corrosion activity of metals after deformation is noted in numerous publications. The mechanisms of intercrystalline and stress corrosion are closely linked with the appearance and existence of defects in the metal lattice. The influence of adsorption of surface-active substances on the mechanical strength of metals has also been widely investigated. Rebinder and his school [89, 90] have shown that adsorption of surface-active substances facilitates deformation and destruction of metals considerably. This effect is due to lowering of the free surface energy and to decrease of the work of formation of a new surface during deformation in presence of surface-active substances.

On the other hand, the effects of mechanical and thermal treatments on the adsorption properties of metals have been studied little, and only fragmentary information on the subject is available at present. Experimental material relating to various metals is collected in this section.

Conway et al. [91] noted the influence of annealing on adsorption of acridine on copper from aqueous solutions. Less acridine was adsorbed on the metal after it had been annealed. The authors attributed the decrease of adsorption after annealing to fusion of surface projections. Balashova and Zhmakin [92] found that deformation and heat treatment of platinum influenced adsorption of sulfuric acid. Adsorption of sulfuric acid on platinum plates subjected to bending deformation or on stretched platinum wire was greater by a factor of 10-20 than on the annealed metal. It was found that the adsorption activity of platinum is reduced considerably by annealing at 700-800°C. At such temperatures recrystallization of platinum proceeds at a considerable rate. The influence of preliminary treatment on adsorption has also been noted by Erbacher [93] and by Tompkins [94]. Campbell [86] found that up to 15% deformation of copper, iron, and aluminum leads to increased adsorption of stearic acid from cyclohexane. The amount of substance adsorbed increases linearly with the percentage deformation. If the deformed specimen was annealed at 600°C, adsorption decreased and was the same as on the undeformed metal. The rate of adsorption was also higher on the deformed than on the annealed metal. The increase in the adsorption of stearic acid was attributed to increased concentration of dislocation emergence points, which are active adsorption sites. Increased adsorption of stearic acid was also observed on aluminum powder after deformation [95]. Rein et al. [96] attempted to investigate the influence of deformation on the capacity of the double layer on a silver single crystal. Heat treatment of a metal may lead to a change in its dissolution mechanism [97]. Certain indirect data on the influence of dislocation concentration on the catalytic and adsorptional properties of metals have been obtained by Japanese scientists [98-101]. They studied decomposition of hydrogen peroxide, ethanol, diazonium salts, and other compounds on copper and silver which had been subjected to cold rolling followed by annealing at various temperatures. The catalytic activity of the metals decreased with increase of the annealing temperature. For two narrow temperature ranges the decomposition rates of these substances were found to fall sharply. This was attributed to decrease of the number of vacancies and dislocation emergence points on the metal surface. This occurs at higher temperatures than disappearance of vacancies.

Adsorption of organic compounds also depends on the crystallographic orientation of the crystal faces. Preferential growth of particular crystal faces often occurs during electrodeposition of metals in presence of surface-active organic substances [102, 103], indicating selective adsorption of the organic compounds. Vozdvizhenskii and Zainullina [104] studied adsorption of organic substances on zinc single crystals. They concluded from the different changes of potential on individual faces of a single crystal that surface-active substances are adsorbed better on the basal and prism faces. Anisotropy of the adsorptional properties of individual faces of single crystals was detected in studies of adsorption of maleic acid on silver [105], of thiourea on copper [106], of hydrogen on platinum [107], and of pyridine on gold [108]. Nikulin and co-workers [109-111] found that the reduction rates of a number of organic substances were influenced by the texture of polycrystalline metals and by the nature of the crystallographic faces of zinc, lead, and silver single crystals.

Examples of the influence of definite orientation of crystal faces and of atomic packing density in different crystallographic directions on the rates of catalytic processes are given in a paper by Rubinshtein [112].

Metal surfaces are usually subjected to mechanical cleaning for adsorption and corrosion determinations; such treatment causes some deformation of the surface layers. For example, Caplan and Cohen [113] showed that iron, first annealed at 900°C and then subjected to abrasive treatment, is oxidized more rapidly by oxygen on heating than annealed iron which had not been cleaned; the coarser the abrasive material used for cleaning the iron surface, the greater was the deformation of the surface layer of the metal and the degree of oxidation of the iron.

The influence of abrasive treatment on corrosion of nickel and iron in hydrochloric acid solutions in presence of inhibitors has been noted [114, 115]. The protective effect of the inhibitors increased with increasing particle size of the abrasive material.

Iofa, Batrakov, and Nikiforova [116, 117] studied the adsorptional properties of zone-refined iron* and Armco iron, cold-rolled and then annealed at 600 and 750-800°C for 6 h in hydrogen (200-300 mm). The first temperature corresponds to the start of recrystallization of iron. Recrystallization of iron usually begins at 500-600°C.† The recrystallization rate increases considerably when the temperature is raised to 750-800°C, and therefore it was to be expected that higher annealing temperatures would alter the adsorptional properties of iron appreciably. The influence of heat treatment on the adsorptional properties of iron was studied with adsorption of tribenzylamine (TBzA), allylphenyl-thiourea (APTU), and iodoethylquinoline (IEQ) from acid solutions as examples. Adsorption of TBzA in $1 N$ HCl was determined by ac measurement (at 800Hz) of the capacity of the electric double layer. The TBzA concentration was varied from 10^{-6} to $10^{-3} M$. Comparison of the results showed that TBzA is adsorbed better on the surface of zone-refined iron annealed at 600°C (type I) than on iron annealed at 750°C (type II). At $10^{-4} M$ TBzA concentration limiting coverage of the surface of iron I is reached, as the capacity of the double layer does not decrease appreciably on further increase of the concentration. In the case of type II iron adsorption does not reach the limit at $10^{-3} M$ concentration of TBzA. The decrease of the double-layer capacity at 10^{-3} TBzA concentration is $10 \ \mu F/cm^2$ ($30 \ \mu F/cm^2$ in the pure electrolyte) in the case of iron II, and $20 \ \mu F/cm^2$ ($30-33 \ \mu F/cm^2$ in the pure electrolyte) in the case of iron I.

Adsorption of APTU from $1 N$ H_2SO_4 was also appreciably greater on iron I. In Fig. 81 the capacity of the double layer and the surface coverage, calculated from Eq. (V.18), for irons I and II are plotted against the logarithm of the APTU concentration at $\varphi = -0.35$ V (N.H.E.). The calculated values of θ give a linear plot in these coordinates; therefore the surface coverage conforms to the Temkin isotherm. Whereas in the case of iron I limiting coverage is observed, apparently due to formation of a monolayer, θ for iron II corresponds to about 0.5 of a monolayer. In calculation

---

*The impurity contents were (%): zone-refined iron C < 10^{-3}, Mn < 10^{-5}, P < 10^{-3}, S < 10^{-3}, Si ~ 10^{-4}, N = 10^{-4}; Armco iron C = 0.04, Mn = 0.02, S = 0.03, P = 0.03.
†In general, the temperature at which recrystallization begins depends on the degree of deformation of the metal [118].

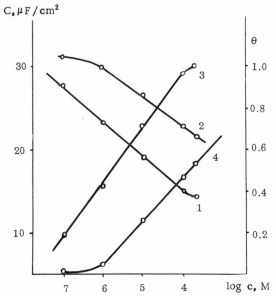

Fig. 81. Decrease of the capacity of the double layer and
calculated isotherms of allylphenylthiourea adsorption from
1 N H_2SO_4 on zone-refined iron annealed at 600°C (1, 3)
and at 750°C (2, 4) (from data in [117]).

of θ for iron II, the capacity found at limiting surface coverage
was taken as the value of C'.*

Similar isotherms were obtained in determination of adsorption
of IEQ from $1\,N$ H_2SO_4. The IEQ concentration was varied from
10^{-8} to $10^{-2}\,M$. The adsorption of IEQ on iron I reaches the limiting value when the IEQ content in the solution is $10^{-4}\,M$, and does
not change when the concentration is raised further to $10^{-2}\,M$. Adsorption of IEQ on iron II is not high even from solutions containing $10^{-2}\,M$ IEQ. Adsorption of the surface-active substances was

---

*Since different electrodes, differing in true surface area, were used for the capacity
measurements for plotting the complete adsorption isotherm, and the capacity values fluctuated in the range of 30-40 $\mu F/cm^2$ in pure 1 N H_2SO_4, for calculation of
θ the capacity values in the pure electrolyte solution were first converted to the
same value (30 $\mu F/cm^2$). It was assumed that the capacity of the double layer does
not differ for the different electrodes and that the capacity fluctuations are due to
differences in surface roughness.

accompanied by increase of the hydrogen overpotential and retardation of anodic dissolution of the iron.

The lowering of the adsorption activity of iron in these experiments could not be attributed to surface oxidation of the specimens during high-temperature annealing, as every precaution was taken to prevent entry of oxygen into the annealing cell. In order to chec the influence of the composition of the atmosphere in which the iron was annealed on the adsorptional properties of the metal, the annealing was performed in hydrogen, in argon, and in vacuo. The composition of the annealing atmosphere did not affect the adsorptional properties of iron. In addition, experiments were carried out in acid solutions, in which the iron surface was etched in order to remove any thin oxide film possibly present on the iron surface. The differences in adsorption activity of the specimens persisted after the acid treatment.

The difference in adsorption activity between iron I and iron II can be attributed to the influence of annealing on the structure of the metal and on the activity of the adsorption sites. Annealing is accompanied by recovery and recrystallization of the iron, leading to removal of residual stresses, changes of crystal orientation, and decreased concentration of dislocations and other lattice imperfections. It was found by x-ray diffraction analysis of iron I and II that some residual stresses persist after 6 h of annealing at 600°C, whereas iron II annealed at 750°C has more equilibrium structure. The interplanar spacing between the iron atoms after annealing at 600°C (d = 2.8643 Å) was appreciably less than the literature value (d = 2.8664-2.8661 Å). The value for iron II (d = 2.8661 Å) is close to literature data.

Nikiforova and Batrakov [120] studied changes in the adsorption activity of zone-refined and Armco iron after annealing over a wider temperature range (annealing time 6 h). The iron was previously cold-rolled. The adsorption activity was estimated from φ vs log i curves determined with the use of $1 N$ KOH containing $10^{-4} N$ Na_2S. The increase of hydrogen overpotential at constant current density (10^{-2} A/cm^2) plotted against the annealing temperature (Fig. 82) is a qualitative indication of changes in the adsorbability of S^{2-} ions on Armco iron. Figure 82 shows that the activity of the iron decreases in a stepwise manner with rise of the annealing temperature. A similar relationship was obtained

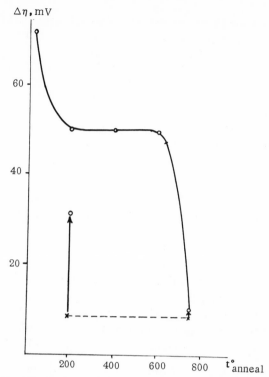

Fig. 82. Increase of the hydrogen overpotential of iron
on introduction of $1 \cdot 10^{-4}$ N Na$_2$S into 1 N KOH as a
function of the annealing temperature. The $\Delta\eta$ values
for oxidized specimens are indicated by crosses, and
for specimens etched after oxidation by arrows (from
data in [120]).

for zone-refined iron. At low annealing temperatures, up to 200°C,
the decrease of adsorption activity is due to decrease of the defect and stress concentration in the metal. The adsorption activity
of iron alters little over the second region of the curve, between
200 and 600°C. The texture of the metal remains unchanged. The
main change in the adsorption activity occurs in the temperature
range from 600 to 750°C, and is caused by recrystallization, which
is accompanied by changes of crystallite orientation and decrease
in the dislocation concentration.

Grain reorientation may change the adsorption properties of
metal significantly. For elucidation of the adsorption activities

TABLE 14. Increase of the Hydrogen
Overpotential (mV) on Introduction
of Surface-Active Anions

Solution	Crystal face		
	(100)	(110)	(111)
$1N$ KOH $+ 1 \cdot 10^{-4}$ N Na$_2$S	50—70	30	~5
$2N$ H$_2$SO$_4 + 1 \cdot 10^{-4}$ N KI	90	—	5—10

of individual faces, Iofa, Batrakov, and Nikiforova [117] determined
polarization curves at the (100), (110), and (111) faces of a steel
single crystal of the following composition (wt.%): C = 0.0093;
Si = 3.04; Mn = 0.13; S = 0.0064; P = 0.0075; Cr = 0.025; Cu = 0.12;
N = 0.012; O = 0.007. The polarization curves were recorded in
$1 N$ KOH and in the same solution with addition of $10^{-4} N$ Na$_2$S. It
was found that introduction of Na$_2$S has different effects on the
hydrogen overpotential (η) at the (100), (110), and (111) faces
(Table 14). The same table includes data on the influence of I$^-$
ions on the hydrogen overpotential in sulfuric acid solutions.

It is seen that the (100) face is the most active in the adsorption process.

Study of the influence of deformation of an annealed metal on
its electrochemical and adsorptional properties is of considerable
interest. It is known that deformation increases the concentration
of defects (in particular, the dislocation density within the metal
and the number of dislocation emergence points on the surface).
Polarization and adsorption measurements were carried out with
steel wire* and with zone-refined iron. The steel wire was deformed by drawing, and some of the specimens were then annealed
at 800°C for 8 h. Determinations of cathodic polarization curves
in $1 N$ H$_2$SO$_4$ with addition of $10^{-4} M$ IEQ showed that adsorption of
the organic compound was greater on the deformed metal than
on annealed specimens. The hydrogen overpotential in $1 N$ H$_2$SO$_4$
was the same for both types of specimens. On addition of $10^{-4} M$
IEQ the hydrogen overpotential increased by 65-70 mV on the de-

*The steel had the following composition (wt.%): C = 0.18; Mn = 0.4; Cr = 0.2; S =
0.025; P = 0.028; Si = 0.14; Ni, trace.

formed wire, and by only 30-35 mV on the annealed specimens. Determinations of the double-layer capacity on annealed and deformed steel wire also showed that adsorption of IEQ was greater on deformed specimens.

In another experiment, zone-refined iron wire previously annealed at 800°C was stretched by 3, 5, and 10%. The extension was performed at the constant rate of 0.03 mm/min in a special apparatus. The stretched specimens were then used for determination of polarization and differential capacity curves. Changes of the hydrogen overpotential ($\Delta \eta$) on addition of 10^{-5} M IEQ to 1 N H_2SO_4 are shown in Table 15.

It follows from the table that the adsorption activity of iron increases with the degree of deformation. This result is confirmed by differential capacity curves. After addition of 10^{-4} M IEQ to 1 N H_2SO_4 the capacity of the double layer (35 $\mu F/cm^2$ in the pure electrolyte) fell to 27 $\mu F/cm^2$ (10% deformation) and to 34 $\mu F/cm^2$ (annealed wire). The greater adsorption of IEQ on annealed steel wire in comparison with annealed zone-refined iron wire is probably attributable to the higher dislocation concentration in the steel, as carbon, phosphorus, and other impurities tend the preserve the dislocations during annealing. The results of selective etching [121] and electron-microscopic studies [122] make it possible to derive an empirical expression for the dependence of dislocation density (m) on the degree of deformation of pure iron (Δl):

$$m = m_0 + k\Delta l^{\alpha}, \tag{VII.5}$$

where m_0 is the dislocation density in the annealed specimen and

TABLE 15. Increase of the Hydrogen Overpotential on Iron after Addition of 10^{-5} M IEQ to 1 N H_2SO_4

Deformation	$\Delta \eta$, mV
0	0
3%	5-10
5%	30
10%	60-70

k and α are constants. This expression is valid in the range of relative deformations $10^{-3} < \Delta l < 10^{-1}$. The values of k and α, taken from the literature [122], where $k = 2.2 \cdot 10^9$ and $\alpha = 0.8$. The dislocation density in annealed metal crystals usually lies in the range of 10^6-10^8 per cm^2. According to Iofa et al. [117], the concentration of dislocation emergence points in zone-refined iron wire annealed at $830°C$ is 10^7 cm^{-2}. This value was taken tentatively as m_0. Calculations with the aid of Eq. (VII.5) gave the following values for the dislocation concentrations at different degrees of deformation: 3%, $1.3 \cdot 10^8$; 5%, $1.9 \cdot 10^8$; 10%, $3.3 \cdot 10^8$.

Increase of the adsorption activity of iron might possibly be attributed to formation of characteristic steps at the crystal glide planes as the result of deformation. Atoms at the edges of the steps may be more active adsorption sites than dislocations. However, annealing of strained iron at 750-800°C leads to a sharp decrease of its adsorption activity whereas the surface microrelief alters to a lesser degree at this temperature. It follows that appearance of steps is not the main cause of the changes in the adsorptional properties of iron. It could also be supposed that deformation by stretching leads to formation of microcracks which tend to increase adsorption and to enlarge the true electrode surface area, as suggested by Foroulis and Uhlig [123, 124] in their explanation of the increase in the rate of iron dissolution after deformation, observed by Green and Saltzman [125]. Our determinations of double-layer capacity showed that the true surface area of the wire specimens is not increased appreciably by deformation up to 10%. The annealed and deformed metal specimens gave similar capacity values.

It is also reported in the literature that deformation of polycrystalline metals by stretching leads to definite predominant crystal orientation on the specimen surface [126]. However, the predominant orientation usually becomes noticeable in a polycrystalline specimen only after decrease of its cross section by 10-20% as the result of deformation. In our case the change in the thickness of the zone-refined iron specimen as the result of stretching deformation was incomparably smaller. Therefore the increase in the adsorption activity of iron after tensile deformation must evidently be attributed to increased dislocation concentration.

On the basis of the foregoing material and literature data, the changes in the surface structure of polycrystalline iron may be

TABLE 16. Influence of Heat Treatment on the Rate of Iron
Corrosion in Presence of Surface-Active
Substances

Solution	Treatment temp.	Corrosion rate, $g/cm^2 \cdot h \cdot 10^5$	γ
1 N H$_2$SO$_4$	600°C	20	—
+ 1 $\cdot$ 10^{-5} N KI	600°C	1.8	11.1
+ 1 $\cdot$ 10^{-6} N KI + + 1 $\cdot$ 10^{-4} M $N(C_6H_5CH_2)_3$	600°C	1.0	20
1 N H$_2$SO$_4$	750°C	8.0	—
+ 1 $\cdot$ 10^{-5} N KI	750°C	6.0	1.3
+ 1 $\cdot$ 10^{-6} N KI + + 1 $\cdot$ 10^{-4} M $N(C_6H_5CH_2)_3$	750°C	2.5	3.2

represented as follows. After cold rolling of iron, its crystals
become predominantly oriented with the (100) face, which is the
most active with respect to adsorption, toward the surface, and
the dislocation concentration increases. This hypothesis is supported by literature data on grain orientation of transformer steel
and iron surfaces as the result of rolling [127-129]. Recrystallization occurs during annealing at 750°C and higher temperatures, and
the crystalline structure of the metal tends to a more equilibrium
state. The concentration of lattice defects decreases, and partial
reorientation of the crystal faces occurs so that faces of lower
surface energy are oriented toward the metal surface. These effects lead to decrease of the adsorption activity of iron after high-
temperature annealing. However, at present it is not possible to
decide on the basis of available data whether crystal orientation
or dislocation concentration has the greater influence on adsorption of surface-active substances.

Changes in the adsorbability of organic substances must lead
to changes in their influence on the rate of iron corrosion. This
is confirmed by the data in Table 16, which gives values of the retardation coefficient of iron corrosion $\gamma = i_0/i$, where i_0 is the
rate of iron corrosion 1 N H$_2$SO$_4$, and i is the corrosion rate in 1 N
H$_2$SO$_4$ in presence of the inhibitor.

A fact of undoubted interest is that zone-refined iron and Armco iron annealed at 600°C and lower temperatures dissolve more rapidly in 1 N H_2SO_4 and more slowly in the same solution with added inhibitor than specimens annealed at 750° and higher temperatures.

The results reported by Trabanelli, Zucchi, and Zucchini [114] and by Talbot [115] can be explained in the light of the foregoing views on adsorption and corrosion data on iron subjected to various mechanical and thermal treatments.

In conclusion, a number of factors which may obscure the influence of texture and surface defects on the adsorptional properties of iron should be mentioned.

As already noted, adsorption of cationic and anionic organic compounds depends on the electrode potential in most cases. There fore at sufficiently negative or positive potentials relative to the potential of zero charge the influence of this factor on adsorption may be more pronounced than the influence of the metal substructure.

In studies of the influence of preliminary treatment on the adsorptional properties of metals, selection of the optimal concentrations is of great importance, because at high enough concentrations considerable adsorption of surface-active substances occurs both on active specimens and on specimens of low activity.

Iron belongs to the group of metals which are easily oxidized in air. The oxides are formed primarily on regions of high free surface energy, and thereby block the most active adsorption sites. Therefore the difference between the adsorption properties of iron specimens annealed at high and low temperatures diminishes after prolonged exposure to air. Figure 82 demonstrates the influence of oxidation as the result of exposure of an iron specimen to air for 10 min at 100°C on its adsorption properties. The specimen was previously annealed in hydrogen at 200°C for 6 h. The adsorption activity fell sharply after oxidation, whereas the activity of specimens previously annealed at 750°C altered little after oxidation. Iron annealed at low temperatures recovers its adsorption activity after treatment of the oxidized electrodes in 1 N H_2SO_4 (indicated by an arrow in the diagram). Electrodes annealed at 750°C retain low adsorption activity even after acid treatment.

5. Dependence of Adsorption
on Structure

The influence of structure of organic compounds on their adsorption from aqueous solutions has not been studied systematically.

Several investigations of the adsorption of organic substances from organic solvents have been reported. Hackerman et al. [68, 130-133] and other investigators [85, 134-138] have shown that in most cases there is a correlation between the adsorbability of surface-active substances from organic solvents and their inhibiting action in aqueous solutions. Adsorption from organic solvents was generally found to be multimolecular. After the adsorbent had been washed with the solvent, chemisorbed particles which occupied a fraction of a monolayer remained on the metal surface. It was found that in the case of compounds of high molecular weight (thiols, amines, alcohols, and nitriles) the amount of chemisorbed substance was independent of the length of the hydrocarbon chain [137].

Investigations of the influence of organic compounds on metal corrosion have been much more numerous. Lowering of the reaction rate (dissolution of the metal, hydrogen evolution, etc.) may give a qualitative indication of the adsorbability of organic substances. However, it is not our purpose here to examine in detail the influence of the structure of organic compounds on the reaction kinetics, as the experimental data accumulated in this field could form the subject of a special monograph, and we will confine ourselves to a concise presentation of certain conclusions.

In publications on this subject [130-150] it is noted that adsorption of surface-active organic compounds increases with their molecular weight and dipole moment. The different classes of aliphatic compounds form the following series in order of decreasing adsorbability: acids > amines > alcohols > esters. Organic molecules containing active electron-donor groups ($-CN$, $-CNS$, $-CNO$, $=CO$, $-CHO$, $-NH_2$) are chemisorbed on the surface of metals having incomplete electron orbitals (e.g., d-orbitals).*

* According to Hayward and Trapnell [151], a sufficient condition for chemical bonding is a high value of the work function for the metal and a low ionization potential of the adsorbate or, conversely, a low value of the work function and a high ionization potential. Thus, the possibility of formation of chemisorption compounds is not necessarily dependent on the existence of incomplete electron orbitals on the one hand and the presence of a free electron pair on the other.

Substitutents in the hydrocarbon chain influence the adsorption activity and inhibiting action of organic compounds. Their influence diminishes with increasing length of the hydrocarbon chain betweer the substituent and the atom which is the adsorption-active center of the molecule. Derivatives of the ethylene [71] and especially of the acetylene [152-154] series have higher adsorption activity owing to interaction of π-electrons with the surface atoms of the adsorbent. This effect was first detected by Gerovich [155, 156] during studies of the adsorption of aromatic hydrocarbons on mercury.

Attempts have been made to establish quantitative relationships between the structure of organic compounds and their adsorbability of inhibiting properties. The relationship was based on the direct connection between the electron density of the adsorption-active atom and the adsorbability of the organic compound at the metal − solution interface.

The Hammett relation has been used by a number of workers [131, 157-165] for this purpose. This semiempirical relation is used in organic chemistry for correlation of chemical reaction rates [166], and may be written in the form

$$\log \frac{k_R}{k_0} = \rho \sigma, \qquad (VII.6)$$

where k_R and k_0 are the rate constants (equilibrium constants) for the reactions of the substituted and unsubstituted compound respectively, and ρ and σ are correlation constants.

The constant σ characterizes the influence of the substituent on the electron density of the reaction center of the molecule, and is independent of the nature of that center and of the structure of the molecule. For electron-acceptor substituents ($-F$, $-Cl$, $-Br$, $-COCH_3$, $-COOH$, etc.) σ is positive, and for electron-donor substituents ($-NH_2$, $-N(CH_3)_2$, $-OCH_3$, etc.) it is negative. The constant ρ represents the influence of the substituent on the electron density of the reaction center, with the molecular structure taken into account.

Donahue and Nobe [159] regarded adsorption as a reversible chemical reaction:

$$RY + M \rightleftarrows RY \ldots M, \qquad (VII.7)$$

where RY is an organic compound with adsorption center Y and substituent R; M is the metal, and RY...M is the chemisorption compound on the metal surface. Then at c = 1 we can write for the adsorption equilibrium

$$k_R = \frac{\theta}{1 - \theta} . \qquad \text{(VII.8)}$$

In the case of an unsubstituted organic compound the surface coverage will be θ_0, and correspondingly k_0 can be represented by the same Eq. (VII.8).

From Eqs. (VII.6) and (VII.8) it is easy to derive the equation

$$\log \frac{\theta (1 - \theta_0)}{\theta_0 (1 - \theta)} = \rho \sigma^1 .^* \qquad \text{(VII.9)}$$

However, Eq. (VII.9) has not been verified experimentally for metals.

With the aid of simple transformations and certain simplifications, Donahue and Nobe [159] correlated the Hammett equation with the corrosion characteristics of metals. The following expression was obtained:

$$\log R = \rho \sigma , \qquad \text{(VII.10)}$$

where R is the ratio of the corrosion rates of the metal in presence of the inhibitor and in the pure solution. Equation (VII.10) is valid for the same concentrations of organic compounds of a given series, and held for methyl-substituted pyridines inhibiting the dissolution of iron in aqueous solutions [167]. However, the inhibiting effect of aniline and its derivatives on corrosion of iron and nickel increases with increasing electronegativity or electropositivity of the substituent [168]. It is possible that in this case the transition from electronegative to electropositive substituents is accompanied by reorientation of the substituted aniline molecules in the double layer. Grigor'ev and Ékilik [164] attribute this type of dependence of the inhibiting action of organic substances on the nature of the substituent, with a minimum corresponding to the unsubstituted organic compound, to participation of two forms of the

---

*Equation (VII.9) can be derived for the general case where the concentration of the substance in solution is c. For this the concentrations of the unsubstituted compound and the compound with substituent R must be equal.

organic compound, molecules (RNH_2) and cations (RNH_3^+) in the adsorption process. Amines are adsorbed in either of these forms, dependent on the influence of the substituent on the electron density of the nitrogen atom. On the one hand, chemisorption of RNH_2 molecules and the strength of their bonds with the surface metal atoms increase with increasing electron-donor properties. On the other hand, increase of the electron density of the nitrogen atom lowers the effective density of the positive charge at the nitrogen atom in the RNH_3^+ ion and diminishes its adsorbability.

The foregoing views on the influence of structure on adsorption of organic compounds are faulty in a number of respects.

1. The role of the double layer is disregarded. The field of the electric double layer should have a considerable influence of the electron density of the adsorption center of the molecule.

2. The nature of the adsorbent and the role of the solvent are not taken into account.

3. It is assumed that adsorption interaction occurs between only one atom in the organic molecule and the metal surface; this is not always true.*

---

*Organic molecules are often adsorbed in flat orientation on the metal surface. This applies primarily to aromatic compounds, which are adsorbed as the result of π-electron interaction of their molecules with the surface [5, 169]. The possibility of π-electron interaction with the metal surface is indicated by the results of determinations of the electronic work function of nickel in presence of adsorbed benzene [170]. Derivatives of unsaturated hydrocarbons and polyfunctional organic molecules can also be adsorbed through several atoms on the metal surface. Chemisorption of trithione methosulfates having the general formula

$$R_1-C=\!\!=C-R_2$$
$$| \qquad |$$
$$S \qquad C=S$$
$$\diagdown \diagup$$
$$S$$

according to Grubitsch and Hilbert [171], occurs by electron interaction of the

$-S-\overset{|}{C}=S$ group with neighboring metal atoms at the (110) and (100) faces, as the distance between the S atoms in this group is equal to the distance between neighboring atoms in the iron lattice.

4. Heterogeneity of the adsorbent surface is ignored.

5. The role of the steric factor in adsorption is not taken into consideration. Steric effects were observed, e.g., in studies of the inhibiting action of phenylimidazoles on the corrosion of iron [163].

6. It is assumed in the derivation of Eq. (VII.10) that inhibition depends only on mechanical shielding of the surface.

7. Equations (VII.9) and (VII.10) are probably valid only if the correlation is performed for organic compounds whose molecules occupy equal areas in adsorption.

Other attempts to establish a connection between the structure and adsorbability of organic compounds have been reported in the literature.

On the basis of NMR spectrum data [172], Cox et al. [172] used the magnitude of the chemical shift of proton resonance frequency in the NH_2 group as a measure of the electron density of the nitrogen atom. They observed fairly good correlation between the chemical shift and the coefficient of inhibition of steel corrosion in presence of methyl- and halosubstituted anilines. However, the NMR spectra were measured in an organic solvent, and the influence of the nature of the solvent on the chemical shift was not taken into account when aqueous solutions were considered.

In a recent publication [173] an attempt was made to correlate changes in the magnetic susceptibility of the adsorbent due to adsorption of organic compounds with the chemical structure of the adsorbed molecules. In distinction from the approach used by Donahue and Nobe, in this investigation account was taken of the degree of polarization of the electron cloud of the adsorbed species, which is determined by the specific nature of the adsorbate—adsorbent interaction.

The influence of the electron structure of the adsorption center in the molecule on the adsorption properties of organic compounds is a problem of undoubted interest, but considerable difficulties of an experimental and theoretical nature arise in its investigation.

6. Influence of Adsorbed Reaction Products and Intermediates on Adsorption of Organic Substances

Electrochemical processes at solid metals are often accompanied by adsorption of intermediate reaction products, or by appearance of the reaction product on the electrode surface owing to its low solubility and slow removal. Coverage of the surface with adsorbate or changes in the nature of the electrode surface. have a considerable influence on adsorption of organic substances. The appearance of an intermediate compound on the electrode surface depends on the ratio of the adsorption and desorption rate constants.

In cathodic polarization of metals, hydrogen evolution comprises several stages. Dependent on the pH of the medium, hydrogen is evolved as the result of discharge either of hydronium ions or of water molecules at the electrode surface. Adsorbed hydrogen atoms, formed by the reaction (VII.11), are partially ionized again, or removed from the surface by an electrochemical (VII.12) or a recombination (VII.13) mechanism:

$$H_3O^+(H_2O) + e \rightarrow H_{ads} + H_2O(OH^-), \qquad (VII.11)$$

$$H_{ads} + H_3O^+(H_2O) + e \rightarrow H_2 + H_2O(OH^-), \qquad (VII.12)$$

$$H_{ads} + H_{ads} \rightarrow H_2. \qquad (VII.13)$$

The amount of atomic hydrogen on the metal surface depends on the ratio of the rate constants of the above processes [174-176].

If discharge of hydronium ions (VII.11) is slow, coverage of the surface with hydrogen is low; this occurs on iron, nickel, and cobalt in acid solutions. In this case the measured differential capacity is close in magnitude to the capacity of the electric double layer and is virtually independent of the potential [25, 28, 29, 76, 77, 177-179].

In alkaline solutions, the slow step on iron, nickel, and cobalt is removal of hydrogen by mechanisms (VII.12) and (VII.13); the differential capacities are substantially higher than the double-layer capacity and depend on the potential [177-181]. Investigations of hydrogen diffusion through an iron membrane in alkaline solutions confirm these views [175, 182, 183]. Adsorption of hydrogen on metals of the iron group in alkaline media begins at potentials on the positive side of the stationary potential of the electrode. At the same time, OH⁻ ions are adsorbed at these potentials on iron, nickel, and cobalt surfaces, since dissolution of metals of the iron group, as was first shown by Kabanov and Leikis [184, 185], proceeds with formation of an intermediate compound MOH_{ads} which is in equilibrium with OH⁻ ions:

$$Fe + OH^- \rightleftarrows FeOH_{ads} + e. \qquad (VII.14)$$

Thus, in a certain range of potentials, H atoms and OH⁻ ions are present on the electrode surface. Figure 83 shows the dependence of the differential capacity on the potential of zone-refined iron in $1\,N$ KOH (ac frequency 800 Hz). The potential is referred to the

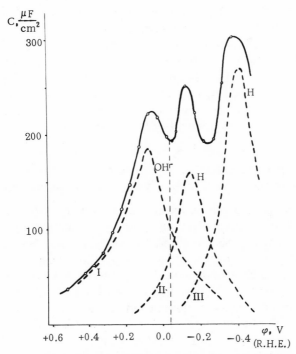

Fig. 83. Differential capacity curves of iron in 1 N KOH. The vertical dash line corresponds to the position of the stationary potential.

potential of a hydrogen electrode in the same solution. The C vs
φ curve can be subdivided into three independent curves, corre- .
sponding to adsorption of OH⁻ ions with formation of FeOH (I) and
to adsorption of hydrogen with high (II) and low (III) bonding energy with the iron surface. A similar curve was obtained earlier
for nickel by Weininger and Breiter [181], but the maxima on the
C vs φ curve were less pronounced. The low adsorbability of organic substances on iron and nickel in alkaline solutions can be
attributed to the presence of adsorbed particles (OH⁻, H), which
saturate the free valence bonds of the surface metal atoms and
lower the energy of adsorption of organic compounds (tribenzylamine, tetrabutylammonium sulfate, naphthalenesulfonic acid,
octyl alcohol, caproic acid) [180].

Other explanations of the low adsorbability of organic substances have also been offered. Turapov and Murtazaev [186] attribute it to a high negative surface charge, when alkali-metal
cations and water molecules displace organic molecules from the
electric double layer. It is assumed that the potential of zero
charge of iron determined in acid solutions ($\varphi_n = -0.37$ to -0.40 V)
remains the same in alkaline solutions, i.e., the potential difference between φ_{st} and φ_n is about -0.5 V. However, as is shown in
Chapter IX, adsorbed hydrogen atoms and OH⁻ ions may influence
the position of the point of zero charge, and the φ_n values in acid
and alkaline solutions probably do not coincide.

Other investigators [180, 187, 188] consider that low adsorbability of organic compounds is due to the presence of oxides on
the metal surface, which intensify its hydrophilic properties.

Only compounds which form chemical bonds with the metal
or its oxidation products during anodic polarization are adsorbed
on iron in alkaline solutions. For example, it has been shown that
organic substances (phenol, tannin, alizarin) which form chemical
compounds with iron ions are adsorbed well on iron under these
conditions [180, 188]. Losev and Kabanov [180] attributed adsorption of these substances to formation of chemical compounds with
hydrated iron oxides. In particular, they suggested that alizarin
forms an iron alizarinate of the composition

$$\text{Fe} \overset{\displaystyle \diagup \text{OH}}{\underset{\displaystyle \diagdown \text{Az}}{-\text{OH}}}$$

Organic substances are adsorbed somewhat better on nickel than on iron in alkaline solutions [189].

In accordance with the mechanisms postulated by Heusler [190] and Bockris [191], dissolution of iron in sulfuric acid solutions is accompanied by formation of an intermediate adsorption product MOH, which is in equilibrium with OH⁻ ions. Therefore a pseudo-capacity appears on the anodic region of the C vs φ curve (see Fig. 71), due to the reversibility of the same process as that occurring in alkaline solutions (VII.11) [28, 77, 179, 192, 193] or to the reaction

$$Fe + 2H_2O \rightleftarrows FeOH^-_{ads} + H_3O^+ + e. \qquad (VII.15)$$

The same dissolution mechanisms as for iron have been proposed for dissolution of cobalt [194, 195]. According to these mechanisms, the exchange currents of the first stage of reaction (VII.14) or (VII.15) are high in comparison with the rates of the subsequent dissolution stages, and, as in the case of iron, increase of the pseudocapacity is observed in the anodic region of the differential capacity curve [77]. In contrast to iron and cobalt, stage (VII.14) on nickel is virtually irreversible and is the limiting stage of the dissolution process. No appreciable accumulation of the intermediate compound NiOH occurs on the metal surface. In this case there is no increase of capacity in the anodic region of the C vs φ curve [77]. It has recently been suggested [197] that the first stage (VII.14) of nickel dissolution is also a quasi-equilibrium stage. However, this model is inconsistent with the results of differential capacity measurements.

According to Heusler [190], the coverage of the surface by FeOH particles, which catalyze dissolution of iron, is not high, and therefore they should not have any direct influence on adsorption or desorption of organic compounds. According to Bockris [191], coverage of the surface by the intermediate compound FeOH approaches unity if this compound is removed slowly during further dissolution of the iron. In this case the adsorbed OH⁻ ions may displace the adsorbed molecules from the metal surface. However, the available experimental material is as yet insufficient for deciding which mechanism is to be preferred.

Oxides which are not removed completely during cathodic polarization are formed on the surface of metals of the iron group during contact with atmospheric oxygen. The presence of oxides on the metal surface influences its adsorption properties. In aqueous solutions oxides render the metal surface more hydrophilic. The wettability of nickel and some other metals by water increases sharply in presence of oxides [198]. Increase of the energy of adsorption of water molecules diminishes adsorption of organic compounds.

In general, oxides also decrease adsorption of surface-active substances from organic solvents. For example, the magnitude and rate of adsorption of 4-ethylpyridine from cyclohexane and of stearic acid from benzene on iron powder previously oxidized by a small amount of oxygen are lower than on reduced iron powder [83, 135]. Organic substances (n-propylamine, benzene) are also adsorbed less well from the vapor phase on oxidized metal surfaces [199].

Schwabe [200] notes that appearance of oxides on a metal surface may lead to a change in the character of adsorption of organic substances. Dibenzyl sulfoxide and di(cyclohexyl)ammonium nitrite are adsorbed on Armco iron and retard its dissolution in acid and neutral solutions. Organic molecules are retained on an oxidized iron surface only by van der Waals forces. Therefore in aqueous solutions the above-named inhibitors are displaced from the metal surface by solvent molecules and do not exert an inhibiting effect.

Balezin et al. [201] found that benzoate ions are adsorbed better on iron than on magnetite from aqueous solutions. On the other hand, di(cyclohexyl)ammonium nitrite was adsorbed better on magnetite.

The action of pickling inhibitors depends on the different adsorbabilities of surface-active substances on clean and oxidized metal surfaces. Machu [141] showed in the case of gelatin that the selective dissolution of oxides from the iron surface is due to lower adsorbability of the inhibitor on the oxidized surface than on the clean metal surface.

7. Chemical Changes of Adsorbed Substances on Electrode Surfaces

Adsorption of a number of organic compounds on metals of the iron group is accompanied by chemical changes of the ad-

sorbed particles. These changes depend on the electrode potential and are often associated with electrochemical processes
occurring on the electrode. Appearance of new species on the
electrode surface alters the inhibiting influence of organic compounds on electrode reactions (secondary inhibition). In the general case, molecules of the original substance and their conversion products are both present on the electrode surface when chemical changes occur.

Horner et al. [202] found that certain onium salts undergo
reduction during cathodic polarization of iron. For example, tri-
phenylbenzylphosphonium and triphenylbenzyllarsonium ions are
converted into triphenylphosphine and triphenylarsine respectively:

$$[(C_6H_5)_3P-CH_2-C_6H_5]^+ \to (C_6H_5)_3P, \qquad \text{(VII.16)}$$

$$[(C_6H_5)_3As-CH_2-C_6H_5]^+ \to (C_6H_5)_3As. \qquad \text{(VII.17)}$$

The reduction rates of these compounds are higher at more
negative potentials. Lorenz and Fischer [203] showed with the aid
of galvanostatic pulse measurements and steady state polarization
curves that the degree of adsorption and inhibiting action of these
onium compounds and of their reduction products on iron in $1\,N$
HCl are approximately equal.

The degree of adsorption of a reduction product rises sharply
with decrease of its solubility relative to the original substance.
Dibenzyl sulfoxide is adsorbed weakly on iron from acid solutions,
and raises the overpotential of electrochemical processes but
slightly. Hydrogen evolved at the electrode surface reduces the
sulfoxide to the less soluble sulfide. According to Schwabe [200],
iron catalyzes this process. Adsorption of dibenzyl sulfide is considerably greater than that of the sulfoxide and increases rapidly
with time, leading to formation of multimolecular layers. Adsorption of dibenzyl sulfoxide on nickel was studied by Smialowski and
Szklarska—Smialowska [204].

Aldehydes and unsaturated hydrocarbons are also reduced by
the hydrogen evolved during cathodic polarization and at the stationary potential [205]. The alcohols formed by reduction of aldehydes are adsorbed less on metals, and the inhibiting effect of
aldehydes gradually diminishes in consequence.

It is reported in the literature [153, 206] that, in addition to
reduction, unsaturated compounds (acrylonitrile, acetylene, etc.)

tend to undergo polymerization on iron. Saturated organic compounds containing oxygen were found in the polymeric coatings formed [207, 208], and it was established that hydrogen evolved during dissolution of iron in the acid or during cathodic polarization is an essential component of the polymerization reaction. When polymer molecules are formed, chemical bonding occurs between their functional groups and metal oxides (with formation of iron alcoholates and carboxylates) [208]; the source of oxygen in formation of such bonds may be water molecules, as in adsorption of, e.g., acetylene. It is interesting to note that chemical conversions of different organic compounds (acetylene, propargyl alcohol, ethenylcyclohexane) adsorbed on iron lead to formation of polymers containing the same functional groups.

Chemical conversions of organic substances used in electroplating technology for production of smooth bright deposits (thiourea, coumarin, etc.) have been studied more fully (e.g., see [209]).

As already noted in Section 2 of this chapter, adsorption of thiourea on metals of the iron group is accompanied by decomposition with formation of hydrogen sulfide [37, 54-73]. According to Iofa [37], decomposition of thiourea in acid solutions proceeds by the following scheme:

$$(NH_2)_2CS + 8H^+ + 6e \rightarrow H_2S + CH_3NH_3^+ + NH_4^+ \dots . \qquad (VII.18)$$

Decomposition of thiourea increases with increasing rate of iron corrosion [54] or with increasing rate of hydrogen evolution during cathodic polarization [55]. In the latter case the thiourea concentration on the iron surface decreases as the electrode potential is shifted in the negative direction.

Thiourea derivatives also decompose on metals of the iron group with formation of hydrogen sulfide. However, the decomposition mechanism of these compounds has been studied less fully. The following scheme has been proposed for phenylthiourea [29]:

$$C_6H_5NCSNH_2 + 8H + 2H^+ \rightarrow C_6H_5NH_3^+ + CH_4 + H_2S + NH_4^+ \dots . \qquad (VII.19)$$

Rogers and Taylor [210] found that coumarin, cinnamic alcohol, propargyl alcohol, and butyne-2-diol-1,4 are reduced during electrodeposition of nickel. The experiments were conducted in a solution containing $NiSO_4$, $NiCl_2$, and H_3BO_3 (Watts bath). The reduction products were analyzed by chromatographic, IR spectro-

scopic, and NMR methods. The reduction of coumarin (I) proceeds
by two parallel reactions. One of these yields melilotic acid (II),
and the other, hydroxyphenylpropanol (III), which is subsequently
reduced to o-propylphenol (IV):

$$(VII.20)$$

Cinnamic alcohol is reduced to phenylpropane, and propargyl alcohol to propane. Reduction of butyne-2-diol-1,4 is accompanied
by formation of several products (n-butane, cis- and trans-
butene-2). The strong influence of the solution pH on the yield
and nature of the products is noted. The reduction is more effective at pH = 1.5 than at pH = 4. At high pH values reduction of OH
groups almost ceases.

Kandler, Roemer, and Häusler [189] found that adsorption of
n-butanol, isobutanol, and glycol from 1 N KOH on nickel is accompanied by chemical changes of these substances. At $\varphi > +0.2$ V
(vs a reversible hydrogen electrode in the same solution) glycol
is oxidized, and its oxidation products are adsorbed on the metal
surface.

8. Influence of Dissolution and Electrodeposition of Metals on the Concentration of Adsorbed Substances

Up to now, our examination of the influence of electrochemical
processes on adsorption of organic substances on metals of the
iron group has been confined to two cases: 1) influence of adsorbed reaction products and intermediates on adsorption of or-

ganic compounds; 2) influence of electrochemical processes on chemical conversions of the adsorbed substances.

Apart from these, another important factor influencing the concentration of an adsorbed surface-active substance is surface renewal by dissolution and electrodeposition of the metal. The rates of these processes depend on the electrode potential. With fairly rapid renewal of the metal surface the rate of adsorption of the organic substance becomes comparable with the rate of formation of the new surface, and the adsorbate concentration falls.

a. Dissolution of Metals. The slope of the anodic branches of the φ vs log i polarization curves for iron, cobalt, and nickel decreases sharply in presence of adsorbed inorganic cations (S^{2-}, I^-, Br^-), organic substances of the cationic type, and other surface-active substances [3, 27-30, 211-213] (see Fig. 84). At considerable anodic currents of metal ionization the adsorbed particles are removed rapidly from the electrode surface together with metal atoms. The overpotential of the anodic process diminishes and the polarization curves recorded in the pure acid solution and in presence of the surface-active substance come closer together. The dissolution process was examined in the first approximation by Heusler and Cartledge [211] for I^- and CO adsorbed on iron. The relationships found can probably be used for explaining the desorption of organic substances in general. The authors cited [211] made the following assumptions:

1) the metal surface is uniform, and adsorption can be represented by the Langmuir isotherm;

2) the ψ_1-potential resulting from adsorption of the particles is small and may be neglected;

3) the OH$^-$ ion activity a_{OH^-} is approximately equal to c_{OH^-}.

According to Heusler [190], the metal dissolves in two stages:

$$Fe + OH^- \rightleftarrows FeOH_{cat} + e,$$
$$FeOH_{cat} + Fe + OH^- \rightarrow FeOH_{cat} + FeOH^+ + 2e. \qquad (VII.21)$$

At the first reaction stage a catalyst $FeOH_{cat}$ is formed on the iron surface, and accelerates the second (potential-determining) stage. Assuming that in the steady state

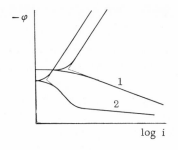

Fig. 84. Schematic polarization curves: 1) in pure acid solution; 2) in solution with added organic substance.

$$i_1 = i_2 + i_3 \,, \qquad\qquad\qquad \text{(VII.22)}$$

where i_1 is the adsorption flux, i_2 is the desorption flux under equilibrium conditions, and i_3 is the desorption flow due to dissolution of the metal, Heusler and Cartledge [211] derived a relationship between φ and the degree of surface coverage by the organic substance, θ:

$$\varphi = \frac{RT}{2\beta F} \ln\left[\left(1 - \frac{k_2}{k_1}\cdot\frac{1}{c_R}\cdot\frac{\theta}{1-\theta}\right)\frac{k_1}{k_3}\cdot\frac{c_R}{c_{OH^-}}\cdot\frac{1}{\theta}\right]. \qquad \text{(VII.23)}$$

In Eq. (VII.23) β is the transfer coefficient on the covered surface, k_1 and k_2 are the equilibrium adsorption and desorption constants of the organic substance, and c_R is the concentration of the organic substance in the bulk solution.

If the rate of desorption of the organic substance due to dissolution of the metal is low, i.e., $i_2 \gg i_3$, the surface coverage is independent of the potential and is determined by the Langmuir isotherm:*

$$\theta = \frac{\dfrac{k_1}{k_2}\,c_R}{1 + \dfrac{k_1}{k_2}\,c_R}. \qquad\qquad\qquad \text{(VII.24)}$$

At more positive potentials desorption is due mainly to dissolution of the metal, $i_2 \ll i_3$. Under these conditions the dependence of coverage on potential is given by the relation

---

*In general, θ also depends on the potential φ under equilibrium conditions. In the present instance it was assumed as a simplification that θ is independent of φ in absence of an electrochemical process.

$$\theta = \frac{k_1}{k_3} \cdot \frac{c_R}{c_{OH^-}} \exp\left[-\frac{2\beta'\varphi F}{RT}\right]. \tag{VII.25}$$

It follows from Eq. (VII.25) that the surface coverage by the inhibitor decreases rapidly with increase of potential. This corresponds to a sharp decrease in the slope of the polarization curves. Figure 85 shows the dependence of surface coverage on potential for three concentrations of the organic substance, 10^{-2}, 10^{-3}, and 10^{-4} N, calculated from Eq. (VII.23). These concentrations correspond to $\theta = 0.91$, 0.5, and 0.091 at negative potentials high enough for i_3 to be neglected. The following values were taken for the calculations: $k_1 = 1$, $k_2 = 10^{-3}$ C/cm^2 · sec · g-equiv, and $k_3 \cdot c_{OH^-} = 10^{-2}$ C/cm^2 · sec. The figure shows that θ varies sharply with φ only at high values of θ. Therefore at low surface coverages, corresponding to a sharp decrease in the slope of the φ vs log i curves; this was found experimentally to be the case [29, 30, 211].

For $\theta =$ const, an expression for the potential can be obtained in explicit form from Eq. (VII.25):

$$\varphi = \frac{2.3RT}{2\beta'F}(\ln c_{H^+} + \log c_R + \text{const}). \tag{VII.26}$$

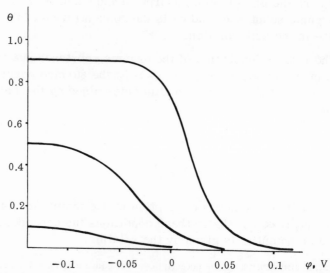

Fig. 85. Dependence of surface coverage on the potential, calculated from Eq. (VII.23).

TABLE 17. Dependence of the Desorption Potential on pH and
Concentration of Organic Compounds in Hydrochloric
Acid Solution

Compound	$(\partial\varphi/\partial \lg c_{opr})_{c_{H^+},\,\theta}$, mV	$(\partial\varphi/\partial \lg c_{H^+})_{c_{opr},\,\theta}$, mV
Phenylthiourea [29]	55-80	50-100
Phenylthiourea [78]	—	40
o-Phenanthroline [29]	30-50	40-50
Propargyl alcohol [214]	—	50
Tribenzylmethyl- ammonium sulfate	40-50	—

If $\beta = 0.5$, the potential φ at which the slope of the polarization curve changes sharply should shift toward more positive poten-
tials with increase of the inhibitor concentration, $(\partial\varphi/\partial \log c_R)_{c_{H^+},\theta} =$
58 mV, and with decrease of the solution pH, $(\partial\varphi/\partial \log c_{H^+})_{c_R,\theta} =$
58 mV. Data reported by Heusler and Cartledge [211] on adsorp-
tion of I⁻ ions and CO are in good agreement with theory. Data
on the influence of the concentration of organic inhibitors and of
solution pH on the desorption potential φ are presented in Table 17.

The coefficients given in Table 17 differ from the theoretical
values. The deviations are due to several causes: 1) difficulty
in determining the desorption potential because, in general, desorption of an organic substance occurs over a certain range of
potentials rather than at some definite potential; 2) difficulty in
fulfilling the conditions $\theta = $ const and $i_2 \ll i_3$ for determination of
$(\partial\varphi/\partial \log c_R)_{c_{H^+},\theta}$ and $(\partial\varphi/\partial \log c_{H^+})_{c_R,\theta}$ from Eq. (VII.26) with
the aid of polarization curves only; 3) the simplified model of the
desorption process.

The Bockris mechanism [191] can also be used for a detailed
examination of the dependence of θ on φ during anodic dissolution of a metal. However, because of the difficulty in exact de-
termination of the coefficients $(\partial\varphi/\partial \log c_R)_{c_{H^+},\theta}$ and $(\partial\varphi/\partial \log c_{H^+})_{c_R,\theta}$
from polarization curves, it is at present impossible to compare
experimental results with calculations based on the Heusler [190]
and Bockris [191] mechanisms.

It should be noted in conclusion that low adsorbability of organic substances may be observed in absence of external polarization of the electrode in the case of metals with high exchange currents at the equilibrium potential if the exchange current is greater than the adsorption rate [215]. For example, in the case of the $Ag - AgClO_4$ system it is difficult to find a surface-active substance which is adsorbed on silver because, according to Gerischer [216, 217], the exchange current at $1M$ silver ion concentration is 24 ± 5 A/cm^2. The same result is to be expected at the stationary potential if the spontaneous solution current of the metal is greater than the rate of adsorption of the surface-active substance.

b. Electrodeposition of Metals. During cathodic deposition of metals the concentration of an organic substance on the electrode surface becomes less than the equilibrium concentration of the adsorbed substance in absence of electrodeposition. This deviation increases with decreasing rate of adsorption of the organic substance and with increasing rate of electrodeposition.

The decrease in the concentration of the adsorbed substance during cathodic polarization may be attributed to:

1) displacement of its molecules when fresh metal atoms appear on the surface in the course of electrodeposition;

2) trapping of organic molecules during formation of the electrolytic deposit.

Surface-active substances themselves influence crystal formation and growth; this is used in plating technology for improving the properties of electrolytic coatings (brighteners, leveling agents). The mutual influence of adsorption of organic substances and the kinetics of crystal growth has been studied by Samartsev [218], Vagramyan [219], Gorbunova [220], Loshkarev [221, 222], Matulis [223], and others.

Extensive material has now been accumulated on electrodeposition of metals in presence of surface-active organic substances (e.g., see the monographs [102, 209, 224-226]). However, there have been hardly any quantitative determinations of adsorption of organic substances under electrodeposition conditions.

The ratio of the rates of processes leading to decreasing concentration of the adsorbed substance varies in accordance with

the electrodeposition conditions (current density, concentration
of surface-active substances, pH and temperature of the solution).

Volk and Fischer [82] showed that in electrodeposition of nickel
from $1 M$ $NiCl_2$ + 20 g H_3BO_3 per liter (pH = 2) adsorption of butyne-
2-diol-1,4 and propargyl alcohol decreases with increase of the
deposition current. The results of adsorption determinations for
butyne-2-diol-1,4 are shown in Fig. 86. The adsorption rate increases with the alcohol concentration, and current density has
less influence on the surface coverage.

The amount of substance incorporated in the electrolytic
deposit increases with increase of the cathodic current density.
According to the concepts of Vermilyea et al. [227], at a certain
critical current density the adsorbed substance is captured completely by the electrolytic deposit. This occurs before the substance can be desorbed, and the surface coverage tends to zero.
The critical current density diminishes with increase of the energy of adsorption of the surface-active substance.

However, it was shown by Schneider et al. [228] that incorporation of o-phenanthroline and of a number of other substances
during electrodeposition of nickel alters little with increase of
the current density from 5 to 40 mA/cm^2. The concentration of the
incorporated substance can be determined by various methods
[227-229]. In the investigation reported by Schneider et al. [228]
the amount of substance incorporated was 10% or less of the degree
of surface coverage.

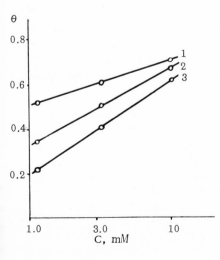

Fig. 86. Dependence of surface cover on
the concentration of butyne-2-diol-1,4
during electrodeposition of nickel at: 1)
$1 \cdot 10^{-3}$; 2) $5 \cdot 10^{-3}$; 3) $1 \cdot 10^{-2}$ A/cm^2
(from data in [82]).

At least two types of surface-active substances are now used jointly in the production of electrolytic coatings: substances which have a leveling effect on the deposits (coumarin, thiourea), and substances which act as brighteners (p-toluenesulfonamide, saccharin, etc.).

Whereas brighteners are adsorbed fairly uniformly over the entire surface, retarding crystal growth and increasing the number of crystallization sites, leveling agents are adsorbed mainly on projections and retard metal deposition on these regions. Brighteners are consumed to a considerably lesser degree than leveling agents during electrolysis, as the latter are reduced or decomposed with relative ease at the electrode surface. It has been shown in a number of investigations [56, 210, 230, 231] that consumption of leveling agents during electrolysis is determined by diffusion. Because of the smaller thickness of the diffusion layer at projections on the metal surface, the organic substance diffuses at a higher rate to these regions than to various surface depressions, and the metal is therefore deposited mainly in the depressions. Some workers consider that nonuniform distribution of inhibitors over the surface is due to preferential adsorption on microprojections [232, 233].

Kruglikov, Kudryavtsev, and co-workers [234, 235] succeeded in demonstrating, by the potential decay method and with the aid of polarization measurements at a rotating disk electrode, that coumarin exerts a leveling effect when its adsorption is determined by diffusion.

Owing to the complexity of electrodeposition processes and to lack of sufficient experimental data on adsorption of organic compounds, mathematical analysis of these phenomena is difficult, although some steps have already been made in this direction [211, 212, 227, 234, 236].

9. Influence of Adsorption of Organic Substances on Metal Corrosion

Adsorption of organic substances on metals of the iron group is closely associated with the use of such substances for protecting metals against corrosion (see reviews in [142, 149, 215, 237-240]).

Frumkin [241, 242] showed that the influence of adsorbed molecules on electrochemical reactions, including dissolution of metals, is determined mainly by two factors: surface coverage θ of the electrode by ions or molecules of the organic substance, and changes of the ψ_1 potential accompanying adsorption.

By the slow discharge theory [243], in presence of an adsorbed substance on the electrode surface we can write the following expression for the rates of cathodic evolution of hydrogen and anodic dissolution of metals in acid solutions:

$$i_c = k_1 [H_3O^+] (1 - \theta) \exp\left\{- \frac{F}{RT} [\alpha\varphi + (1 - \alpha)\psi_1]\right\}, \quad \text{(VII.27)}$$

$$i_a = k_2 (1 - \theta) \exp\left\{\frac{F}{RT} 2\beta (\varphi - \psi_1)\right\}^*, \quad \text{(VII.28)}$$

where ψ_1 depends on θ. In the region of medium surface coverages it may be assumed in the first approximation that $\psi_1 = \psi_1^0 + \Delta\psi_1\theta$, where ψ_1^0 is the value of the ψ_1-potential when $\theta = 0$. Electrochemical reactions on covered regions of the surface and the low reaction rates of hydrogen ionization and electrodeposition of metal cations on the uncovered surface are disregarded.

Thus, in solutions which do not contain oxygen or other oxidizing agents the corrosion of a metal in the active state is determined by the course of the two coupled processes discussed above. Retardation of the anodic and cathodic processes, separately or simultaneously, in presence of organic inhibitors leads to retardation of metal corrosion. However, adsorption of organic compounds does not always slow down metal dissolution, as the two factors θ and ψ_1 may act in opposite directions. It follows from Eqs. (VII.27) and (VII.28) that in adsorption of substances of the cationic type the effects of the two factors are additive and the rate of metal dissolution falls. Adsorption of organic compounds of the anionic type may raise the corrosion rate if the processes are accelerated to a greater extent by the change of ψ_1-potential than they are retarded by increase of θ.† This effect was observed by Iofa et al. [1, 37,

---

*Equation (VII.28) is written for the reaction $M - 2e \rightarrow M^{2+}$.

†This reasoning is valid only for acid solutions. In neutral and alkaline solutions evolution of hydrogen is determined by discharge of water molecules, and the term containing the ψ_1 potential in the expression for the current of hydrogen evolution will have the opposite sign.

45, 46] for iron in presence of various sulfonic acids. The relativ
effects of the ψ_1-potential and θ on iron corrosion changed with in
creasing length of the hydrocarbon chain in the sulfonic acid molecules, and the influence of sulfonic acids changed from stimulatior
to inhibition.

Increase of the rate of metal corrosion in presence of organic
substances may be attributed to other factors in addition to change
of the ψ_1-potential. In many cases corrosion is stimulated by othe
factors: 1) lowering of the overpotential of the cathodic process
owing to reduction of the organic substance; 2) the catalytic action
of the adsorbed substance on hydrogen evolution; 3) formation of
readily soluble complexes by metal ions with the adsorbed substance, which lower the overpotential of the anodic process.

Under practical conditions electrolyte solutions contain oxygen or other oxidizing agents; this complicates the electrochemica
and adsorption picture. Antropov [244] examined the influence
of inhibitors on metal corrosion under conditions of oxygen and
hydrogen depolarization, in relation to the ψ_1-potential.

In presence of oxygen or other oxidizing agents, shift of the
stationary electrode potential may lead to desorption of the organic substance owing to a change of the electrode charge or to
removal of the adsorbate together with metal ions during rapid
dissolution of the metal. At high concentrations of oxidizing agent
having high oxidation—reduction potentials the metal passes from
the active into the passive state. * In this case the influence of
the inhibitor on metal corrosion should be assessed by examination of the complete anodic potentiostatic curve, which has, in addi
tion to an active region, a region of passivity and transpassivity
or anodic activation (breakdown). This approach was first used
by Batrakov [245, 246] and later by Tomashov and Chernova [247].
Assessment of inhibitor efficiency from changes in the form of the
φ vs log i potentiostatic curve has also been proposed in other
publications [248, 249].

Inhibitors are classified as cathodic, anodic, and mixed, in
accordance with their influence on electrochemical reactions.
Schematic potentiostatic curves explaining the action of inhibitors

---

*At more positive potentials possible oxidation of the organic inhibitor is not ex-
cluded.

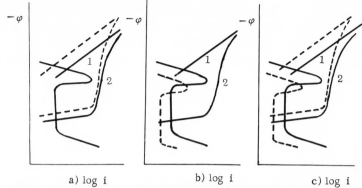

a) log i   b) log i   c) log i

Fig. 87. Schematic polarization curves explaining the action of a) cathodic,
b) anodic, and c) mixed inhibitors. The continuous curves correspond to
pure solutions and the dash curves to solutions containing inhibitors. Cathodic curve 1 corresponds to hydrogen evolution, and curve 2 to reduction of
the oxidizing agent.

of different types are shown in Fig. 87. Current densities in potentiostatic diagrams are equivalent to the metal solution rates.

Cathodic inhibitors influence the overpotential of the cathodic
process (Fig. 87a). Increase of the overpotential of the cathodic
process shifts the stationary potential into the negative direction
and retards corrosion in the active region and in the region of
anodic activation and transpassivation. In the passive region increase of the overpotential of the cathodic process does not affect
dissolution of the metal. It should be noted that if the stationary
potential of the metal is at the boundary between the active and
passive regions, increase of the overpotential of the cathodic process takes the metal from the passive into the active region and
results in a sharp rise of the corrosion rate; conversely, lowering
of the overpotential of the cathodic process results in transition
from the active to the passive state and lowers the corrosion rate
sharply.

Anodic inhibitors (see Fig. 87b) generally increase the overpotential of the anodic process. The adsorbed molecules may or
may not take part in formation of a passivating oxide. In the former case the passivation potential of the metal and the critical
passivation current change in presence of inhibitors (benzoates,
cinnamates). In the latter case the inhibitor merely lowers the

critical anodic current but does not alter the position of the passivation potential (e.g., in the case of adsorption of various aliphat acids on steel in 5 N NH_4NO_3 [250]).

Cases are possible where a surface-active organic substance which has a stimulating effect in a certain potential range may act as an inhibitor in another range. For example, lauryl sulfate accelerates corrosion of stainless steel in the active region, but prevents breakdown of the protective film in the passive state in 0.02 N HCl over a wide potential range [251] (see Fig. 88). Betti et al. [251] regard this effect as the result of competitive adsorptic of Cl⁻ ions and lauryl sulfate on the passive film. Increase of the lauryl sulfate concentration to 1% leads to complete displacement of Cl⁻ ions from the oxidized steel surface, and film breakdown does not occur even at very high positive potentials. Increase of the current at potentials of 1.7-1.8 V is due to oxygen evolution.

Inhibitors of the mixed type (Fig. 87c) raise the overpotentials of the cathodic and anodic processes. Derivatives of thiourea and quinoline are examples of such inhibitors.

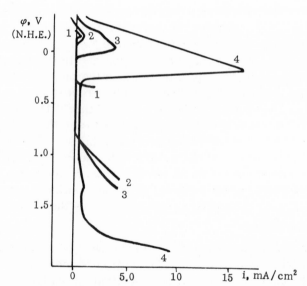

Fig. 88. Potentiostatic curves for stainless steel (13% Cr) in 0.02 N HCl: 1) pure solution; in presence of lauryl sulfate: 2) 0.01; 3) 0.1; 4) 1% (from data in [251]).

Substances whose molecules consist of an organic base and an acid anion having oxidizing properties [di(cyclohexyl)ammonium chromate, isoamyl chromate, etc.] are widely used as inhibitors of the mixed type. In aqueous solutions these substances dissociate or undergo hydrolysis with formation of anions having oxidizing properties and organic cations. The anions are adsorbed on the metal surface and facilitate formation of a protective oxide film. The organic cations are also adsorbed on the oxidized metal surface. Joint adsorption of anions and cations, discussed in detail in Section 2 of this chapter, probably occurs, as the range of influence of NO_2^- ions on the anodic process is widened appreciably in presence of dicyclohexylammonium cations [252]. Infrared spectroscopic data on adsorption of butyl nitrite from the vapor phase [253] also confirm these views.

Potentiostatic curves for iron in sodium sulfate solutions in presence of di(cyclohexyl)ammonium nitrite [254] are shown in Fig. 89. Lowering of the overpotential of the cathodic process in presence of the inhibitor is due to partial reduction of NO_2^- ions.

The slope of polarization curves often changes in presence of surface-active substances. Increase of the slope may be attributed to:

1) increased adsorption of the organic substance increasing the overpotential of the process (with decreased adsorption of the substance lowering the overpotential), due to a shift of the electrode potential from the stationary value into the cathodic or anodic region;

2) a change in the mechanism of the process (e.g., in presence of the adsorbed substance dissolution of metals of the iron group may proceed directly by the reaction $M - 2e \rightarrow M^{2+}$ rather than by formation of the intermediate product MOH);

3) formation of oxides on the metal surface in presence of oxidizing inhibitors;

4) gradual formation of compounds which are more effective inhibitors than the compounds initially adsorbed;

5) formation of multimolecular layers, layers of colloidal particles, and polymeric coatings, having high ohmic resistance, on the electrode surface.

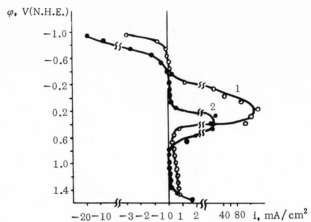

Fig. 89. Potentiostatic curves for iron in pure 1 N Na_2SO_4 (1) and with addition of 2% of di(cyclohexyl)ammonium nitrite (2) (from data in [254]).

Decrease of the slope of the polarization curves can be attributed to:

1) desorption of the organic substance which raises the overpotential of the process (electrostatic expulsion from the double layer, desorption during dissolution of the metal, reduction or oxidation accompanied by removal, adsorption of other compounds)

2) increased adsorption of an organic substance which lowers the overpotential of the process;

3) change of the dissolution mechanism (e.g., formation of complexes);

4) a substantial rate of reduction (or oxidation) of the organic substance, varying with the electrode potential.

The corrosion rates of many metals and alloys in various media can be regulated over wide ranges with the aid of organic inhibitors, but the mechanism of their action may vary considerably in accordance with the reaction conditions and the nature of the inhibitor.

References

1.   Z.A. Iofa, V.V. Batrakov, and Cho Ngok Ba, Zashchita Metallov, 1:56 (1965).

2. P. J. Hillson, J. Chim. Phys., 49:88 (1952); Trans. Faraday Soc., 48:462 (1952).
3. É. I. Mikhailova and Z. A. Iofa, Élektrokhimiya, 1:107 (1965).
4. J. O'M. Bockris and D. A. J. Swinkels, J. Electrochem. Soc., 111:736 (1964).
5. J. O'M. Bockris, M. Green, and D. A. J. Swinkels, J. Electrochem. Soc., 111:743 (1964).
6. T. Hurlen, Acta Chem. Scand., 14:1555 (1960).
7. P. D. Lukovtsev and S. D. Levina, Zh. Fiz. Khim., 21:599 (1947).
8. Z. A. Iofa, V. V. Batrakov, and Cho Ngok Ba, in: Corrosion Inhibitors [in Russian], Izd. "Radyans'ka Ukraina," Kiev (1965), p. 11.
9. A. N. Frumkin, Dokl. Akad. Nauk SSSR, 154:1432 (1964).
10. T. N. Voropaeva, B. V. Deryagin, and B. N. Kabanov, Izv. Akad. Nauk SSSR, Otd. Khim. Nauk, 257 (1963).
11. E. O. Ayazyan, Dokl. Akad. Nauk SSSR, 100:473 (1955).
12. L. I. Antropov, Zh. Fiz. Khim., 37:965 (1963).
13. R. M. Vasenin, Zh. Fiz. Khim., 27:878 (1953).
14. V. L. Kheifets and B. S. Krasikov, Zh. Fiz. Khim., 31:1992 (1957).
15. P. D. Lukovtsev, S. D. Levina, and A. N. Frumkin, Acta Physicochim. URSS, 11:21 (1939); Zh. Fiz. Khim., 13:916 (1939).
16. V. L. Kheifets and L. S. Reishakhrit, Uch. Zap. Leningrd. Gos. Univ., No. 169, 173 (1953).
17. A. N. Frumkin, J. Chem. Phys., 7:552 (1939).
18. V. Fomenko, Emission Properties of Elements and Chemical Compounds [in Russian], Izd. Akad. Nauk Ukr.SSR, Kiev (1961).
19. A. N. Frumkin, Svensk. Kem. Tidskr., 77:300 (1965).
20. A. N. Frumkin, N. S. Polianovskaya, N. B. Grigor'ev, and I. A. Bagotskaya, Electrochim. Acta, 10:793 (1965).
21. W. J. Lorenz and H. Fischer, Electrochim. Acta, 11:1597 (1966).
22. Z. A. Iofa, E. Lyakhovetskaya, and K. Sharifov, Dokl. Akad. Nauk SSSR, 84:543 (1952).
23. A. N. Frumkin, Vestn. Mosk. Gos. Univ., No. 9, 37 (1952).
24. Z. A. Iofa, Vestn. Mosk. Gos. Univ., No. 2, 139 (1956).
25. V. V. Losev, Dokl. Akad. Nauk SSSR, 88:499 (1953).
26. W. J. Lorenz, K. Sarropoulos, and H. Fischer, Extended Abstracts of the 19th CITCE Meeting, Detroit, USA (1968), p. 65.
27. Z. A. Iofa and L. A. Medvedeva, Dokl. Akad. Nauk SSSR, 69:213 (1949).
28. Z. A. Iofa (Jofa), V. V. Batrakov, and Cho Ngok Ba, Electrochim. Acta, 9:1645 (1964).
29. H. Jamaoka and H. Fischer, Electrochim. Acta, 10:679 (1965).
30. H. Jamaoka, W. Lorenz, and H. Fischer, Fundamental Problems of Modern Theoretical Electrochemistry [Russian translation], Proceedings of the 14th CITCE Meeting, Izd. "Mir," Moscow (1965), p. 464.
31. H. Fischer and H. Jamaoka, Chem. Ber., 94:1477 (1961).
32. S. D. Beskov, Uch. Zap. Mosk. Gos. Ped. Inst., 63:129 (1951).
33. G. Walpert, Z. Phys. Chem., A151:219 (1930).
34. V. A. Kuznetsov and Z. A. Iofa, Zh. Fiz. Khim., 21:201 (1947).

35. N. Hackerman, E. S. Snavely, Jr., and J. S. Payne, Jr., J. Electrochem. Soc., 113:677 (1966).

36. T. Murakawa and N. Hackerman, Corrosion Sci., 4:387 (1964).

37. Z. A. Jofa (Iofa), Proceedings of the Second European Symposium on Corrosion Inhibitors, Ferrara, Ann. Univ. Ferrara, Ser. V, 151 (1965).

38. A. Shlygin, A. Frumkin, and V. Medvedovskii, Acta Physicochim. URSS, 4:911 (1936).

39. N. A. Balashova and A. N. Frumkin, Dokl. Akad. Nauk SSSR, 20:449 (1938).

40. A. N. Frumkin, O. A. Petrii, A. M. Kossaya, V. S. Éntina, and V. V. Topolev, J. Electroanal. Chem., 16:175 (1968).

41. V. S. Éntina and O. A. Petrii, Élektrokhimiya, 4:457 (1968).

42. R. Burshtein, M. Surova, and I. Zaideman, Zh. Fiz. Khim., 24:214 (1950).

43. M. A. Vorsina and A. N. Frumkin, Acta Physicochim. URSS, 18:341 (1943).

44. Z. A. Iofa, Dokl. Akad. Nauk SSSR, 119:971 (1958).

45. Z. A. Iofa and G. N. Tomashova, Zh. Fiz. Khim., 34:1036 (1960); Corrosion, 19:26t (1963).

46. Z. A. Iofa, Cho Ngok Ba, and M. K. Vasil'eva, Zh. Fiz. Khim., 39:2182

47. T. P. Hoar and R. P. Kehra, European Symposium on Corrosion Inhibitors, Ferrara, 1960, Compt. Rend. Univ. Ferrara, 73 (1961).

48. B. B. Damaskin and N. V. Nikolaeva-Fedorovich, Zh. Fiz. Khim., 35:1279 (1961).

49. S. I. Zhdanov and B. A. Kiselev, Zh. Fiz. Khim., 40:484 (1966).

50. T. Murakawa, S. Nagaura, and N. Hackerman, Corrosion Sci., 7:79 (1967).

51. T. Murakawa, T. Kato, S. Nagaura, and N. Hackerman, Corrosion Sci., 8:341, 483 (1968).

52. T. P. Hoar and R. D. Holliday, J. Appl. Chem., 3:502 (1953).

53. L. Cavallaro, L. Felloni, and G. Trabanelli, European Symposium on Corrosion Inhibitors, Ferrara, 1960, Compt. Rend. Univ. Ferrara, 111 (1961).

54. A. I. Fil'ko, Uch. Zap. Mosk. Gos. Ped. Inst., 146:62 (1960).

55. T. K. Ross and D. H. Jones, European Symposium on Corrosion Inhibitors, Ferrara, 1960, Compt. Rend. Univ. Ferrara, 163 (1961).

56. Yu. K. Vyagis, A. I. Bodnevas, and Yu. Yu. Matulis, Zashchita Metallov., 1:359 (1965); 2:201 (1966).

57. C. Roth and H. Leidheiser, J. Electrochem. Soc., 100:553 (1953).

58. J. Elze, Metall., 9:104 (1955).

59. A. A. Sutyagina and K. M. Gorbunova, Zh. Fiz. Khim., 35:2515 (1961).

60. J. Edwards, Trans. Inst. Metal Finishing, 39:33 (1962).

61. O. K. Galdikiene, in: Theory and Practice of Bright Electrolytic Coatings [in Russian], Gospolitnauchizdat, Vilnius (1963), p. 65.

62. D. J. Brown, J. Appl. Chem., 2:202 (1952).

63. Yu. K. Vyagis and A. I. Bodnevas, Tr. Akad. Nauk Lit.SSR, B3(38):3 (1964).

64. Yu. K. Vyagis, A. I. Bodnevas, and Yu. Yu. Matulis, Zashchita Metallov., 2:471 (1966).

65. R. M. Morgart and O. É. Panchuk, Zh. Prikl. Khim., 32:833 (1959).

66. A. V. Pamfilov, O. É. Panchuk, and G. G. Kossyi, Zh. Prikl. Khim., 32:1399 (1959).

67. H. Fischer, Chem.-Ingr. Tech., 36:582 (1964).
68. A. C. Mackrides and N. Hackerman, Ind. Eng. Chem., 47:1773 (1955).
69. J. Bougault, E. Cattelain, and P. Chabrier, Bull. Soc. Chim. France, 7:781 (1940).
70. S. A. Balezin and S. K. Novikov, Uch. Zap. Mosk. Gos. Ped. Inst., 45:62 (1947).
71. L. Cavallaro and G. P. Bolognesi, Atti Acad. Sci., Ferrara, 24:235 (1946-1947).
72. L. Cavallaro and L. Felloni, Atti Acad. Sci., Ferrara, 25:139 (1947-1948).
73. L. Cavallaro, L. Felloni, G. Trabanelli, and F. Pulidori, Electrochim. Acta, 8:521 (1963).
74. A. Yamaguchi, R. B. Penland, S. Muzushima, T. J. Lane, Columba Curran, and J. V. Quagliano, J. Am. Chem. Soc., 80:527 (1958).
75. T. J. Lane and A. Yamaguchi, J. Am. Chem. Soc., 81:3824 (1959).
76. V. V. Batrakov, Gamil Hanna Avad, and Z. A. Iofa, Zashchita Metallov (in press).
77. V. V. Batrakov, Gamil Hanna Avad, É. I. Mikhailova, and Z. A. Iofa, Élektrokhimiya, 4:601 (1968).
78. J. Heidemeyer and H. Kaesche, Corrosion Sci., 8:377 (1968).
79. W. Machu and V. K. Gouda, Werkstoffe Korrosion, 13:745 (1962).
80. F. I. Khaibullin, in: Physicochemical and Technological Investigations of Mineral Ores [in Russian], Izd. "Nauka," Tashkent (1965), p. 158.
81. H. C. Gatos, European Symposium on Corrosion Inhibitors, Ferrara, 1960; Compt. Rend. Univ. Ferrara, 257 (1961).
82. O. Volk and H. Fischer, Electrochim. Acta, 4:251 (1961); 5:112 (1961).
83. R. Annand, R. M. Hurd, and N. Hackerman, J. Electrochem. Soc., 112:138, 144 (1965).
84. Z. A. Iofa and V. V. Batrakov, Abstracts of Papers at the Interuniversity Electrochemical Conference [in Russian], Novocherkassk (1965), p. 21.
85. R. J. Adams, H. L. Weisbecker, and W. J. McDonald, J. Electrochem. Soc., 111:774 (1964).
86. R. B. Campbell, Nature, 197:374 (1963).
87. S. Z. Roginskii and Ya. B. Zel'dovich, Acta Physicochim. URSS, 1:595 (1934).
88. S. M. Kochergin, Usp. Khim., 24:779 (1955).
89. V. I. Likhtman, E. D. Shchukin, and P. A. Rebinder, Physicochemical Mechanics of Metals [in Russian], Izd. Akad. Nauk SSSR, Moscow (1962).
90. Yu. V. Goryunov, N. V. Pertsov, and B. D. Sums, The Rebinder Effect [in Russian], Izd. "Nauka," Moscow (1966).
91. B. E. Conway, R. G. Barradas, and T. Zawiozkii, J. Phys. Chem., 62:676 (1958).
92. N. A. Balashova and G. G. Zhmakin, Dokl. Akad. Nauk SSSR, 143:358 (1962).
93. O. Erbacher, Z. Phys. Chem., 156:135 (1931).
94. F. Tompkins, Ind. Eng. Chem., 42:1469 (1950).
95. M. S. Beletskii, Tr. Vsesoyuzn. Alyumin.-Magnievo Inst., No. 44, 193 (1960).
96. R. G. Rein, C. M. Sliepcevich, and R. D. Daniels, J. Electrochem. Soc., 112:739 (1965).
97. W. J. Lorenz and G. Eichkorn, Ber. Bunsenges, 70:99 (1966).

98.   I. Uhara, Nature, 192:867 (1961).
99.   I. Uhara, S. Yanagimoto, K. Tani, G. Rachi, and S. Teratani, J. Phys. Chem.,
       66:2691 (1962).
100.   S. Kishimoto, J. Phys. Chem., 66:2694 (1962).
101.   I. Uhara, Sh. Kishimoto, J. Jashida, and T. Hikino, J. Phys. Chem., 69:880
       (1965).
102.   A. T. Vagramyan and Z. A. Solov'eva, Methods for Investigating Electro-
       deposition of Metals [in Russian], Izd. Akad. Nauk SSSR, Moscow (1960).
103.   S. M. Kochergin, Texture of Electrodeposited Metals [in Russian], Metal-
       lurgizdat, Moscow (1960).
104.   G. S. Vozdvizhenskii and F. Kh. Zainullina, Proceedings of Conference on the
       Influence of Surface-Active Substances on Electrodeposition of Metals [in
       Russian], Gospolitnauchizdat, Vilnius, Lit.SSR (1957), p. 103.
105.   K. J. Bachmann and K. J. Vetter, Z. Phys. Chem. (N.F.), 51:98 (1966).
106.   A. I. Levin, V. M. Rudoi, L. A. Kryuchkov, and L. M. Gryaznukhina, Élek-
       trokhimiya, 2:914 (1966).
107.   F. G. Will, J. Electrochem. Soc., 112:451 (1965).
108.   A. Hamelin and G. Valette, Compt. Rend., C267:127, 211 (1968).
109.   V. N. Nikulin and M. Z. Tsypin, Zh. Fiz. Khim., 35:58 (1961).
110.   V. N. Nikulin and P. V. Zamuragin, Zh. Fiz. Khim., 35:287 (1961).
111.   V. N. Nikulin, Zh. Fiz. Khim., 38:1103 (1964).
112.   A. M. Rubinshtein, Usp. Khim., 21:1287 (1952).
113.   D. Caplan and M. Cohen, Corrosion Sci., 6:321 (1966).
114.   G. Trabanelli, F. Zucchi, and G. L. Zucchini, Corrosion et Anticorrosion,
       14:375 (1966).
115.   J. Talbot, Corrosion et Anticorrosion, 15:43 (1967).
116.   Z. A. Iofa, Yu. A. Nikiforova, and V. V. Batrakov, Third International Con-
       gress on Metal Corrosion (abstracts of papers) [in Russian], Moscow (1966),
       p. 129.
117.   Z. A. Iofa (Jofa), V. V. Batrakov, and Yu. A. Nikiforova, Extended Abstracts
       of the 17th CITCE Meeting, Tokyo (1966), p. 60; Vestn. Mosk. Gos. Univ.
       (Khimiya), No. 6, 11 (1967); Corrosion Sci., 8:573 (1968).
118.   V. I. Ivanov and K. A. Osipov, Recovery and Recrystallization in Metals on
       Rapid Heating [in Russian], Izd. "Nauka," Moscow (1964).
119.   W. B. Pearson, Handbook of Lattice Spacings and Structures of Metals and
       Alloys, Pergamon Press, London (1958).
120.   V. V. Batrakov and Yu. A. Nikiforova, The Double Layer and Adsorption on
       Solid Electrodes [in Russian], Tartu, Est.SSR (1968), p. 29.
121.   W. G. Johnston and I. Gilman, J. Appl. Phys., 30:129 (1959).
122.   G. T. Hahn, Acta Metallurgica, 10:727 (1962).
123.   Z. Foroulis and H. Uhlig, J. Electrochem. Soc., 111:522 (1964); 112:1177
       (1965).
124.   Z. Foroulis, J. Electrochem. Soc., 113:532 (1966).
125.   N. Green and I. Saltzman, Corrosion, 20:293t (1964).
126.   M. C. Smith, Principles of Physical Metallurgy, Harper and Bros., New York
       (1956).

127.   C. S. Barrett, Structure of Metals, McGraw-Hill, New York (1952).
128.   I. P. Kudryavtsev, Texture in Metals and Alloys [in Russian], Izd. "Metal-
       lurgiya," Moscow (1965).
129.   S. S. Gorelik, Recrystallization of Metals and Alloys [in Russian], Izd.
       "Metallurgiya," Moscow (1967).
130.   E. L. Cook and N. Hackerman, J. Phys. Colloid Chem., 55:549 (1951).
131.   N. Hackerman and A. C. Makrides, Ind. Eng. Chem., 46:523 (1954).
132.   N. Hackerman and A. H. Roebuck, Ind. Eng. Chem., 46:1481 (1954).
133.   J. J. Bordeaux and N. Hackerman, J. Phys. Chem., 61:1323 (1957).
134.   C. C. Nathan, Corrosion, 12:161t (1956).
135.   E. B. Greenhill, Trans. Faraday Soc., 45:625 (1949).
136.   F. P. Bowden and A. C. Moore, Trans. Faraday Soc., 47:900 (1951).
137.   F. E. Matsen, A. C. Makrides, and N. Hackerman, J. Chim. Phys., 22:1800 (1954).
138.   J. W. Shepard and J. P. Ryan, J. Phys. Chem., 60:127 (1956).
139.   C. Mann, B. Lauer, and C. Hultin, Ind. Eng. Chem., 28:1049 (1936).
140.   F. Rhodes and W. Kuhn, Ind. Eng. Chem., 21:1066 (1929).
141.   W. Machu, Korrosion Metallschutz, 13:20 (1958).
142.   I. N. Putilova, S. A. Balezin, and V. P. Barannik, Inhibitors of Metal Cor-
       rosion [in Russian], Goskhimizdat, Moscow (1958).
143.   R. H. Cardwell and L. H. Eilers, Ind. Eng. Chem., 40(10):1951 (1948).
144.   A. C. Makrides and N. Hackerman, J. Phys. Chem., 56(8):707 (1955).
145.   H. R. Baker and W. A. Zisman, Ind. Eng. Chem., 40(12):2338 (1948).
146.   O. L. Riggs and R. L. Every, Corrosion, 18:262 (1962).
147.   Z. Ostrowski, Acta Chim. Hung., 20:215 (1959).
148.   L. Cavallaro and L. Felloni, Ann. Chimica, 49:579 (1959).
149.   I. L. Rozenfel'd and V. P. Persiantseva, Zashchita Metallov., 2:5 (1966).
150.   L. I. Antropov, Corrosion Sci., 7:607 (1967).
151.   D. O. Hayward and B. M. W. Trapnell, Chemisorption, Plenum Press, New York
       (1964).
152.   G. L. Foster, B. D. Oakes, and C. H. Kucera, Ind. Eng. Chem., 51:825 (1959).
153.   I. N. Putilova and E. N. Chislova, in: Corrosion Inhibitors [in Russian], Izd.
       "Radyans'ka Ukraina," Kiev (1965), p. 35.
154.   I. N. Putilova, A. M. Lolua, I. I. Suponitskaya, and G. M. Maslova, Zashchita
       Metallov., 4:392 (1968).
155.   M. A. Gerovich and O. G. Ol'man, Zh. Fiz. Khim., 28:19 (1954).
156.   M. A. Gerovich, Dokl. Akad. Nauk SSSR, 96:543 (1954); 105:1278 (1955).
157.   N. Hackerman, European Symposium on Metal Corrosion, Ferrara, 1960;
       Compt. Rend. Univ. Ferrara, p. 99 (1961).
158.   V. P. Grigor'ev and O. A. Osipov, Third International Congress on Metal
       Corrosion (abstracts of papers) [in Russian], Moscow (1966), p. 158.
159.   F. M. Donahue and K. Nobe, J. Electrochem. Soc., 112:886 (1965).
160.   L. I. Antropov, G. G. Vrzhosek, and V. F. Panasenko, in: Corrosion Inhibitors
       [in Russian], Izd. "Radyan'ska Ukraina," Kiev (1965), p. 23.
161.   F. M. Donahue, A. Akiyama, and K. Nobe, J. Electrochem. Soc., 114:1006
       (1967).
162.   F. M. Donahue and K. Nobe, J. Electrochem. Soc., 114:1012 (1967).

163. V. P. Grigor'ev and V. V. Kuznetsov, Zashchita Metallov., 3:178 (1967).

164. V. P. Grigor'ev and V. V. Ékilik, Zashchita Metallov., 4:31 (1968).

165. V. P. Grigor'ev and V. K. Ékilik, Zashchita Metallov., 4:582 (1968).

166. Yu. A. Zhdanov and V. I. Minkin, Correlation Analysis in Organic Chemistry [in Russian], Izd. Rostovsk. Univ. (1966).

167. R. C. Ayers, Jr., and N. Hackerman, J. Electrochem. Soc., 110:507 (1963).

168. N. Hackerman and R. M. Hurd, First International Congress on Metal Corrosion, Butterworths, London (1961), p. 313.

169. A. V. Pamfilov and V. S. Kuzub, Ukr. Khim. Zh., 28:528 (1962).

170. M. Suhrman, Z. Metallkunde, 46:770 (1955).

171. H. Grubitsch and F. Hilbert, Werkstoffe Korrosion, 17:289 (1966).

172. P. F. Cox, R. L. Every, and O. L. Riggs, Jr., Corrosion, 20:299t (1964).

173. V. P. Grigor'ev, O. A. Osipov, V. V. Ékilik, and V. V. Kuznetsov, The Double Layer and Adsorption on Solid Electrodes [in Russian], Tartu, Est.SSR (1968), p. 50.

174. A. N. Frumkin, Zh. Fiz. Khim., 10:568 (1937).

175. A. N. Frumkin, Zh. Fiz. Khim., 31:1875 (1957).

176. J. O'M. Bockris, Modern Aspects of Electrochemistry, Vol. 3, Plenum Press, New York (1964).

177. V. É. Past and Z. A. Iofa, Dokl. Akad. Nauk SSSR, 106:1050 (1956); Zh. Fiz. Khim., 33:1230 (1959).

178. V. V. Sobol' and Z. A. Iofa, Dokl. Akad. Nauk SSSR, 138:1151 (1961).

179. V. V. Batrakov and Z. A. Iofa, Élektrokhimiya, 1:123 (1965).

180. V. V. Losev and B. N. Kabanov, Izv. Akad. Nauk SSSR, Otd. Khim. Nauk, 414 (1957).

181. J. L. Weininger and M. W. Breiter, J. Electrochem. Soc., 111:707 (1964).

182. I. A. Bagotskaya, Dokl. Akad. Nauk SSSR, 110:397 (1956).

183. I. A. Bagotskaya and A. I. Oshe, Proceedings of the Fourth Conference on Electrochemistry [in Russian], Izd. Akad. Nauk SSSR, Moscow (1959), p. 82.

184. B. N. Kabanov and D. I. Leikis, Dokl. Akad. Nauk SSSR, 58:1685 (1947); Zh. Fiz. Khim., 20:995 (1946).

185. B. Kabanov, R. Burshtein, and A. Frumkin, Discussions Faraday Soc., 1:259 (1947).

186. M. K. Turapov and A. M. Murtazaev, Izv. Akad. Nauk Uz.SSR (Khim. Ser.), 2:53 (1957).

187. W. Machu, Korrosion, Metallschutz, 10:277 (1934).

188. S. D. Levina, Zh. Prikl. Khim., 29:1353 (1958).

189. L. Kandler, D. Roemer, and E. Häusler, Electrochim. Acta, 8:233 (1963).

190. K. E. Heusler, Z. Elektrochem., 62:582 (1958).

191. J. O'M. Bockris, D. Drazic, and A. R. Despic, Electrochim. Acta, 4:325 (1961).

192. J. O'M. Bockris and H. Kita, J. Electrochem. Soc., 108:676 (1961).

193. H. Jamaoka, Einfluss von organischen Inhibitoren auf die Säurekorrosion des Eisens, Dissertation, Technische Hochschule, Karlsruhe (1963).

194. Z. A. Iofe and Wei Pao-ming, Zh. Fiz. Khim., 36:2558 (1962).

195. M. H. Tikkanen and T. Tuominen, Third International Congress on Metal Corrosion (abstracts of papers) [Russian translation], Moscow (1966), p. 111.

196. N. Sato and G. Okamoto, J. Electrochem. Soc., 111:897 (1964).
197. K. Heusler and L. Gaiser, Electrochim. Acta, 13:59 (1968).
198. R. A. Erb, J. Phys. Chem., 69:1306 (1965).
199. Yung-Fang Yu, J. J. Chessick, and A. C. Zettlemoyer, J. Phys. Chem., 63:1626 (1959).
200. K. Schwabe, Z. Phys. Chem., 226:1 (1964).
201. S. A. Balezin, P. G. Kuznetsov, and I. A. Podol'nyi, in: Corrosion Inhibitors [in Russian], Izd. Sudostroenie, Moscow–Leningrad (1965), p. 18.
202. L. Horner and A. Mentrup, Liebigs Ann. Chem., 65:646 (1961); F. Böttger, H. Fuchs, and L. Horner, Chem. Ber., 96:3141 (1963).
203. W. Lorenz and H. Fischer, Proceedings of the Second European Symposium on Corrosion Inhibitors, Ferrara, Ann. Univ. Ferrara, Sez. V, 109 (1965).
204. M. Smialowski and Z. Szklarska-Smialowska, Bull. Acad. Polon. Sci., 1:155 (1953).
205. V. S. Kemkhadze and S. A. Balezin, Zh. Obshch. Khim., 22:1848 (1952).
206. I. N. Putilova, N. V. Rudenko, and A. N. Terent'ev, Zh. Fiz. Khim., 38:494 (1964).
207. E. J. Duwell, J. M. Todd, and H. C. Butzke, Corrosion Sci., 4:435 (1964).
208. G. W. Poling, J. Electrochem. Soc., 114:1209 (1967).
209. V. I. Lainer, Modern Electroplating Technology [in Russian], Izd. "Metallurgiya," Moscow (1967).
210. G. T. Rogers and K. J. Taylor, Electrochim. Acta, 8:887 (1963); 11:1685 (1966); Trans. Inst. Metal Finishing, 43:75, 120 (1965).
211. K. E. Heusler and G. H. Cartledge, J. Electrochem. Soc., 108:732 (1961).
212. K. E. Heusler, Fundamental Problems of Modern Theoretical Electrochemistry, Proceedings of the 14th CITCE Meeting [Russian translation], Izd. "Mir," Moscow (1965), p. 453.
213. Z. A. Iofa and Wei Pao-ming, Zh. Fiz. Khim., 37:2300 (1963).
214. H. Fischer and H. Jamaoka, Chem. Ber., 94:1477 (1961).
215. H. Fischer, European Symposium on Corrosion Inhibitors, Ferrara, Compt. Rend. Univ. Ferrara, 1 (1961).
216. H. Gerischer and R. P. Tischer, Z. Elektrochem., 61:1159 (1957).
217. H. Gerischer, Z. Elektrochem., 62:256 (1958).
218. A. G. Samartsev, Dokl. Akad. Nauk SSSR, 2:478 (1935).
219. A. T. Vagramyan, Electrodeposition of Metals [in Russian], Izd. Akad. Nauk SSSR, Moscow (1950); Dokl. Akad. Nauk SSSR, 22:238 (1939); Acta Physicochim. URSS, 19:148 (1944).
220. K. M. Gorbunova and A. I. Zhukov, Zh. Fiz. Khim., 22:1097 (1948); 23:605 (1949).
221. M. A. Loshkarev, O. A. Esin, and V. I. Sotnikova, Zh. Obshch. Khim., 9:1412 (1939).
222. M. A. Loshkarev and A. A. Kryukova, Dokl. Akad. Nauk SSSR, 62:97 (1948); 77:919 (1950); Zh. Fiz. Khim., 23:209, 221, 1457 (1949).
223. Yu. Yu. Matulis and A. I. Bodnevas, Izv. Akad. Nauk SSSR, Otd. Khim. Nauk, No. 4, 577 (1954).
224. V. I. Lainer and N. T. Kudryavtsev, Principles of Electroplating [in Russian], Vol. 1, Metallurgizdat, Moscow (1953).

225.   H. Fischer, Elektrolytische Abscheidung und Elektrokristallization von Metallen, Springer Verlag, Berlin (1954).

226.   F. A. Lowenheim (editor), Modern Electroplating, 2nd ed., John Wiley and Sons, Inc., New York—London—Sydney (1963).

227.   P. B. Price, D. A. Vermilyea, and M. B. Webb, Acta Metallurgica, 6:524 (1958).

228.   R. Schneider, L. Albert, and H. Fischer, Metalloberfläsche, 21:65 (1967).

229.   J. Edwards, Trans. Inst. Metal Finishing, 41:140, 147, 157 (1964).

230.   O. Kardos and D. G. Foulke, Electrochemistry and Electrochemical Engineering (edited by P. Delahay and C. W. Tobias), Vol. 2, Interscience, New York (1961), p. 145.

231.   S. S. Kruglikov, N. T. Kudryavtsev, and U. I. Sinyakov, Trans. Inst. Metal Finishing, 44:152 (1966).

232.   S. E. Beacom and R. J. Riley, J. Electrochem. Soc., 106:309 (1959).

233.   J. D. Thomas, Proc. Am. Electropol. Soc., 43:60 (1956).

234.   S. S. Kruglikov, Yu. D. Gamburg, and N. T. Kudryavtsev, Electrochim. Acta, 12:1129 (1967).

235.   S. S. Kruglikov, N. T. Kudryavtsev, and R. P. Sobolev, Electrochim. Acta, 12:1263 (1967).

236.   S. I. Krichmar, Élektrokhimiya, 1:858 (1965).

237.   T. P. Hoar, Proceedings of the Second International Congress of Surface Activity, Vol. 3, Butterworths, London (1957), p. 81.

238.   J. Elze, Gesundh.-Ingr., 76:195 (1955).

239.   A. Bukowiecki, Schweiz, Arch. Angew. Wiss. Tech., 20:169 (1954).

240.   H. Fischer, J. Electrochem. Soc. Japan, 35:169 (1967).

241.   A. N. Frumkin, Electrochim. Acta, 9:465 (1964).

242.   A. N. Frumkin, Fundamental Problems of Modern Theoretical Electrochemistry, Proceedings of the 14th CITCE Meeting [in Russian], Izd. "Mir," Moscow (1965), p. 302.

243.   A. N. Frumkin, V. S. Bagotskii, Z. A. Iofa, and B. N. Kabanov, Kinetics of Electrode Processes [in Russian], Izd. Mosk. Gos. Univ., Moscow (1952).

244.   L. I. Antropov, Zashchita Metallov, 2:279 (1966).

245.   V. P. Batrakov, in: Corrosion and Metal Protection [in Russian], Oborongiz, Moscow (1962), p. 33.

246.   V. P. Batrakov, Extended Abstracts of the 14th CITCE Meeting [in Russian], Moscow (1963).

247.   N. D. Tomashov and G. P. Chernova, Passivity and Protection of Metals Against Corrosion [in Russian], Izd. "Nauka," Moscow (1965).

248.   R. D. Weed, Corrosion, 20:321 (1964).

249.   Corrosion Technol., 12:28 (1965).

250.   M. Smialowski and Z. Sklarska-Smialowska, Proceedings of the Second European Symposium on Corrosion Inhibitors, Ferrara, Ann. Univ. Ferrara, Sez. V, 183 (1965).

251.   A. Betti, L. Cavallaro, G. Trabanelli, and F. Zucchi, Corrosion, 18:351t (1965).

252.   K. S. Rajagopalan, European Symposium on Corrosion Inhibitors, Ferrara, 1960; Compt. Rend. Univ. Ferrara, 685 (1961).

253.   G. W. Poling and R. P. Eischens, J. Electrochem. Soc., 113:218 (1966).
254.   I. L. Rozenfel'd and V. P. Persiantseva, Dokl. Akad. Nauk SSSR, 156:162 (1964).

Chapter 8

Mechanism of Adsorption of Organic Compounds on Electrodes of Metals of the Platinum Group

In examination of the adsorption of organic compounds on metals of the platinum group, the question of the mechanism of the adsorption process is of primary importance. This question has been considered in numerous publications. As far as can be judged from the treatment of the experimental results, it was assumed in earlier investigations that, as on a mercury electrode, the organic molecules are physically adsorbed or reversibly chemisorbed [1-8]. However, further accumulation of experimental material led to the conclusion that adsorption of organic substances on platinum electrodes may be partially accompanied by extensive changes in the adsorbed molecules. The possibility of chemical conversions of organic compounds adsorbed on platinum electrodes has been mentioned by Sokol'skii [9]. This is a natural conclusion in the light of modern views on adsorption of organic compounds from the gas phase on catalytically active materials [10, 11]. Dissociative chemisorption of gaseous hydrocarbons and alcohols has been studied in detail in a number of investigations relating to catalysis (e.g., see [12-20]). Piersma [21] has published a brief survey of views on the mechanism of adsorption of organic substances from electrolyte solutions on metals of the platinum group.

The views presented below on the adsorption mechanism of organic substances and on the nature of adsorbed particles on metals of the platinum group are based mainly on evaluation of the results of electrochemical determinations, gas chromatographic analysis, and experiments with C^{14}, H^2, and H^3 isotopes. Determinations of the volumes of hydrocarbons adsorbed by platinum black have also been carried out on a small scale [22]. As yet there have been no publications relating to studies of adsorption of organic substances under comparable conditions with the aid of other physicochemical methods, although there are references to the use of such methods, e.g., infrared spectroscopy [21, 23].

The hypothesis that organic substances undergo chemical changes during adsorption on metals of the platinum group originally arose from comparison of charging curves and potentiodynamic curves for electrochemical oxidation of various adsorbed compounds. Giner [24] found that CO_2 is adsorbed on a platinum electrode covered with adsorbed hydrogen, forming a product which is oxidized in the same potential range as the adsorption products of methanol and formic acid [25]. This product, given the name of "reduced CO_2," $(CO_2)_r$, can apparently be detected after chemisorption of certain hydrocarbons on platinum [26]. Giner suggested that $(CO_2)_r$ may have the composition CO or COOH. At the same time, certain conclusions regarding the adsorption mechanism of organic substances on platinum were put forward by Frumkin et al. [27-29] for alcohols, and by Schlatter [30], Grubb [31], Niedrach [22], and Bianchi [32] for hydrocarbons.

Various groups of investigations have now compiled extensive experimental material on the nature of adsorption of various classes of organic compounds on platinum metals. The idea of dissociative adsorption has gained fairly wide acceptance with remarkable rapidity, so that the present discussions are concerned mainly with the composition of the adsorbed particles.

Elucidation of the mechanism of adsorption of organic substances has been based on a combination of the following experimental results.

1. Detection of adsorbed hydrogen, produced by dehydrogenation of organic compounds, among the adsorption products.

2. Establishment of a stoichiometric ratio between the amount of hydrogen formed during dehydrogenation of an organic compound and the amount of carbon compound adsorbed.

3. Determination of the number of sites occupied by the adsorbed organic compound on the electrode surface.

4. Analysis of the products appearing in solution or in the gas phase during adsorption of various organic substances on metal surfaces.

5. Analysis of the products formed during cathodic reduction of adsorbed organic compounds.

6. Comparison of potentiodynamic curves or charging curves determined during electrochemical oxidation of various adsorbed organic compounds.

7. Determination of the number of electrons released during formation of a CO_2 molecule from an adsorbed particle.

8. Determinations of the relationships of the kinetics of adsorption of organic substances on platinum metals. Inferences drawn from studies of adsorption kinetics with regard to adsorption of organic substances will be examined separately in Chapter X.

It should be pointed out that determination of the nature of particles adsorbed on catalysts is a complicated problem even in investigations of adsorption from the gas phase [10, 11]; this is largely due to the surface heterogeneity of solid metals. The difficulties increase in passing to electrolyte solutions. It is therefore not surprising that in some cases there are considerable discrepancies between the experimental results and conclusions obtained by different investigators for the same systems. In discussions of these discrepancies it is necessary to take into consideration the differences between the properties of the electrodes used (the state of the electrode surface, method of preparation, preliminary treatment, nature of the support upon which the catalyst is deposited, etc.) and between the experimental conditions (composition and temperature of the electrolyte, time of adsorption of the organic compound, the electrode potential), and the potentialities of the methods used for studying adsorption. Thorough purification of the solutions is necessary for the determinations; this is especially important in work with smooth electrodes.

In the subsequent discussion we use the term "adsorbed substance" (methanol, methane, etc.) to describe the adsorption prod-

ucts formed on the surface during contact of the electrode with a
solution of the particular adsorbate. In reality, these products
may differ greatly in nature from the original substance. Therefore, in contrast to the adsorption effects on a mercury electrode
discussed earlier in this book, description of the adsorbed layers
of organic substances on metals of the platinum group requires the
use of certain new terms and definitions, as proposed by Gilman
[33]. The stoichiometry of the adsorbed layer of the
organic substance denotes the relative amounts of different elements (carbon, hydrogen, oxygen, etc.) on the surface. The
stoichiometry of an adsorbed particle denotes the
relative amounts of elements in any individual particle. The
composition of the adsorbed particles, or their
chemical formula, indicates the number of atoms of each element
in the adsorbed particle. The structure of an adsorbed
particle includes its composition, the nature of the valency
bonds between its atoms, and the nature of the valency bonds with
the surface. Indication of bond orientation between the atoms in
the adsorbed particle and between the particle and the surface
(the overall structure) is not sufficient for description of the
structure of adsorbed particles. On a heterogeneous surface a
number of "fine" structures differing in bond lengths and strengths,
etc., may be expected with the same orientations. Finally, the
structure of the adsorbed layer comprises all the
information on the structure of the individual particles and of the
adsorbent surface.

The existing data on adsorption of organic substances from
electrolyte solutions cannot as yet provide a complete description
of the structure of the adsorbed layer, and at best can only lead to
limited conclusions with regard to the stoichiometry of the adsorbed layer and of the adsorbed particles.

1.   Adsorption of Methanol on

Platinum Metals

Adsorption of methyl alcohol on the surface of a platinum
electrode is among the processes which have been studied most
fully. This is because the combination of adsorption processes
occurring during contact of a platinum electrode with methanol
solutions is simpler in certain respects than in the case of other
organic compounds.

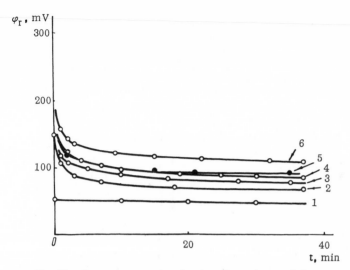

Fig. 90. Curves representing the time dependence of the potential of a Pt/Pt electrode after addition of methanol as a function of the initial potential: 1) 53; 2) 151; 3) 425; 4) 492;
5) 745; 6) 992 mV. Supporting electrolyte 0.1 N H_2SO_4. Methanol concentration 0.7 M (from data in [34, 35]).

The mechanism of methanol adsorption on a platinum electrode has been studied by Petrii, Podlovchenko, et al. [29, 34-36].
A potential* $\varphi_r = 0.4$-0.5 V was applied to a platinum electrode
in a solution of the supporting electrolyte. Methyl alcohol was
then added to the solution on open circuit. This was accompanied
by a shift of the potential in the cathodic direction (Fig. 90). The
shift of potential in the cathodic direction continues even after a
considerable time after addition of methanol.[†]

The charging curve or the potentiodynamic curve determined
after the Pt/Pt electrode has been washed free from methanol
(Fig. 91) shows two plateaus or waves, one of which (a) corresponds to oxidation of adsorbed hydrogen, and the other (b) to
oxidation of the adsorbed organic compound [25, 29, 34-36]. At
potentials acquired by the platinum electrode in methanol solu-

---

*In this chapter φ_r denotes the potential measured against a reversible hydrogen elec-
trode in the same solution, and φ is the potential relative to the standard hydrogen
electrode.
[†]The increase of the potential of a platinum electrode in methanol solutions after
an initial shift in the cathodic direction, reported in certain publications [37-39],
is apparently due to the presence of dissolved oxygen in the solutions used.

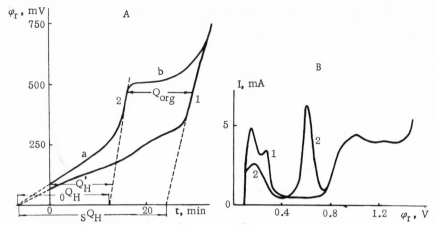

Fig. 91. Anodic charging curves (A) and potentiodynamic curves (B) for a Pt/Pt electrode in 0.1 N H_2SO_4 (1) and in presence of methanol chemisorbed on open circuit (2). Methanol added at 500 mV; methanol concentration 0.5 M. Current density in determination of the charging curves $1 \cdot 10^{-4}$ A/cm^2. Potential sweep rate in determination of potentiodynamic curves 5 mV/sec (from data in [35]).

tions ("rest potentials") adsorbed hydrogen is also detected by pulse potentiodynamic or galvanostatic methods in presence of methanol [7].

The shifts of potential observed on addition of methanol are qualitatively similar to the shifts of potential caused by adsorption of Cl$^-$, Br$^-$, and I$^-$ anions, which had been studied by Obrucheva [40]. However, the mechanisms of establishment of potentials in methanol solutions and of adsorption potentials are different:

a) the $\varphi_r^!$ values established after addition of methanol are virtually independent of the initial φ_r^0 as long as the latter is in the double-layer region, whereas the shifts of potential due to adsorption are strongly dependent on the original φ_r^0;

b) the final $\varphi_r^!$ in methanol solutions is practically independent of the solution pH, whereas the adsorption shifts are greater in acid than in alkaline solutions;

c) the rate of shift of φ_r depends little on the initial potential as long as the latter is in the range $\varphi_r \sim 0.35\text{-}0.7$ V;

d) the initial rate of shift of φ_r increases linearly with the concentration of the organic compound and depends little on the solution pH with a given initial φ_r;

e) determinations of polarization curves indicate that oxidation of methanol occurs in the double-layer region of potentials (see Chapter V) at the platinum electrode.

The shift of the electrode potential on open circuit in the cathodic direction is equivalent, on closed circuit and at a constant potential, to flow of anodic current across the electrode – solution interface. The value of this current i can be calculated from the equation

$$i = C \frac{d\varphi_r}{dt} \, , \qquad\qquad (VIII.1)$$

where C is the capacity of the electrode, determined from the slope of the charging curve in the supporting electrolyte; $d\varphi_r/dt$ is the initial rate of shift of the potential after addition of the organic substance. The anodic current, which diminishes with time, can be recorded directly if the electrode potential is maintained constant (e.g., at 0.4 V) with the aid of a potentiostat [41] during addition of methanol to the solution in which the platinum electrode is immersed. The experimentally determined anodic current at the instant of methanol addition virtually coincides with the value calculated with the aid of Eq. (VIII.1), from the potential shift curves on open circuit.

The shift of the electrode potential in the cathodic direction, the occurrence of an anodic current at constant electrode potential, and the appearance of adsorbed hydrogen during adsorption of methanol can be explained in two ways. It may be supposed that the shift of potential or appearance of an anodic current are due to transfer of electrons from the methanol molecules to the metal (electrochemical mechanism) [42]. In this case, at potentials of the hydrogen region, adsorbed hydrogen atoms appear as the result of the following reactions:

$$H^+ + \bar{e} \rightarrow H_{ads} \text{ (acid solution) } , \qquad\qquad (VIII.2)$$

$$H_2O + \bar{e} \rightarrow H_{ads} + OH^- \text{ (alkaline solution) } . \qquad (VIII.2a)$$

On the other hand, it is possible that the adsorption process itself causes hydrogen atoms to split off and leads to formation of a methanol dehydrogenation product on the surface (chemical mechanism) [27, 29, 34, 35, 43-46]. In this case the anodic current flowing during adsorption of methanol at $\varphi_r = 0.4$ V is due to ionization of hydrogen atoms split off from the alcohol molecules. The possibility of dehydrogenation of various organic compounds on contact with platinum metals has been suggested in numerous publications (e.g., [2, 9, 47-50]). However, adequate experimental evidence of the dehydrogenation mechanism was not given in these publications.

It was shown by Frumkin and Podlovchenko [27] that a choice between these hypotheses can be made on the basis of determinations of the dependence of the initial methanol oxidation current (or of the rate of shift of the potential on addition of methanol) on the solution pH.

In the case of the electrochemical mechanism, it is to be expected from the laws of electrochemical kinetics that, regardless of pH (in the pH region far from pH of methanol dissociation), the oxidation rate i should be constant at a constant electrode potential φ, measured against a constant reference electrode. On the other hand, if we assume that the slow step is dehydrogenation of the organic compound RH, e.g., by the scheme

$$RH \xrightarrow{\text{slow}} R_{ads} + H_{ads} \qquad (VIII.3)$$

then the initial oxidation rate should, in the first approximation, remain constant at constant φ_r.

Taking into account that the energy of hydrogen bonding varies with the surface coverage of the electrode [51, 52], let us assume that the dehydrogenation kinetics is determined by the maximum bonding energy at regions not covered with adsorbed hydrogen. Then it follows from the kinetics of reactions on heterogeneous surfaces [52-54] that we can write the following expression for the initial rate of the process after addition of the organic compound:

$$i = k_1 [RH] \exp\left(-\frac{\beta\mu_H}{RT}\right), \qquad (VIII.4)$$

where μ_H is the chemical potential of adsorbed hydrogen, expressed in electrical units (V); [RH] is the volume concentration of the organic substance; k_1 and β are constants.

Since

$$\mu_H = -F\varphi_r + \text{const}, \qquad (VIII.5)$$

it follows from Eq. (VIII.4) that

$$i = k \, [\text{RH}] \exp\left(\frac{\beta F \varphi_r}{RT}\right) ; \qquad \text{(VIII.6)}$$

$$\left[\frac{\partial i}{\partial (\text{pH})}\right]_{\varphi_r, \, [\text{RH}]} = 0. \qquad \text{(VIII.7)}$$

The experimental data presented above on the influence of pH on the potential shift curves in methanol solutions, and direct determinations of the rate of methanol oxidation during the initial instant of contact of the electrode with methanol [35, 41, 45, 46] show that the oxidation rate is approximately constant at constant φ_r and support the dehydrogenation mechanism of adsorption. If the electrode surface is free from adsorbed hydrogen and organic compound, the rate of reaction (VIII.3) should be independent of the potential. Accordingly, at potentials of the double-layer region, when the amounts of adsorbed gases present are small, the rate of methanol oxidation during the first instant of contact with the surface depends little on the potential.* For example, with change of φ_r from 350 to 750 mV the reaction rate is increased only by a factor of about 8-10 [35, 41], whereas with an electrochemical mechanism an increase of the rate by a factor of 10^3 could be expected in the same range of potentials.

Thus, one of the products of methanol adsorption in platinum is adsorbed hydrogen, formed by dehydrogenation of the organic molecules. Because of the higher exchange current of reactions (VIII.2) and (VIII.2a) in comparison with the rate of methanol oxidation during establishment of the stationary potential, the latter should be close to the equilibrium potential of reactions (VIII.2) and (VIII.2a) [7, 27]. Rapid exchange between hydrogen formed by dehydrogenation of methanol and hydrogen ions in solution was confirmed by Smith et al. [56] in experiments with methyl alcohol labeled with tritium.

---

*It could be concluded [55] from the results reported by Frumkin and Podlovchenko [27] that a strong dependence of the dehydrogenation rate on the potential should also be observed at φ_r of the double-layer region, although the amounts of hydrogen are small in this φ_r region. In reality, as was subsequently shown [41], the experimental data in [27] on oxidation at φ_r of the double-layer region referred not to the dehydrogenation reaction but to oxidation of the chemisorption products of organic compounds, which has a different mechanism (see Chap. X).

TABLE 18

Methanol concentration, M	Potential of methanol addition φ_r^0, mV	Time of contact of electrode with methanol t_{ads}, min	Potential φ_r' after t_{ads}, mV	$\kappa = \dfrac{Q_{org}}{sQ_H}$	$m = \dfrac{Q_H'}{sQ_H}$	$\dfrac{k}{m} = \dfrac{Q_{org}}{Q_H'}$
3,0	509	46	53	0.43	0.454	0.95
0,5	511	70	80	0.44	0.44	1.0
0,1	515	78	97	0.393	0.414	0.96
0,01	510	3	231	0.177	0.165	1.07
0,01	510	113	128	0.360	0.324	1.11
0.001	509	221	171	0.294	0.260	1.13

Certain conclusions on the nature of the organic particles adsorbed on the surface of Pt/Pt electrodes in methanol solutions were drawn by Podlovchenko and Gorgonova [29] from the results of determinations by the method of electrochemical oxidation in the adsorbed layer. The experimental results are presented in Table 18.

In Table 18, k is the ratio of the amount of organic substance adsorbed, Q_{org} (in coulombs) to the amount of hydrogen sQ_H adsorbed at $\varphi_r = 0$ in the pure electrolyte solution, while m is the ratio of the amount of hydrogen Q_H' adsorbed at φ_r' to sQ_H (Fig. 91).

Regardless of the methanol concentration and of the potentials attained by the electrode, the amounts of adsorbed organic compound and of adsorbed hydrogen, expressed in electrical units, are always approximately equal. The values of k and m are in practice determined only by the values of φ_r', to an increasing extent with decrease of the latter. The simplest explanation of these results is based on the assumption that particles of the stoichiometric composition HCO, formed as the result of dissociative chemisorption of methanol

$$CH_3OH \rightarrow HCO_{ads} + 3H_{ads} \tag{VIII.8}$$

are adsorbed on the platinum surface.

The sum of the amount of organic substance adsorbed on the electrode and the amount of hydrogen $_0Q_H$ which can be adsorbed in its presence at $\varphi_r = 0$ is approximately $_sQ_H$. This means that the HCO particle occupies three adsorption sites on the surface,

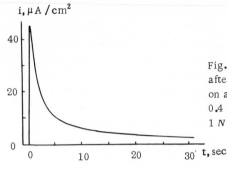

Fig. 92. Variation of current with time after the start of adsorption of methanol on a smooth platinum electrode at $\varphi_r = 0.4$ V in a solution of 0.1 M CH$_3$OH + 1 N H$_2$SO$_4$ (from data in [45, 46]).

i.e., three surface platinum atoms. Therefore the area corresponding to one adsorbed HCO particle is ~23 $\overset{\circ}{A}^2$.

Similar conclusions were reached by Bagotskii et al. [44-46] in investigations of the adsorption of methanol on a smooth platinum electrode with the aid of pulse methods. The electrode potential was changed abruptly from $\varphi_r = 1.2$ V, at which methanol is not adsorbed to any appreciable extent, to $\varphi_r = 0.4$ V, and the variation of current with the time after this change was measured (Fig. 92). The amount Q'_H of hydrogen split off during adsorption of methanol on the electrode can be determined by integration of the resultant i vs t curve. The amount Q_{org} of chemisorbed organic substance was then determined by the anodic pulse method, and the decrease $\Delta_S Q_H$ of the amount of adsorbed hydrogen in its presence was found by the cathodic pulse method. Table 19 shows that all three quantities are approximately equal at any instant during adsorption; this may be attributed to the occurrence of reaction (VIII.8) during adsorption of methanol.

TABLE 19

Adsorption time t_{ads}, sec	Q'_H, mC/cm^2	$\Delta_S Q_H$, mC/cm^2	Q_{org}, mC/cm^2
0.7	0.027	0.023	0.03
1	0.097	0.082	0.090
15	0.120	0.111	0.119
120	0.150	0.149	0.154

Similar experiments were carried out at various potentials of the platinum electrode, and at various temperatures (up to 80°C) [44-46, 57]. It was shown [35, 57] that the chemisorption processes become more complicated at elevated temperatures in the range of low potentials of the platinum electrode in concentrated acid solutions of methanol. Under these conditions methane is evolved; it is formed by hydrogenation of methanol by adsorbed hydrogen. Molecular hydrogen was evolved in analogous experiments with alkaline solutions at anodic potentials close to $\varphi_r = 0$ [57]. The nature of this effect was not investigated in detail, but some suggestions on its causes will be discussed below. Despite the increased complexity of the chemisorption processes it was concluded that, regardless of the experimental conditions, the stoichiometric composition of the adsorbed layer corresponds to HCO as before. * As in the experiments described above, it was assumed that the stoichiometry and composition of the adsorbed particles again correspond to the formula HCO.

In subsequent publications doubts were cast on the view that the stoichiometric composition of adsorbed methanol corresponds to HCO.† Biegler and Koch [58], using pulse methods of measurement, concluded that CO is adsorbed on smooth platinum in methanol solutions. However, this conclusion is not completely convincing, because their data on the amounts of substance adsorbed were taken from publications by other investigators [7, 41]. The

---

* The following structures were considered as possible "over-all" structures of chemisorbed methanol [29, 45, 46]:

$$
\begin{array}{cc}
\mathrm{C{-}O{-}H} & \mathrm{O{-\!-}C{-}H} \\
\diagup \;|\; \diagdown & |\;\;\diagup \\
\mathrm{Pt \;\; Pt \;\; Pt} & \mathrm{Pt \;\; Pt \;\;\; Pt} \\
\mathrm{I} & \mathrm{II}
\end{array}
$$

On the basis of indirect results (in particular, the possibility of dissociation of the adsorbed particles at high pH and the low adsorbability of tertiary alcohols on platinum), Bagotskii, Vasil'ev, et al. [45, 46] consider that structure I is the more probable. However, it must be pointed out that deuterium exchange data for adsorption of alcohols on catalysts [10] indicate, on the contrary, that the activity of the hydrogen in the hydroxyl group is high; i.e., they provide evidence that predominant formation of structure II is more probable.

† The suggestion that glycolaldehyde is chemisorbed in methanol solutions [63] is not considered in detail, because the arguments in its favor are not entirely reliable.

conditions of preliminary treatment of the electrodes in the studies cited [7, 41, 58] were different (moreover, the experiments described in [41] were carried out with the use of platinized platinum). In a later investigation |59], in which the quantity of electricity consumed for oxidation of the adsorption product was determined after the electrode had been washed with a solution of the pure electrolyte, it was found that the composition of the adsorbed methanol depended on the electrode potential. According to [59], at $\varphi_r \geq 0.4$ V the adsorbed particle requires two electrons for oxidation to CO_2, while at $\varphi_r < 0.4$ V the presence of a more highly reduced form is not excluded. At the same time, Biegler [59] observed a number of substantial deviations, the cause of which are not clear, from data reported in [44-46].

Breiter [60, 61] investigated the nature of adsorbed methanol in acid solution by a combination of methods: anodic charging curves for electrodes of extended surface area, and gas-chromatographic determination of CO_2. Since the reaction of anodic oxidation of an adsorbed substance having the composition $C_sH_pO_q$ may be represented by the equation

$$C_sH_pO_q + (2s - q)H_2O = sCO_2 + (4s + p - 2q)H^+ + (4s + p - 2q)\bar{e} ,$$

the number R of electrons required for formation of a CO_2 molecule from an adsorbed particle is given by the expression

$$R = 4 + (p - 2q)/s. \qquad (VIII.9)$$

The average R for formation of a CO_2 molecule from the adsorbed methanol was found to be somewhat greater than 2. This corresponds to an average stoichiometric composition $H_2C_2O_3$ of the adsorbed particles, or to simultaneous adsorption of HCO and COOH on the electrode. However, Breiter observed considerable scatter of the R values in different experiments.

Because of the different views which have arisen, the mechanism of methanol adsorption has recently been studied more fully with the aid of certain new techniques. Podlovchenko, Frumkin, and Stenin [62] accompanied electrochemical measurements by analysis of the liquid phase for methanol, formaldehyde, and formic acid. Formation of HCOOH (or HCOOK) and CH_2O was not observed during electrochemical oxidation of the substance chemisorbed from methanol solutions in 1 N H_2SO_4 and 1 N KOH at a

Pt/Pt electrode at potentials in the double-layer region. It was thus demonstrated by direct determinations that the stably chemisorbed substance is oxidized almost completely to CO_2.

Podlovchenko et al. [64] used methyl alcohol containing labeled carbon (C^{14}) for determining the composition of the chemisorbed substance. After adsorption from a solution of the labeled alcohol the solution was replaced by $1\,N$ KOH, in which the chemisorbed residue was electrochemically oxidized. The quantity of electricity required for oxidation of the adsorbate was compared with the amount of carbonate formed. The accuracy attained was higher than in Breiter's experiments [60, 61]. The determinations showed that in oxidation of the adsorbate accumulated either on open circuit or during anodic polarization in acid or alkali the number of electrons (R) corresponding to formation of one CO_2 molecule is 2.8 ± 0.2. This value corresponds to the stoichiometric composition HCO. The somewhat low value may be due either to the presence of other species, e.g., HCOO, in the chemisorbed layer or to some constant error in determination of the quantity of electricity, arising from difficulty in applying corrections for changes of the structure of the ionic double layer and of the amount of oxygen adsorbed (see Chapter V); this error is greater in alkaline than in acid solutions.

The mechanism of chemisorption of methanol from dilute solutions on platinum − ruthenium and rhodium electrodes was investigated by comparison of the amount of hydrogen formed by dehydrogenation of methanol with the amount of organic substance adsorbed [65, 66]. In the case of a platinum − ruthenium electrode, when methanol is brought into contact with the electrode on open circuit at $\varphi_r > 0.3$ V, it must be taken into account that a considerable proportion of the hydrogen formed by dehydrogenation is expended for charging the electrode because of the high capacities in the 300-700 mV range. The determinations show that the values of Q'_H and Q_{org} are approximately equal on rhodium and platinum − ruthenium electrodes; therefore chemisorption of particles of the stoichiometric composition HCO occurs.*

---

*However, in distinction from the Pt electrode, on rhodium-coated and platinum − ruthenium electrodes the sum of Q_{org} and the amount of hydrogen adsorbed at $\varphi_r = 0$ in presence of chemisorbed methanol exceeds sQ_H by about 20-30%. Some possible causes of this result have been discussed by Petrii et al. [65, 66].

Gaseous products were evolved after methanol in fairly high concentrations ($>0.5\ M$) had been brought into contact with rhodium-coated or platinum—ruthenium electrodes on open circuit in the region of anodic φ_r close to 0. In the case of the rhodium electrode the gas contained $\sim 7\%$ methane, 11% hydrogen, and 82% argon (which had been used for clearing oxygen from the system). The gas formed in alkaline solutions also contained methane (3%), hydrogen (14%), and argon (83%). In the case of the platinum—ruthenium electrode in $1\ N\ H_2SO_4$, the gas phase was found to contain methane (7%), hydrogen (92%), and CO_2 (1%). The high hydrogen content in the gas phase may be attributed to the low values of $\varphi_r^!$ attained in experiments with concentrated methanol solutions (from ~ 0 to 30 mV). The formation of methane is due to reduction of methanol by adsorbed hydrogen. As these processes occur even at normal temperatures, it may be concluded that rhodium and platinum—ruthenium catalysts are more active than platinized platinum in reduction of methanol. Finally, the formation of CO_2 indicates that complete oxidation of methanol accompanies hydrogenation at a platinum—ruthenium electrode in concentrated solutions. In distinction from acid solutions, in the case of alkaline methanol solutions in contact with a platinum—ruthenium electrode at $\varphi_r > 0$ mainly gaseous hydrogen was evolved, containing only traces of methane (the possible causes of this effect are discussed below).

2. Identification of the Adsorption Products of Saturated Alcohols, Ethylene Glycol, Aldehyde, and Acids

The interaction of alcohols containing more than one carbon atom in the molecule with the surface of a platinum electrode is found to be more complex than the chemisorption of methanol. When these alcohols are brought into contact with a platinum electrode at potentials in the double-layer region, the potential is again shifted in the cathodic direction. However, at high concentrations ($>0.1\ M$) of these compounds in acid solutions the minimum potential is reached fairly rapidly, and the potential then either remains constant or rises to more anodic values, giving a minimum on the φ_r vs t curve (Fig. 93) [28, 67, 68]. The $\varphi_r^!$ values which become established in C_2H_5OH, C_3H_7OH, and C_4H_9OH solu-

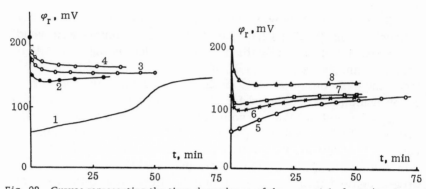

Fig. 93. Curves representing the time dependence of the potential of a Pt/Pt electrode after addition of ethanol (1-4) and butanol (5-8) as a function of the initial potential: 1, 5) 60; 2) 240; 3) 500; 4) 1045; 6) 109; 7) 196; 8) 505 mV. Supporting electrolyte 0.1 N H_2SO_4. Ethanol and butanol concentrations 0.5 M (from data in [28]).

tions are appreciably higher than the potentials of the Pt/Pt electrode in methanol solutions under the same conditions. The difference in behavior between these compounds and methanol becomes more evident from a comparison of the potential – time curves after addition of the alcohols at φ_r^0 close to 0. In this case alcohols containing more than one carbon atom in the molecule shift the potential to more positive values; this is accompanied by evolution of gaseous products. Chromatographic analysis (Table 20) showed that the gas evolved consists of hydrogenation products of the original substances.

In illustration, Fig. 93 shows curves representing the shift of potential after contact of ethanol and butanol with a Pt/Pt electrode at various potentials. It follows from Fig. 93 that hydrogenation of ethanol occurs after a certain delay; i.e., it has a certain induction period. No such delay is observed with higher alcohols.

Thus, the complex course of the change of potential in the case of these alcohols is due to transition from dehydrogenation, causing a shift of potential in the cathodic direction, to predominant hydrogenation which shifts φ_r in the anodic direction; the latter process begins after some delay.

Evolution of gaseous products is also observed when alcohols are brought into contact with a Pt/Pt electrode polarized to φ_r in the double-layer region (Table 20).

TABLE 20

Alcohol	Concentration, M	φ_r^0, mV	φ_r', mV	Gas composition, vol.%			
				CH_4	C_2H_6	C_3H_8	C_4H_{10}
C_2H_5OH	2	57	159	13.4	86.6	—	—
C_2H_5OH	2	499	157	68.2	31.8	—	—
C_3H_7OH	1	54	140	—	19 3	76.3	4.4
C_4H_9OH	Satd.	41	131	1	1	23	76
C_4H_9OH	"	511	138	—	2.8	69.4	27.8

After an alcohol has been brought into contact with an electrode saturated with hydrogen, the hydrocarbon with the same number of carbon atoms in the molecule as the original alcohol predominates in the gas evolved, whereas contact of a "degassed" electrode (the term sometimes used to describe an electrode polarized to potentials in the double-layer region) with alcohols results in predominant cleavage of the molecules at the C_1-C_2 bond. Butanol also breaks down to a small extent at the C_2-C_3 bond. Thus, a "degassed" Pt/Pt electrode surface has a stronger destructive effect on alcohols than a surface covered with adsorbed hydrogen.

Formation of hydrogenation products when alcohols are introduced in the double-layer φ_r region apparently indicates that an autohydrogenation reaction occurs. This is accompanied by oxidation of some of the alcohol molecules and its decomposition products. Therefore a reaction of the Cannizzaro type [69] occurs at the electrode surface. Examination of thermodynamic data on saturated alcohols [67, 70] suggests that hydrogenation of saturated alcohols is to be expected on thermodynamic grounds.

A fact of special interest is that when a Pt/Pt electrode is brought into contact with propanol solutions butane appears in the gas phase. It should be regarded as the product of partial recombination of C_2H_5 radicals formed by rupture of C_1-C_2 bonds.

When alcohols are introduced at φ_r of the hydrogen region, dehydrogenation, autohydrogenation, and splitting processes begin after removal of adsorbed hydrogen from a part of the surface. Formation of gaseous products continues for a long time even after the potential has become virtually constant, and the amount of gases formed is greater than could be obtained by complete exhaustion of all the hydrogen adsorbed on the surface. The composition of the evolving gas varies during the experiment, with a progressive increase of the relative content of the alcohol breakdown products. This is the result of the gradual decrease of coverage of the surface by adsorbed hydrogen.

Charging curves determined after the electrodes have been washed to remove organic substances reveal the presence of small amounts of adsorbed hydrogen in all cases,* and have arrest corresponding to oxidation of chemisorbed substances. These arrests are considerably longer than the corresponding arrests observed in the case of methanol (Fig. 94). In the case of the compounds under consideration the ratio of the amount of organic substance adsorbed at $\varphi_r^!$ to the amount of hydrogen adsorbed is greater than unity, and increases with increasing surface coverage and length of the hydrocarbon chain. While oxidation of chemisorbed methanol is complete at $\varphi_r \sim 800$ mV (at low current densities or with slow potentiodynamic sweeps), considerable amounts of the chemisorption products of higher alcohols are oxidized at potentials of the oxygen region.

Accurate determination of the composition of the chemisorbed substance presents a complicated problem in this case. As the result of dehydrogenation, hydrogenation, and autohydrogenation reactions and splitting of the molecules, adsorption of particles such as HCO, RHCOH, C_nH_{2n+1}, C_nH_{2n}, etc., on the electrode surface is to be expected. The experimental results which will be discussed below appear to indicate the presence of HCO_{ads} and various species with larger numbers of carbon atoms on the surface. Indeed, methane is not formed in appreciable amounts during hydrogenation of propanol and butanol although C_1-C_2 bond rupture occurs. It may be supposed that methanol is formed in solution or that particles containing C_1 atoms are chemisorbed on the surface. The latter suggestion appears more probable. Oxidation of the chemisorbed substances begins at approximately the same φ_r which, e.g., in 0.1 N H_2SO_4 at i = 10^{-4} A/cm^2 is 500 ± 20 mV; this can also be attributed to presence of particles of the

---

* During the first instant of ethanol chemisorption on open circuit, when dehydrogenation predominates, a considerable amount of adsorbed hydrogen may be detected; this amount gradually decreases [67].

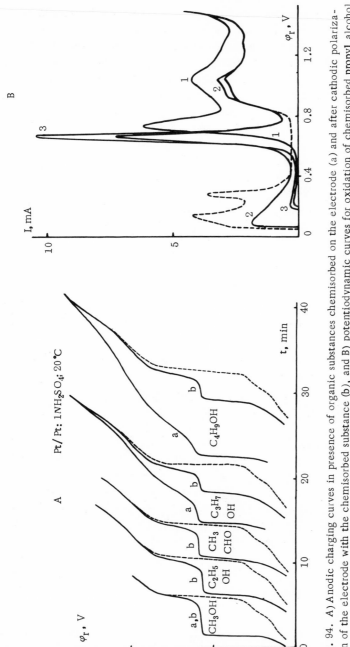

Fig. 94. A) Anodic charging curves in presence of organic substances chemisorbed on the electrode (a) and after cathodic polarization of the electrode with the chemisorbed substance (b), and B) potentiodynamic curves for oxidation of chemisorbed propyl alcohol [before (1) and after (2) cathodic polarization of the electrode] and chemisorbed methyl alcohol (3) (from data in [71]). Curves recorded in the supporting electrolyte are dashed. Methanol was adsorbed at $\varphi_r = 480$ mV, and the other alcohols on open circuit (added at 500 mV); concentration of alcohols 0.5 M. Current density in recording the charging curves $3 \cdot 10^{-4}$ A/cm^2. Sweep rate in recording the potentiodynamic curves 4.9 mV/sec.

same composition in the chemisorbed layers. Experiments on cathodic reduction of the adsorption products [71] * confirm that more than one type of adsorbed particle is present and that one of these has the same composition in the products of all alcohol chemisorption. Whereas in the case of chemisorbed methanol cathodic polarization of the electrode has virtually no influence on the amount of chemisorption products and the character of their oxidation,[†] the amount of products of chemisorption of other alcohols decreases substantially.[‡] Comparison (Fig. 94) of galvanostatic and potentiodynamic curves for oxidation of chemisorbed methanol with those for the products remaining after cathodic polarization of the electrode in the case of chemisorption of other saturated alcohols suggests that a substance of the same composition is adsorbed on the electrode in all cases; its composition is the same as that chemisorbed from methanol solution. Not only do the oxidation potentials coincide, but the ratios of the quantities of electricity required for removal of adsorbed hydrogen and of the organic substance after cathodic polarization of the electrode to $_S Q_H$ is the same and fairly close to unity (~1.15) in all cases. Table 18 shows that the same ratio is characteristic of particles chemisorbed in methanol solutions.

It follows from Fig. 94 that desorption of "difficultly oxidizable" particles from the electrode surface occurs during cathodic polarization of the electrode. Chromatographic analysis of the gases evolved during cathodic polarization revealed that the main hydrogenation products were ethane in the case of chemisorbed ethanol, and ethane and propane (C_2H_6 > C_3H_8, with traces of butane and isobutane) in the case of propanol. Methane was not present in the gas phase in amounts exceeding 0.05 vol.% (this was the sensitivity of the chromatographic determination of methane in [71]). Thus, the "difficultly oxidizable" particles appear to be particles containing more than one carbon atom.

---

*This method for elucidation of the nature of the chemisorption products of organic compounds was proposed by Niedrach [22].

[†]A deeper study demonstrated that cathodic polarization has some influence on chemisorbed methanol and on its reduction to methane. However, the observed effects are small [62].

[‡]It will be shown below that in this respect the behavior of alcohols and hydrocarbons (with more than two carbon atoms in the molecule) on contact with platinum surfaces is analogous.

Chemisorption of ethylene glycol on platinum has been studied by Podlovchenko and Stenin [71] and by Weber et al. [72]. According to [72], at a temperature of 70°C in 1 N H_2SO_4 the quantity of electricity produced by ionization of the hydrogen formed as the result of dehydrogenation of ethylene glycol at $\varphi_r =$ const is always equal to the quantity of electricity expended for oxidation of the carbon-containing residues formed in the process. Assuming that at low temperatures oxidation of ethylene glycol proceeds with formation of oxalic acid,* Weber et al. [72] represented the chemisorption process on a smooth platinum electrode as follows:

$$\text{HOCH}_2-\text{CH}_2\text{OH} \rightarrow [\text{HO}-\underset{\underset{\text{Pt}}{\diagup}\underset{\text{Pt}}{\diagdown}}{\text{C}}\text{------}\underset{\underset{\text{Pt}}{\diagup}\underset{\text{Pt}}{\diagdown}}{\text{C}}-\text{OH}]_{\text{ads}} + 4\text{H}_{\text{ads}} \qquad (\text{VIII.10})$$

Analogous determinations at higher temperatures, at which oxidation of adsorbed ethylene glycol to CO_2 was presumed, led to the conclusion that $C-C$ bond rupture occurs in the molecule with formation of two COH particles, which are also formed during chemisorption of methanol:

$$\text{HO}-\text{CH}_2-\text{CH}_2\text{OH} \rightarrow 2[\underset{\underset{\text{Pt Pt Pt}}{\diagup\mid\diagdown}}{\text{COH}}]_{\text{ads}} + 4\text{H}_{\text{ads}} \qquad (\text{VIII.11})$$

Somewhat different conclusions were drawn [71] in a study of chemisorption of ethylene glycol at a Pt/Pt electrode in 1 N H_2SO_4, by the method of electrochemical oxidation in the adsorbed layer. The curve representing the shift of potential after ethylene glycol had been brought into contact with an electrode previously polarized to $\varphi_r = 500$ mV had a minimum. This suggests that dehydrogenation is accompanied by reduction of ethylene glycol by adsorbed hydrogen. However, evolution of gaseous products was not observed. Ethylene glycol chemisorbed under these conditions is partially reduced to ethane during cathodic polarization. The electrochemical oxidation curve of the product remaining after reduction of chemisorbed ethylene glycol is similar to the corresponding curve for chemisorbed methanol. Therefore ethylene glycol chemisorbed on a Pt/Pt electrode probably consists both

---

*The conclusions regarding the oxidation products of adsorbed ethylene glycol were drawn from the results of analysis of the products of ethylene glycol oxidation under stationary conditions [73]; it will be shown below that this is not always valid.

of HCO particles and of particles containing two carbon atoms, which may undergo cathodic reduction.

Adsorption of aldehydes (formaldehyde and acetaldehyde) on platinum electrodes has been studied by Petrii, Podlovchenko, et al. [28, 67, 68, 71]. Dehydrogenation and hydrogenation reactions occur when formaldehyde comes into contact with platinum; the final hydrogenation product appears to be methanol.* A peculiarity in the behavior of formaldehyde is that, despite the occurrence of hydrogenation, the potentials established in its presence are nearly as low as in methanol solutions. In alkaline formaldehyde solutions at elevated temperatures even more negative potentials than that of the reversible hydrogen electrode are reached, and evolution of hydrogen occurs. After contact with formaldehyde, a small amount of adsorbed hydrogen and a considerable amount of chemisorbed substance is found on the electrode surface. This substance is oxidized in the same range of potentials as chemisorbed methanol (Fig. 95). The total amount of chemisorbed substance and of hydrogen which can be adsorbed in its presence is close to $_sQ_H$. The organic product is not desorbed during cathodic polarization of the electrode and therefore behaves similarly to the product of methanol chemisorption.

Acetaldehyde undergoes dehydrogenation and destructive hydrogenation during chemisorption on a Pt/Pt electrode. These processes have higher rates than the corresponding processes in the case of saturated alcohols. The products of acetaldehyde chemisorption are hydrogen (in small amounts) and an organic substance of heterogeneous composition, containing both HCO particles and particles with two carbon atoms (Fig. 94). The latter are removed in the form of ethane during cathodic polarization of the electrode [71].

Trassatti and Formaro [76, 77] used the cathodic and anodic galvanostatic pulse method for studying chemisorption of glycoaldehyde (CH_2OHCHO) on smooth platinum. At $\varphi_r = 0.4$ V and θ ≤0.7 adsorption apparently proceeds in accordance with the scheme

$$CH_2OH - CHO \rightarrow \underset{\underset{Pt\ \ \ Pt}{\diagdown}}{C} - OH - \underset{\underset{Pt}{|}}{CO} + 3H_{ads} \qquad (VIII.12)$$

_____

*According to Johnson and Kuhn [74] a polymeric product is formed when platinum comes into contact with concentrated formaldehyde solutions. Polymerization of formaldehyde during adsorption was also presumed by Koch [75].

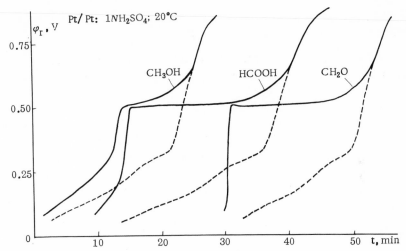

Fig. 95. Anodic charging curves in presence of methanol, formaldehyde, and formic acid chemisorbed on open circuit. Organic substances added at 500 mV; concentration 0.5 M. Curves for the pure electrolyte are indicated by dash lines. Current density $1 \cdot 10^{-4}$ A/cm^2 (from data in [68]).

At $\theta > 0.7$ the molecule loses only two hydrogen atoms during adsorption. The nature of the adsorbed particles also depends on the electrode potential.

Much attention is devoted in the literature to studies of chemisorption of formic acid [2, 8, 25, 45, 46, 49, 68, 74, 78-96]. In certain cases the interest in this process was apparently due to its presumed simplicity. In reality, however, experimental investigations led to the opposite conclusion, and the views put forward on the mechanisms of HCOOH adsorption are contradictory.

Sufficiently reliable and unambiguous conclusions were obtained only in studies of the chemisorption of HCOOH on palladium electrodes, owing to the use of different experimental methods. For example, an elegant demonstration of the dehydrogenation of formic acid during adsorption was given by Bagotskii, Vasil'ev, and Polyak [45, 46, 78] with the aid of a palladium membrane electrode which divided the cell into two parts (Fig. 96). Both parts of the cell were filled with $1\,N$ H$_2$SO$_4$. Each part contained an auxiliary electrode and a reference electrode. A potential of 0.5 V was applied to the membrane surface in the left-hand side of the cell, and 0.6 V in the right-hand side, with the aid of potentiostats.

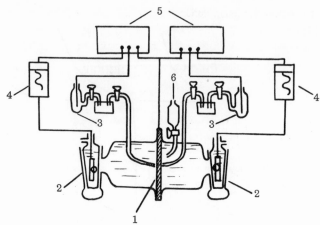

Fig. 96. Cell with a palladium membrane and two potentiostatic circuits for investigation of dehydrogenation of organic substances: 1) palladium membrane; 2) auxiliary electrodes; 3) reference electrodes; 4) current recorders; 5) potentiostats; 6) vessel for introduction of the organic substance (from data in [78]).

No current flowed through the external circuits. When formic acid was added to the right-hand side, an anodic current due to oxidation of the organic compound at the palladium electrode arose in the right-hand external circuit. After some time an anodic current also appeared in the left-hand circuit. This current can be attributed only to ionization of hydrogen atoms which diffused through the membrane from right to left. As formation of measurable amounts of hydrogen owing to discharge of hydrogen ions is excluded at 0.6 V, the only possible source of hydrogen atoms is dehydrogenation of formic acid on palladium.*

Adsorption of formic acid at an active palladium electrode on open circuit is accompanied by a shift of potential in the negative direction, to values on the negative side of φ_r of an equilibrium

---

*Similar experiments with methanol solutions are difficult to carry out, because dehydrogenation of methanol on palladium in acid solution proceeds at only a low rate [81]. Palladium membranes have been used for investigating dissociative adsorption [97, 98]. However, in these studies appearance of hydrogen on the opposite side of the membrane was observed only after the potential of the side in contact with methanol reached values at which formation of adsorbed hydrogen by reactions (VIII.2) or (VIII.2a) is also possible. Therefore the results reported in [97, 98] do not provide a direct proof of the dehydrogenation mechanism.

hydrogen electrode in the same solution [94]. At $\varphi_r < 0.8$ V, a gas consisting of CO_2 and H_2 is evolved. For elucidation of the mechanism of formation of molecular hydrogen, Polyak et al. [94] determined the isotope composition of gases evolved during decomposition of DCOOH in light water and of HCOOK in heavy water. The analytical results are given in Table 21.

The content of the tracer in the gas is appreciably higher than the equilibrium value corresponding to complete isotope exchange. Hydrogen is apparently formed immediately in molecular form in direct proximity to the electrode surface by the reactions:

$$2HCOOH \rightarrow H_2 + 2HCOO_{ads} \qquad (VIII.13)$$

or

$$HCOOH \rightarrow H_2 + CO_2 \qquad (VIII.13a)$$

It was shown in [68] that the potential of a Pt/Pt electrode is shifted in the cathodic direction in presence of HCOOH, although under certain conditions a minimum on the φ_r vs t curve can also be observed [79]. Considerable amounts of chemisorption products are detected at φ_r^0 close to 0. The ratio of the amounts of adsorbed hydrogen and chemisorbed substance does not remain constant with variation of the HCOOH concentration or of the adsorption time. In the light of these results it must be assumed that at $\varphi_r > 0$ hydrogenation of formic acid occurs in addition to dehydrogenation. Despite this, the sum of Q_H and Q_{org} after adsorption of HCOOH at different φ_r^0 always remains approximately constant and close to $_sQ_H$, as in the case of methanol (Fig. 95). Brummer and Makrides [86] reached a similar conclusion. The oxidation regions of the products of chemisorption of HCOOH and of CH_3OH

TABLE 21

Composition of solution	Isotope composition of gas evolved, %		
	H_2	HD	D_2
0.25 N DCOOH + 1 N H_2SO_4 + H_2O	98.37	1.63	—
0.25 N DCOOH + H_2O	92.5	7.5	—
0.25 N DCOOK + 1 N KOH + H_2O	57.2	33.6	9.2
0.25 N HCOOK + 1 N KOD + D_2O	15.15	35.85	49

coincide. Apparently in HCOOH solutions HCO particles formed
by hydrogenation of formic acid are chemisorbed, or else simultaneous chemisorption of HCO and COOH occurs. The latter al-
ternative corresponds stoichiometrically to adsorption of
$$\underset{}{HC}^{\!\!\!\!O}_{\!\!\!\!/\!\!/} -COOH$$
molecules [68, 86]. According to Binder et al. [96] the
firmly chemisorbed product formed in presence of HCOOH and
$HCOO^-$ on a Raney platinum electrode requires 1.6 electrons per
site occupied for oxidation to CO_2 in acid solution, and 0.4-0.5
electron in alkaline solution. Simultaneous adsorption of HCO and
COOH on the electrode was postulated. The formation of HCO
possibly occurs as the result of reaction (VIII.14) [79]:

$$HCOOH + H_{ads} \rightarrow HCO_{ads} + H_2O . \qquad (VIII.14)$$

After addition of HCOOH at $\varphi_r \leq 100$ mV the potential of a
Pt/Pt electrode in 0.1 and $1 N$ H_2SO_4 reaches values close to 0
(4-20 mV). In such cases at $\varphi_r \leq 30$ mV a gas is evolved, consisting only of hydrogen and the inert gas (nitrogen or argon) used
for removing oxygen from the solution [68]. Bagotskii et al. [93]
attributed the evolution of hydrogen to a reaction between formic
acid and adsorbed hydrogen:

$$H_{ads} + HCOOH \rightarrow H_2 + COOH_{ads} . \qquad (VIII.15)$$

The occurrence of this reaction was demonstrated in adsorption of HCOOH from the gas phase on catalysts covered with H_{ads}
[10].

Adsorption determinations showed [93] that at $\varphi_r \gtrsim 0.4$ V adsorption of HCOOH on smooth platinum is accompanied by dehydrogenation of the formic acid molecule, with formation of a chemisorbed COOH particle which occupies one adsorption site. At lower potentials the amount of hydrogen formed during adsorption
of HCOOH and ionized is considerably less than the amount of
chemisorbed substance [88, 93]. Breiter [88] interpreted this
result as proof of adsorption of HCOOH molecules on the electrode surface. However, it can probably be explained by the
occurrence of a hydrogenation reaction, which leads to nonelectrochemical removal of hydrogen from the surface, during adsorption of HCOOH [68, 79, 92, 93].

The foregoing conclusions regarding adsorption of organic
compounds on platinum are of general character. This is evident

TABLE 22

Compound, conc.	Electrolyte	φ_r^0, mV	Δt, h	φ_r', mV	Gas composition,[*] vol. %			
					H_2	CH_4	C_2H_6	C_3H_8
C_2H_5OH	$1N$ H_2SO_4	504	16	51	—	92	8	—
	$1N$ NaOH	500	2	36	3	96	1	—
	$1N$ H_2SO_4	60	2.5	28	8	55	37	—
	$1N$ NaOH	60	1.5	7	9	90	Trace	—
C_3H_7OH	$1N$ H_2SO_4	400	0.75	76	—	35	50	15
	$1N$ NaOH	500	12	33	—	31	68	Trace
	$1N$ H_2SO_4	60	0.75	35	—	30	40	30
	$1N$ NaOH	60	1.5	7.8	--	11	88	Trace
CH_3CHO	$1N$ H_2SO_4	500	3.5	187	—	95	5	—
		60	12	188	—	89	11	—

[*]Gases not analyzed for butane

from Tables 22 and 23, which contain the results of analysis of the gaseous products evolved during contact of ethanol, propanol, and acetaldehyde with platinum – ruthenium (Table 22) and rhodium (Table 23) catalysts [66, 99].

In all cases of chemisorption of organic compounds on these metals, as on a platinum electrode, formation of adsorbed hydrogen can be detected. The quantity of electricity consumed for

TABLE 23

Compound, conc.	Electrolyte	φ_r^0, mV	Δt, h	φ_r', mV	Gas composition,[*] vol. %				
					H_2	CH_4	C_2H_6	C_3H_8	C_4H_{10}
C_2H_5OH	$1N$ H_2SO_4	46	19	46	3	95	2	—	—
		400	58	6	2	96	2	—	—
	$1N$ NaOH	270	5	20	14	85	1	—	—
C_3H_7OH	$1N$ H_2SO_4	79	19	22	6	9	53	30	2
		453	49	26	1	16	77	5	1
	$1N$ NaOH	61	5	19	33	10	53	1	—
		453	64	2	5	6	86	2	1
CH_3CHO	$1N$ H_2SO_4	77	47	2 5	5	94	1	—	—
		500	95	2	2	98	—	—	—

[*]Gases not analyzed for butane.

oxidation of the difficultly oxidizable particles formed during
adsorption of alcohols increases with the length of the carbon
chain. The difficultly oxidizable particles are removed during
cathodic reduction of the adsorbed substances. The curves representing the variation of potential after contact of CH_3CHO with
Pt/Ru and Rh electrodes are similar to the corresponding curves
for Pt/Pt both in shape and in the values of the established potentials.

At the same time, there are characteristic peculiarities and
distinctions in the behavior of different metals on the platinum
group in solutions of organic compounds. After propanol and
ethanol have been brought into contact with rhodium and platinum—ruthenium electrodes, regardless of the potential at which
the addition is made, φ_r shifts in the negative direction only and
φ_r' values close to the reversible hydrogen potential become established. This result, in conjunction with the accompanying rapid
evolution of products of alcohol reduction, indicates that the hydrogenation rate of organic compounds at low anodic potentials is
fairly high, and exceeds over lengthy periods of time the rate of
their reduction by adsorbed hydrogen. *

Comparison of the data in Tables 20, 22, and 23, and the considerably higher rate of gas evolution on platinum—ruthenium and
rhodium catalysts than on a Pt/Pt electrode, lead to the conclusion that these metals exert a stronger dehydrogenating, hydro-
genating, and destructive action on aldehyde and alcohols. For
example, it is noteworthy that reduction of acetaldehyde on rhodium is accompanied by complete breakdown of the molecule, and
only very small amounts of ethane are found in the gas phase
(Table 23). Reactions involving rupture of C_2-C_3 as well as of
C_1-C_2 bonds are also characteristic of rhodium and platinum—
ruthenium catalysts. The processes of HCOOH dehydrogenation
and hydrogenation, which were also observed on platinum, proceed very intensively on a rhodium electrode, and CO_2 and CH_4
are found in the gas phase.

---

*At a Pt/Pt electrode in alkaline solutions in presence of ethanol and propanol hy-
drogenation products are evolved, but φ_r is again shifted only in the cathodic direction.

3. Adsorption of Carbon Monoxide and Carbon Dioxide

Carbon monoxide, CO, is of interest as a compound of simple composition, the adsorption characteristics of which have much in common with those of organic compounds, and which can be used in identification of chemisorption products.* Some interesting results relating to the mechanism of adsorption of CO from aqueous solutions on platinum metals were obtained a relatively long time ago (e.g., [100, 101, 102]). For example, Wieland [100] observed oxidation of CO to CO_2 by oxygen in water in presence of metallic palladium, which became saturated with hydrogen in the process. Bruns and Zarubina [102], in the course of a study of adsorption of electrolytes on platinized carbon in presence of CO, found that the potential of carbon in solutions containing CO is shifted somewhat in the oxygen direction relative to the hydrogen potential. Platinized carbon adsorbs Na^+ ions in a CO atmosphere and therefore becomes negatively charged in alkaline solution. At the same time an equivalent amount of CO_2 appears in the solution. Two mechanisms were suggested for this adsorption:

$$CO + H_2O \rightarrow CO_2 + H_2 \qquad \text{(VIII.16)}$$

or

$$CO + 2Na^+ + 2OH^- \rightarrow CO_2 + H_2O + 2Na^+ + 2\bar{e} . \qquad \text{(VIII.17)}$$

Adsorption of CO on platinum has recently been studied by galvanostatic [103-113] and potentiodynamic [87, 114-116] methods.

According to Sokol'skii, Fasman, and co-workers [104, 107], when CO is brought into contact with a Pt/Pt electrode or platinum black polarized to $\varphi_r \sim 0.5$ V in $1 N H_2SO_4$ the potential initially shifts slowly in the cathodic direction. However, the subsequent course of the potential strongly depends on the temperature. At $t^0 \geq 30°C$ the shift of φ_r in the cathodic direction is

---

*Interest in studies of adsorption of CO on platinum metals is also due to attempts to use it, and hydrogen made by conversion of hydrocarbons and containing an admixture of carbon monoxide, in fuel cells.

followed by increase of the potential toward more anodic values. The anodic charging curves obtained after removal of dissolved CO have a region corresponding to ionization of adsorbed hydrogen, while in the region $\varphi_r \sim 0.55$ V, an arrest appears, corresponding to oxidation of the chemisorbed product formed during interaction of CO with platinum. Analysis of the gas phase after saturation of a platinum electrode with carbon monoxide showed the presence of approximately equimolecular amounts of carbon dioxide and hydrogen. These results were attributed to conversion of carbon monoxide on platinum in aqueous solutions in accordance with Eq. (VIII.16). Veselovskii et al. [117] also noted the possibility of carbon monoxide conversion on Pt (and on Pd). According to Binder et al. [96], the reaction between carbon monoxide and water in presence of Raney platinum may also proceed in accordance with the scheme

$$2CO + H_2O \rightarrow COOH_{ads} + HCO_{ads} \qquad \text{(VIII.18)}$$

so that the same species as are adsorbed on platinum from methanol and formic acid solutions appear on the surface.

At 50-70°C, predominant formation of platinum dicarbonyl, $Pt(CO)_2$, on the surface is presumed [107]. However, it is possible that under these conditions reactions between carbon monoxide and adsorbed hydrogen [28], i.e., processes of the Fischer–Tropsch synthesis type [118], leading to a shift of φ_r in the anodic direction, begin. For example, Binder et al. [96] postulate the reaction

$$H_{ads} + CO \rightarrow HCO_{ads} \ . \qquad \text{(VIII.19)}$$

The data reported by Sokol'skii, Fasman et al. [107-111] indicate that the processes occurring in presence of carbon monoxide are strongly dependent on the nature of the metal (Pt, Rh, Pd, Ru, Os) and on the composition of the solution (H_2SO_4, $HClO_4$, HCl, KOH). The most stable chemisorption of CO is observed on Pd electrodes. Heat treatment is needed to decompose the surface compounds formed in this case. The influence of the nature of a catalyst on its behavior in presence of CO has been noted by Niedrach, McKee, et al. [119-121].

From a comparison of the quantities of electricity required for oxidation of adsorbed CO and for formation of an oxygen monolayer on platinum, Warner and Schuldiner [105] presumed the existence of physically adsorbed CO on a monolayer of chemisorbed carbon monoxide. Brummer and Ford [106], Breiter [112], and Gilman [114] also concluded that, in addition to firmly chemisorbed carbon monoxide, relatively weakly bonded particles, which are desorbed when CO is removed from the solution, are present.*

Gilman [114-116, 122] used the MPP method for detailed study of adsorption of CO on platinum and rhodium. Maximum coverage of the platinum surface with CO in 1 N HClO$_4$ is attained in the φ_r range from 0.4 to 0.9 V, when about 25% of the surface platinum atoms still remain free. Gilman concluded that two structures of adsorbed CO are possible: linear or one-site, and bridged or two-site. Two electrons correspond to each occupied adsorption site in oxidation of the first form, and one electron in oxidation of the second. The quantity of electricity Q_{CO}^t involved in oxidation of the whole adsorbed layer can be represented by the equation

$$Q_{CO}^t = Q_{CO}^b + Q_{CO}^l = 2\Delta_S Q_H - Q_{CO}^b, \qquad \text{(VIII.20)}$$

where $\Delta_S Q_H$ is the decrease of the amount of adsorbed hydrogen when CO is adsorbed on the platinum surface; Q_{CO}^b is the quantity of electricity corresponding to two-site adsorption; Q_{CO}^l is the quantity of electricity corresponding to one-site adsorption. It was found that at 30° only two-site adsorption occurs initially, and only after Q_{CO}^b has reached 0.44Q_{CO}^t does one-site adsorption begin and then Q_{CO}^l may increase to 0.46Q_{CO}^t. These data on two types of adsorbed CO on the platinum surface are consistent with the results of infrared spectroscopic studies of adsorption of CO from the gas phase [125, 126], and of simultaneous electrochemical and gas-chromatographic determinations of CO adsorption [112].

Potentiodynamic curves for oxidation of carbon monoxide adsorbed on rhodium at various φ_r in the range from -0.2 to 0.5 V in 4 N H$_2$SO$_4$ at 80°C are almost identical [122]. This suggests that the structure of the adsorbed species is constant in this φ_r range. According to Gilman, one CO molecule occupies one hydrogen adsorption site on rhodium, in distinction from platinum; i.e., only the linear structure of adsorbed CO is present.

---

*Very different values are given in various publications [105, 106, 114-116] for the quantity of electricity required for oxidation of adsorbed CO at maximum adsorption. The causes of the discrepancies, which are apparently due to the peculiarities of the experimental techniques and methods used in evaluation of the results, have been discussed in the literature [116, 123, 124].

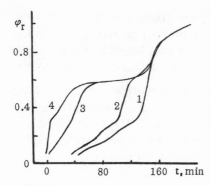

Fig. 97. Charging curves of a Pt/Pt electrode in 0.5 M H_2SO_4 (1), and in presence of carbon monoxide adsorbed on open circuit after different time intervals from the start of adsorption: 2) 100 sec; 3) 500 sec; 4) 2000 sec. Current density 1 mA/cm² (from data in [113]).

According to Breiter [113], carbon monoxide is adsorbed on a Pt/Pt electrode in $1\,N$ H_2SO_4 and $1\,N$ KOH on open circuit (φ_r^0 = 0.1 V) with formation of two types of adsorbed particles, which are oxidized in different potential ranges. The more easily oxidized particles appear when coverage of the surface by adsorbed carbon dioxide approaches completion (Fig. 97). Oxidation of these particles begins at lower φ_r with increase of the solution pH. On the other hand, the oxidation φ_r of particles of the second type is independent of pH. The amount of the more difficultly oxidized species is ~0.79 of the total amount of adsorbed particles, expressed in electrical units.* The ratio of $2\Delta_sQ_H$ to the maximum amount of CO adsorbed is, on the average, 1.37 in acid solution; i.e., the number of platinum atoms per adsorbed molecule is somewhat greater than unity. From this Breiter concluded that the linear and bridged forms of CO_{ads} are present on the surface.

Adsorption of CO_2, first reported by Giner [24], was subsequently studied by a number of investigators. Johnson and Kuhn [74] showed that no products of CO_2 adsorption pass into solution. They correlated the quantity of electricity expended for oxidation of $(CO_2)_r$ in alkaline solution with the amount of carbonate formed, and found that a product formed by reduction of CO_2 with addition of two electrons (possibly carbon monoxide) is adsorbed on the sur-

---

*A similar result was obtained by McKee [127] for weakly and strongly bonded CO adsorbed at room temperature from the gas phase on platinum powder.

faces. Vielstich and Vogel [84] suggested that formate radicals are adsorbed in presence of CO_2.*

Breiter [112] found with the aid of charging curves and gas chromatography that two electrons are required for formation of one CO_2 molecule by oxidation of each adsorbed particle. The charging curves determined in presence of this product and in presence of CO_{ads} virtually coincide in the potential region of the arrest. However, in these cases the oxidation potentials were somewhat higher than the oxidation potentials of the products of CH_3OH, $HCOOH$, and CH_2O adsorption.

This last result for $(CO_2)_r$ was not confirmed in later publications [79, 129], the authors of which consider that the particles adsorbed on a Pt/Pt electrode as the result of interaction of CO_2 with H_{ads} are largely the same as during anodic polarization of the electrode in CH_3OH, CH_2O, and $HCOOH$ solutions. The potentiodynamic and charging curves in presence of these substances are very similar, while the polarization curves for their oxidation almost coincide (Fig. 98). However, according to Breiter [130], the kinetic relationships in oxidation of adsorbed CO, CO_2, CH_3OH, and $HCOOH$, which are very similar at low temperatures, begin to differ with rise of temperature. According to Brummer and Turner [131] $(CO_2)_r$ in $12\,M\,H_3PO_4$ at 130° already consists of two different species. The greater proportion of the adsorption product requires about 1.2 e per platinum atom occupied by it for oxidation to CO_2, and is oxidized at relatively low potentials.

To conclude this section, concerned with the mechanism of adsorption of compounds of the $C_xH_yO_z$ type on platinum metals, it is of interest to compare the results of electrochemical studies of the chemisorption of these compounds on catalysts with data on chemisorption from the gas phase. There are very few direct experimental proofs of the structure of particles adsorbed under these conditions, and in certain instances the conclusions drawn by different investigators are contradictory. Infrared spectro-

---

*Piersma et al. [128] explained the results obtained in adsorption of CO_2 on platinum with the assumption that the adsorbed carbon dioxide shields H_{ads} and hinders its oxidation. This hypothesis seems improbable in the light of the data presented below.

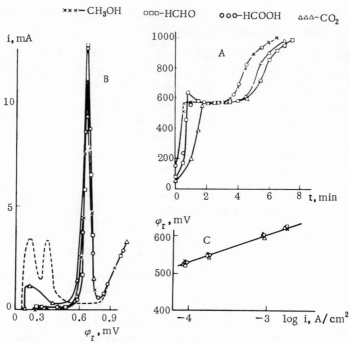

Fig. 98. Curves for electrochemical oxidation of $(CO_2)_r$ and of CH_3OH, HCHO, and HCOOH chemisorbed during anodic polarization: A) Galvanostatic curves, i = 0.36 mA/cm^2; B) potentiodynamic curves, potential sweep rate 5 mV/sec; the dashed curve corresponds to the pure electrolyte, 1 N H$_2$SO$_4$; C) polarization curves for oxidation of the chemisorbed products (from data in [129]).

scopy is one of the most powerful methods for determining the structures of adsorbed particles. The IR spectra of adsorbed HCOOH have been studied in detail by a number of investigators [132]. The presence of CO, formate ions, covalently bonded formate, and other less well-defined species on catalyst surfaces has been reported. Blyholder et al. [17] obtained IR spectra of various nickel (and also Fe and Co) catalysts after contact with certain organic compounds. The IR spectra of adsorbed saturated alcohols of normal structure (from methyl to n-butyl), acetaldehyde, methyl vinyl ether, and ethylene oxide contain bands corresponding to chemisorbed CO and to acyl structures (H−$\overset{\mid}{C}$=O), which are stable surface species. Absorption bands corresponding to CH$_3$, CH$_2$, and certain other hydrogen-containing species are also present. Secondary and tertiary alcohols, acetone, methyl ethyl ketone, and diethyl ether do not form adsorbed particles on nickel which can be detected with the aid of IR spectra. From a comparison of these results with the data discussed earlier it can be concluded that adsorption of organic compounds from solutions has certain features in common with adsorption from the gas phase.*

4. Behavior of Platinum-Metal Electrodes in Solutions of Hydrocarbons

Attempts to use hydrocarbons as electrochemical fuel in fuel cells have resulted in the appearance of numerous publications concerned with the mechanism of adsorption of hydrocarbons on the platinum metals.[†] Niedrach [22] showed that the behavior of hydrocarbons depends on their structure; methane, higher saturated hydrocarbons, and unsaturated hydrocarbons differ in this respect. Volumetric measurements [22] showed that the volume of different gases chemisorbed on platinum increases in the following series[‡]:

$$CH_4 \ll C_2H_6, \ C_3H_8 < C_2H_4, \ C_3H_6 < H_2.$$

Saturated hydrocarbons are adsorbed on platinum only from acid solutions; the rate of adsorption from alkaline solutions is very low.*

---

*In a discussion of the nature of the bonds between adsorbed particles and metallic surfaces [17] an analogy is presumed between the chemistry of metallic surface compounds and the chemistry of organometallic compounds. Direct confirmation of this analogy can be obtained from the properties of isolated organometallic complexes. It has been found, for example, that ethanol reacts with certain complex compounds of ruthenium to form CO-containing complexes and CH_4 [133]. According to Blyholder et al. [17], the formation of CO and acyl structures on catalytically active surfaces implies that the general principles of stability of organometallic compounds are directly applicable to adsorption on metallic surfaces.

†The development of American research in this field was stimulated by the work of Heath and Worsham [134], who were the first to find that saturated hydrocarbons can be oxidized at low temperatures in electrolyte solutions [23, 135].

‡This is also valid for adsorption from the gas phase [10].

§The rate of adsorption of saturated hydrocarbons from alkaline solutions seems to become appreciable [30] when a platinum catalyst deposited on carbon is used.

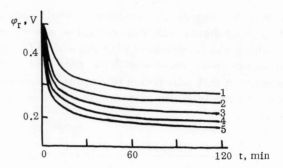

Fig. 99. Shift of the potential of a Pt/Pt electrode in 1 N H_2SO_4 in presence of methane at
various temperatures (°C): 1) 20; 2) 40; 3) 60;
4) 80; 5) 95. Methane partial pressure 1 atm.
Initial potential 500 mV (from data in [138]).

Some hypotheses regarding the mechanism of chemisorption
of the simplest hydrocarbon, methane, can be put forward on the
basis of determinations reported in the literature [136-139]. As
the adsorbability of methane on platinum is lower than that of
other hydrocarbons, adsorption measurements are possible only
with the use of highly pure methane and of electrolyte solutions
carefully freed from surface-active substances. When methane is
brought into contact with a platinum electrode on open circuit, the
potential shifts in the cathodic direction. Even after a considerable
time $d\varphi_r/dt < 0$ (Fig. 99). Charging curves determined after removal of dissolved methane by a current of argon show the presence of hydrogen and of a chemisorbed organic substance (Fig.
100). Examination of the curves shows that after prolonged contact of the electrode with methane Q_H': $Q_{org} \simeq 1.2$-2 in the temperature range from 20 to 80°. The sum of the amounts of the organic substance and hydrogen at $\varphi_r = 0$ is approximately equal to
$_S Q_H$ [138]. The chemisorbed organic substance is not removed
appreciably during cathodic polarization of the electrode.* These

---

*In measurements with the aid of platinum—Teflon porous semimicroelectrodes
Niedrach [137] obtained results indicating appreciable desorption of adsorbed methane
from the surface during cathodic polarization. The difference from the results reported by Marvet and Petrii [138] may be due to the different experimental conditions; primarily, to differences in the structure and activity of the catalysts. Virtually no desorption was observed by Taylor and Brummer [139] at 130° or by Marvet
and Petrii [138] at lower temperatures.

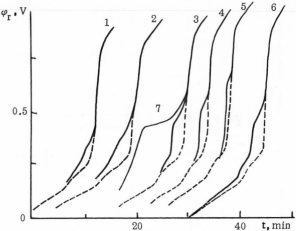

Fig. 100. Charging curves of a Pt/Pt electrode in 1 N H_2SO_4 in presence of methane chemisorbed on open circuit at various temperatures (°C): 1) 95; 2) 80; 3, 6) 60; 4) 40; 5) 20. Curve 6 was obtained after cathodic polarization of the electrode with chemisorbed methane. 7) Charging curve in presence of chemisorbed methane at 60°C. Current density $1 \cdot 10^{-4}$ A/cm² (from data in [138]).

results mean that dehydrogenation of methane and interaction of the particles formed with water and its oxidation products occur on platinum in methane solutions. As the result of these processes predominant chemisorption of oxygen–containing particles, possibly of the HCO type, also formed during chemisorption of methanol, occurs on the surface. This hypothesis provides an explanation of the value of the ratio Q'_H: $Q_{org} \simeq 1.2\text{-}2$, and is consistent with the results of a comparison of the potential range and rates of oxidation of particles chemisorbed in methane and methanol solutions (Fig. 100). Similar conclusions were reached by Taylor and Brummer [139] in a study of adsorption of methane on platinum in 80% H_3PO_4 at 130°C.

Adsorption of ethane, propane, and butane on a platinum electrode on open circuit is accompanied by a shift of the potential in the cathodic direction (Fig. 101) and by evolution of gaseous products: methane in the adsorption of ethane, and methane and ethane in the adsorption of propane [22]. According to Bianchi and Longhi [32], propylene is also formed during the adsorption of propane. If propane is chemisorbed at a constant electrode

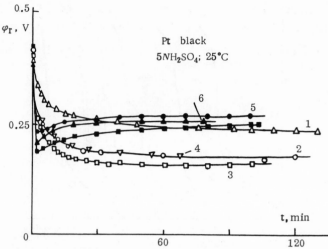

Fig. 101. Shift of the potential of a platinum-black electrode in presence of methane (1), ethane (2), propane (3), isobutane (4), ethylene (5), propylene (6), and cyclopropane (7). Electrolyte 5 N H_2SO_4, 25°C (from data in [22]).

potential, the amount of methane in the gas formed decreases with increase of the φ_r of adsorption; this is due to decrease of the amount of hydrogen adsorbed on platinum with increasing φ_r [31].

The nature of the substances adsorbed on the electrode was elucidated by cathodic reduction of the adsorbed products [22]. Reduction of chemisorbed ethane resulted in predominant evolution of ethane and methane with small amounts of propane and butane. The gas formed during reduction of chemisorbed propane contained $C_3H_8 > C_2H_6 > CH_4$ and a small amount of butane. According to Shropshire and Horowitz [140], in desorption of chemisorbed butane during cathodic polarization of the electrode the gas phase consists mainly of butane, with small amounts of propane, ethane, and methane. These results were interpreted as the consequence of dehydrogenation, cracking, and autopolymerization of the organic compounds during adsorption. Determination of the CH_4:C_2H_6 ratio in the products of cathodic desorption of adsorbed propane in 85% H_3PO_4 at 100°C showed that breakdown of the propane molecule at the $C-C$ bond is most pronounced during adsorption at potentials in the double-layer region [141]. After cathodic polarization a certain amount of an undesorbable product of hydrocarbon chemisorption remains on the electrode surface [22, 141].

Charging curves determined after contact of ethane, propane, and isobutane [22, 142-145] with a platinum electrode have arrests due to ionization of adsorbed hydrogen. The sharp rise of potential in the double-layer region at $\varphi_r \sim 0.5$-0.6 V (25°) is followed in every case by a new arrest corresponding to oxidation of carbon-containing particles (Fig. 102). The arrest shifts to lower potentials with rise of temperature. Although the curves for all the hydrocarbons are similar in shape, the curve length increases in the series $C_2H_6 < C_3H_8 < C_4H_{10}$, i.e., in order of increasing molecular weight of the adsorbate. When the plateau potential is reached, carbon dioxide is evolved; this is the final oxidation product of the adsorbed substances. The presence of adsorbed hydrogen on the electrode in presence of saturated hydrocarbons suggests that the rest potentials in these cases are determined by reaction (VIII.2).

There are indications of slow changes in the nature of the adsorbed species with increase of the adsorption time. For example, the amount of ethane in the gas collected during cathodic reduction of the adsorbate diminished with increase of the time of contact of ethane with platinum black [22]. This decrease is the result of slow formation of particles which are not desorbed during cathodic polarization. Slow accumulation on the surface of platinum in butane solutions of a chemisorbed product which is not desorbed and which differs in behavior from the main mass of the absorbed butane has been detected by the radioactive tracer method [146].

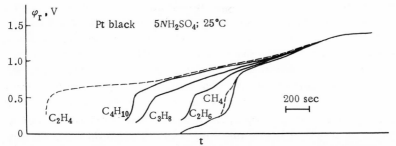

Fig. 102. Galvanostatic curves for oxidation of hydrocarbons adsorbed on platinum black on open circuit in 5 N H_2SO_4 at 25°C (from data in [22]).

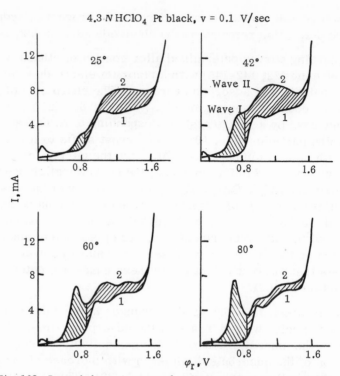

Fig. 103. Potentiodynamic curves for platinum black in 4.3 N HClO$_4$
(1) and in presence of ethane chemisorbed at $\varphi_r = 0.46$ V (2). Potential sweep rate 0.1 V/sec (from data in [136]).

Gilman [33, 147, 148] used the MPP method for a detailed investigation of ethane on a smooth platinum electrode in 1 N HClO$_4$.
The electrode was held in ethane solution at a potential of 0.4 or
0.3 V for various times. The decrease $\Delta_S Q_H$ of hydrogen adsorption was then determined with the aid of a cathodic potentiodynamic pulse, and the charge Q_{org} associated with oxidation of the
adsorbate was found with the aid of an anodic pulse. A linear relationship between $\Delta_S Q_H$ and Q_{org} was obtained over a wide range of
adsorption times. With certain assumptions, this result makes it
possible to establish the average stoichiometric composition of
the adsorbed particles. If it is assumed that the composition of the
adsorbed particle is C_2H_{6-p}, i.e., that it is formed as the result of
rupture of $p - C - H$ bonds and formation of p bonds with the surface
and if it is taken into account that complete oxidation of the ethane
molecule requires 14 electrons, it can be shown that

$$\frac{\Delta_S Q_H}{Q_{org}} = \frac{p}{14 - p} \, . \qquad (VIII.21)$$

The experimental slope $\Delta_S Q_H / Q_{org}$ is 0.375, which gives p = 4; i.e., the average stoichiometric composition of adsorbed ethane is C_2H_2. Although this conclusion regarding the nature of adsorbed ethane is qualitative because of the large number of assumptions made, these experimental data provide adequate evidence of the chemical conversions of ethane molecules during adsorption.

At $\varphi_r > 0.5$ V the ratio $\Delta_S Q_H / Q_{org}$ increases; this should correspond to formation of particles even poorer in hydrogen, and possibly of C_2, i.e., "carbide" structures [33, 148].

Analysis of potentiodynamic curves for oxidation of chemisorbed ethane before and after cathodic reduction shows that in reality at least two different adsorbed species correspond to the average stoichiometric composition C_2H_2. Thus, whereas the initial coverage of the surface with ethane was 0.45, it decreased to 0.30 when a potential of 0.09 V was reached, but the residue was not reduced even at -0.21 V after 100 sec of desorption. The reducible portion of the substance is oxidized at high φ_r, and is represented by the second wave on the potentiodynamic curve for oxidation of the adsorbed ethane (Fig. 103).* The particles which are not removed during cathodic reduction are easily oxidizable, and correspond to wave I, which lies mainly before the potentials of oxygen adsorption. Gilman's data involve the assumption that the ratio between the components of the adsorbed product remains constant at different surface coverages.

Similar results were obtained by Niedrach et al. [136, 137, 149] in studies of adsorption of ethane on a semimicroelectrode made of platinum black. It was shown that the ratio of the charges corresponding to the first and second oxidation waves of the adsorbed ethane is independent of the adsorption time and of the elec-

---

*It should be pointed out that subdivision of the potentiodynamic curve into two waves was somewhat arbitrary. It has also been suggested that the observed waves correspond to intermediate oxidation products of adsorbed ethane, formed during linear variation of φ_r, rather than to the substances initially adsorbed (see discussion in [148]). However, this explanation contradicts other data on chemisorption of ethane, discussed in the main text; in particular, the influence of cathodic reduction on the chemisorption products.

trode potential up to $\varphi_r \sim 0.5$ V. This ratio diminishes at $\varphi_r >$ 0.5 V. The relative number of particles giving rise to the first oxidation wave increases with rise of temperature. For elucidation of the nature of these particles, potentiodynamic curves were determined after contact of the electrode with CO (at 0.3 V), CO_2 (at 0.06 and 0.2 V), and HCOOH. Although the curves for the different adsorbates did not coincide completely, it could nevertheless be concluded that particles of similar nature are adsorbed in all cases.* It was suggested that the first oxidation wave of adsorbed ethane corresponds to oxidation of oxygen-containing particles with one carbon atom, and the second to dehydrogenated "ethylene-like" particles.

Thus, the over-all scheme of ethane chemisorption of platinum may be written as follows:

$$C_2H_6 \rightarrow C_2H_{5\,ads} + H_{ads}$$

(VIII.22)

$$\text{Particles II} \qquad \text{Particles I}$$
$$\text{(from } C_2H_2 \text{ to } C_2) \qquad (C_1H_xO_y)$$

In distinction from [147], some desorption of particles whose oxidation corresponds to wave I was observed during cathodic polarization by Niedrach et al. [137, 126]. This result may be due to the somewhat arbitrary subdivision of the oxidation curve of adsorbed ethane into two waves.

Burshtein, Pshenichnikov, et al. [145] studied the nature of ethane adsorbed at 90° in 1 N H_2SO_4 by comparing the amount of hydrogen formed during dehydrogenation on open circuit with the amount of carbon-containing chemisorption product. With certain assumptions, this ratio shows that for short adsorption times the composition of the adsorbed particles corresponds to C:H = 1:2. With long adsorption times the C:H ratio tends to 1:1. The authors consider that the C:H ratio alters as the result of slow, continued dehydrogenation of the adsorbed particles with simultaneous formation of methane:

$$C_2H_6 \rightarrow a\,CH_4 + b\,(CH_x)_{ads} + z H_{ads}$$

(VIII.23)

---

*The products of chemisorption of hydrocarbons have been compared with CO_{ads}, $(CH_3OH)_{ads}$, and $(CO_2)_r$ in other publications [131, 150].

The adsorption behavior of propane and butane on platinum black is, in general, similar to that of ethane [136, 144, 149].* According to Grubb and Lazarus [151], chemisorption of propane is accompanied by rupture of $C-C$ bonds, with predominant conversion of particles with one carbon atom into oxygen-containing particles which are not reduced during cathodic polarization. This conclusion was based on the results of chromatographic determination of CO_2 during galvanostatic oxidation of propane chemisorbed on platinum at 0.2 V and 65°C. The maximum rate of CO_2 formation was higher than would be observed if four electrons were required for formation of a CO_2 molecule from an adsorbed particle.

Brummer et al. [92, 152-154] carried out a detailed investigation of the chemisorption of propane on smooth platinum in concentrated phosphoric acid at various temperatures (80-140°). The principal conclusion drawn from this series of studies is that the adsorbate consists of several species corresponding to different degrees of oxidation of propane, and that the surface coverage by these species is a function of the potential, the adsorption time, and the experimental temperature. The nature of the adsorbed particles was investigated by the following method. Propane was adsorbed on the electrode at the chosen potential. A potential $\varphi_r = 0.7$ V was then applied abruptly to the electrode; † at this potential the adsorbate underwent partial oxidation, and anodic or cathodic potentiodynamic pulses were applied after various time intervals. The $Q_{org}/\Delta_S Q_H$ ratio determined in this way gives the number of electrons per adsorption site for oxidation of the adsorbate. Experiments on anodic oxidation were also carried out after cathodic reduction (at 0.06 V) of previously adsorbed propane, during which the adsorption product is partially removed. This part of the chemisorbed product was formed as the result of relatively mild dehydrogenation, and was designated the CH-α species. The composition of the CH-α species alters with in-

---

*The possibility of different orientations of propane molecules in the adsorbed layer has been noted [144].

†The so-called "dynamic" hydrogen electrode proposed by Giner [155] was used in these and in certain other investigations. The potential of this electrode, which consists of a cathodically polarized platinum plate, is close to the potential of a reversible hydrogen electrode in the same solution.

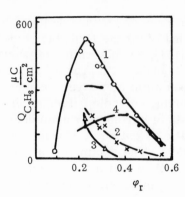

Fig. 104. Dependence of the amount of propane adsorbed on smooth platinum in 13 M H_3PO_4 at 130°C on the potential: 1) total coverage; 2) coverage by CH-α particles; 3) coverage by CH-β particles; 4) coverage by O-type particles (from data in [154]).

creasing potential toward deeper dehydrogenation. The substance which is not desorbed during cathodic reduction consists of two species: the O-type and the CH-β type. The O-type particles occupy about 0.6 of the electrode surface in the potential range of 0.22-0.40 V. They are the more active electrochemically. The number of electrons per site occupied by each O-type particle is 1.28 ± 0.03. This is close to the values found, as stated above, for chemisorbed methanol. The composition of the CH-α and CH-β particles is difficult to determine exactly. CH-β particles are not formed if the electrode potential during adsorption of propane exceeds 0.4 V, and are difficult to oxidize (they may possibly be the products of polymerization of hydrocarbon radicals). The dependence of the total surface coverage by the adsorbate and of the coverage by different types of particles on the electrode potential is shown in Fig. 104.

Formation of different spaces during adsorption of propane has been confirmed [156] by experiments with a Pt/Pt electrode in 3 N H_2SO_4 at 90°C. In addition, Pshenichnikov et al. [156] determined the amount of adsorbed propane with propane present in the solution and after argon had been blown through the solution. In the latter case the amount of adsorption products is smaller (Fig. 105). This is probably due to removal of a portion of the adsorbate which is reversibly adsorbed on the platinum surface.

Data on H−D exchange of C_3H_8 on platinum in 85% D_3PO_4 in D_2O at 100°C [157] indicate that the adsorption of at least a part of the propane molecules is reversible. The electrode was held at a constant potential and C_3H_8 was passed through the solution.

Dissolved propane was then removed, the electrode was polarized to 0.05 V, and the desorption products were investigated by chromatographic and mass-spectrometric methods. It was found that the methane, ethane, and propane formed during desorption are either completely deuterated or contain only one hydrogen atom in the molecule. Comparison of this result with the data presented above leads to the conclusion that the exchange occurs directly at the potential of propane adsorption; i.e., that adsorption of CH-α particles is at least partially reversible.

Cairns et al. [158, 159] consider that adsorption of propane on platinum from hydrofluoric acid solutions at 30 and 90° proceeds in accordance with the scheme

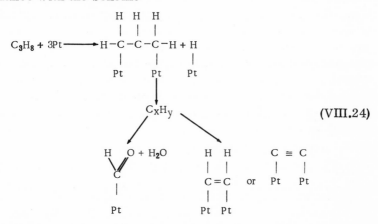

$$\text{(VIII.24)}$$

Particles of the O-, CH-α, and CH-β types were detected by Turner and Brummer [160] in adsorption of n-hexane on smooth

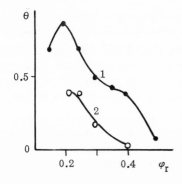

Fig. 105. Dependence of the surface coverage of a Pt/Pt electrode on the potential in 3 N H_2SO_4 at 95°C: 1) propane present in the solution; 2) after argon has been blown through the solution (from data in [156]).

platinum in 12 M H_3PO_4 at 130°C. The amount of CH-α particles increases when propane is replaced by hexane.

Podlovchenko, Kolyadko, and Petukhova [161] investigated adsorption of n-hexane on a Pt/Pt electrode in 1 N H_2SO_4 at 80°C. In presence of hexane, the electrode potential on open circuit and at $\varphi_r^0 > 80$ mV is shifted in the negative direction only, and reaches values close to the final potentials in ethane, propane, and butane solutions. The charging curves determined after the electrode had been washed to remove hexane had an arrest corresponding to oxidation of H_{ads}, and an arrest ~10-30 times as large corresponding to oxidation of the chemisorbed substance. Most of the latter (about 95%) is removed from the surface during hydrogenation. Chromatographic analysis of the gaseous hydrogenation products showed absence of methane, pentane, and unsaturated hydrocarbons, and the presence of only small amounts of ethane, propane, and butane. Evidently, the desorption products consisted mainly of hexane. Therefore it seems that particles containing six carbon atoms and corresponding to a low degree of dehydrogenation of the original molecules are, in the main, chemisorbed on the surface:

$$C_6H_{14} \rightarrow C_6H_{14-x} + xH_{ads} \qquad \text{(VIII.25)}$$

Calculations showed that x may have values from 1 to 5, dependent on the experimental conditions (temperature, activity of the electrode).

When ethylene, acetylene, propylene, and cyclopropane are brought into contact with a platinum electrode polarized to potentials of the double-layer region, the potential first shifts sharply in the cathodic direction, and then increases toward more anodic values [22, 30, 142, 143, 162, 163] (Fig. 101). The final potentials φ_r^i depend on the temperature and are in general higher than φ_r^i established in presence of saturated hydrocarbons. If ethylene is removed by argon at the established φ_r^i from the solution surrounding the platinum electrode, an additional shift of potential in the negative direction occurs. A similar effect is observed if the temperature of the solution containing an electrode covered with chemisorbed ethylene is raised [162]. When unsaturated hydrocarbons are brought into contact with platinum saturated with hydrogen, the electrode potential shifts rapidly in the anodic direction, approaching the values established if the potential at which the hydrocarbons were introduced was in the double-layer region. However, the potentials attained with different initial φ_r^0 do not coincide completely. This is due to dehydrogenation and hydrogenation of the organic substances. As the result of dehydrogenation of ethylene adsorbed hydrogen appears on the electrode surface; the amount present first increases with time and then begins to decrease. An appreciable decrease of the amount of adsorbed hydrogen formed during dehydrogenation of ethylene corresponds to the start of an appreciable shift of the electrode potential in the anodic direction; i.e., it is caused by the start of the hydrogenation reaction [22, 162].

In accordance with these data, the views put forward in [28, 68] regarding the nature of rest potentials in solutions of saturated alcohols with more than two carbon atoms in the molecule, and of aldehydes, can be used for determining the nature of the potentials which become established in solutions of unsaturated hydrocarbons. Adsorbed hydrogen must be of primary significance for the potentials established, because the exchange current of reactions (VIII.2) and (VIII.2a) on platinum, even at low surface coverages with H_{ads}, is apparently greater than the exchange currents of the other electrode processes. An approximate estimate of the exchange current of electrochemical reactions involving C_2H_4 [164] supports this hypothesis.

The shifts of potential during contact of unsaturated hydrocarbons with platinum are accompanied by evolution of gaseous products [22, 165, 166]: $C_2H_6 > CH_4$ in the case of ethylene, C_2H_4 and C_2H_6 in the case of acetylene, and $C_3H_8 > C_2H_6 > CH_4$ in the case of propylene. The surface coverages attained by carbon-containing particles in presence of unsaturated hydrocarbons are considerably greater than in the case of saturated hydrocarbons, as the hydrogenation reaction releases sites for further adsorption.

Careful chromatographic analysis of the products of ethylene chemisorption in $1\,N\,H_2SO_4$ at $100\,°C$ on platinum black oxidized by atmospheric oxygen showed that small amounts of CO_2 were evolved in addition to ethane and methane [167]. For elucidation of the nature of the process of CO_2 formation, platinum black was brought into contact with gaseous ethylene; here again CO_2 was initially formed, apparently, as the result of interaction of adsorbed oxygen with ethylene. However, the CO_2 content of the gas fell as reduction of the platinum black proceeded, and finally only methane and ethane were detected. The platinum black was then brought into contact with 1 N H_2SO_4 or water saturated with ethylene. Carbon dioxide again appeared in the gas phase, probably formed by reaction (VIII.26) catalyzed by platinum:

$$C_2H_4 + 4H_2O \rightarrow 2CO_2 + 12H_{ads} \qquad\qquad (VIII.26)$$

A similar process occurs in the case of ethane, but at only a low rate.

When the products of adsorption of unsaturated hydrocarbons are subjected to cathodic hydrogenation, the gas phase contains $C_2H_6 \gg C_4H_{10} > CH_4 > C_3H_8$ in the case of adsorbed ethylene, and $C_3H_8 \gg CH_4 > C_2H_6$, with traces of hexane, in the case of adsorbed propylene [22]. Thus, considerable cracking and polymerization occur.* A very interesting fact, discovered by Niedrach [22], is that the amount of ethane in the gases formed by hydrogenation of adsorbed ethylene decreases with increasing time of contact between the electrode and C_2H_4 and with shift of the electrode potential in the anodic direction. If ethylene is adsorbed at potentials of 0.7–0.9 V and the adsorbed products are then hydrogenated, the amount of ethane and especially of butane in the resultant gas is greatly decreased and the relative contents of methane and propane increase.

Gilman [33, 168] compared the potentiodynamic curves for oxidation of the adsorption products of ethylene and acetylene at 30–60° in 1 N $HClO_4$. The comparison was made at equal sweep rates and similar surface coverages. In distinction from ethane, the chemisorption products of which give two oxidation waves, in oxidation of adsorbed ethylene and acetylene only one wave was observed, located at potentials of the oxygen region and corresponding to the second oxidation wave of adsorbed ethane. The potentiodynamic curves obtained after contact of the electrode with ethylene and acetylene for various times at 0.2, 0.3, and 0.4 V virtually coincided (Fig. 106). This meant that the adsorbed particles in different adsorbates have identical structures. To explain this result, it was suggested that the composition of the adsorbate corresponds to the formula C_2H_2, i.e., the chemisorption mechanism is nondissociative for acetylene and dissociative for ethylene. In agreement with this, Gilman observed an anodic cur-

---

*According to Bianchi et al. [165, 166] hydrogenation of adsorbed ethylene and acetylene yields mainly butane and propane respectively.

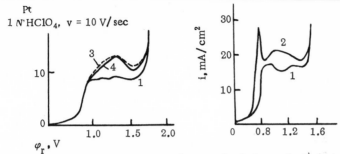

Fig. 106. Potentiodynamic curves of a smooth platinum electrode in 1 N HClO$_4$(1), and in presence of ethane (2), ethylene (3), and acetylene (4) adsorbed at φ_r = 0.4 V (from data in [147, 168]). Potential sweep rate 10 V/sec.

rent due to ionization of hydrogen formed during dehydrogenation of ethylene on platinum at 0.4 V. However, it should be pointed out that the shift of the potential of platinum in the cathodic direction in presence of acetylene [163] also appears to indicate that acetylene undergoes dehydrogenation during adsorption.

The structures of adsorbed ethylene and acetylene change at φ_r > 0.5 V. The potentiodynamic curves for oxidation of the adsorption products are shifted toward more anodic potentials. This evidently indicates formation of "carbide" structures during adsorption. Formation of surface carbides during adsorption of ethylene and acetylene at φ_r > 0.35 V has also been presumed by Bianchi et al. [165, 166].

The situation is also more complex in the case of adsorption of ethylene on platinum in 85% H$_3$PO$_4$ at 120° [33, 148, 168]. In this case the adsorbed layer is oxidized at lower potentials and the potentiodynamic curves have two overlapping oxidation waves. Evidently considerable rupture of C−C bonds occurs during adsorption at high temperatures. Gilman [33, 148] investigated the possibility of conversion of the species corresponding to the second oxidation waves into more easily oxidizable species. For this the Pt/Pt electrode was held at 0.3 V, and the resultant adsorbed layer was then partially oxidized at φ_r ≃ 0.8 V to remove the species of higher electrochemical activity. A potential of 0.3 V was then applied again to the electrode; it was presumed that at this potential the difficultly oxidizable species was converted into the easily oxidizable form. To detect the latter, the difficultly

oxidizable species was desorbed at -0.2 V and the anodic potentiodynamic curve was then recorded. The experiments showed
that the conversion proceeds to only a small extent.

Burshtein, Pshenichnikov, et al. [143, 162, 169-172] investigated the nature of the organic species chemisorbed on platinum
and rhodium in solutions of ethylene in sulfuric acid at 90° by
comparing the amount of hydrogen formed on the electrode with
the amount of substance chemisorbed. The character of the potentiodynamic curves for oxidation of the ethylene chemisorption
products depends on the temperature, the adsorption potential, the
partial pressure of the hydrocarbon, and the electrode material
(Fig. 107). For example, decrease of the ethylene partial pressure
was accompanied by increase of the oxidation rate of the adsorbate
at low φ_r, owing to formation of easily oxidizable particles. It was
concluded that the number of hydrogen atoms per carbon atom in
the chemisorbed product is, on the average, 1.5 on platinum and
~2 on rhodium. The number of sites occupied by one chemisorbed
particle is four on platinum and two on rhodium, while the number
of electrons per site required for oxidation of a particle to CO_2
is 3 and 6 respectively. Thus, the compositions of the adsorbate
on platinum and rhodium are significantly different. Evidently on
rhodium the adsorption process is accompanied mainly by rupture
of one of the double bonds in the molecule, while on platinum $C-H$
bond rupture may also occur. At the same time, reversible adsorption of some of the adsorbed ethylene was detected by comparison of the surface coverage determined by the pulse method on
smooth platinum with the amount of firmly adsorbed substance
(after removal of dissolved ethylene) on a Pt/Pt electrode. The
coverage was appreciably greater in the former case than in the
latter. It was assumed that adsorption of ethylene calculated per
unit surface should be the same on the smooth and the platinized
electrodes.

The publications discussed above contain experimental evidence
indicating that one of the main characteristics of adsorption phenomena on metals of the platinum group is chemisorptive interaction between hydrocarbons and the electrode, leading to considerable changes in the nature of the particles during adsorption.
Chemisorption processes in electrolyte solutions are allied in this
respect to reactions of gaseous hydrocarbons with catalysts, which
have been studied fairly extensively [10-20]. Various examples

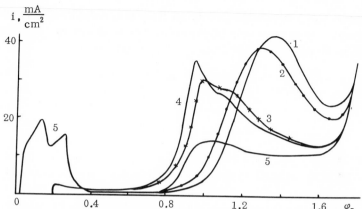

Fig. 107. Potentiodynamic curves for oxidation of ethylene chemisorbed on a Pt/Pt electrode at various partial pressures (atm): 1) 1; 2) 0.1; 3) 0.01; 4) 0.001. 5) Curve in the supporting electrolyte, 1 N H_2SO_4; 90°C. Ethylene adsorbed at $\varphi_r = 0.2$ V (from data in [172]).

can be cited indicating parallelism in the behavior of hydrocarbons during adsorption from the gas phase and from solutions [21, 23, 142, 145, 147, 162, 163, 168], although the presence of water makes it possible for reactions leading to formation of oxygen-containing particles to occur.

It appears that the influence of the solvent is not confined to participation in purely chemical interaction with the adsorbable particles. In discussions of adsorption phenomena on metals of the platinum group, Bockris et al. [21, 23, 173-178] attach primary importance to competition for surface sites between the organic substances and water molecules. This conclusion was based on studies of the energetics of adsorption of various substances (ethylene, naphthalene, benzene) on platinum by analysis of adsorption isotherms determined with the aid of radioactive tracers. It was found that adsorption is usually accompanied by small and sometimes even positive enthalpy and positive entropy of adsorption, although stable bonds are formed between the adsorbate and the surface. Thus, a value of 0.0 ± 4.0 kcal/mole was obtained for the heat of adsorption, ΔH_{ads}, of ethylene on platinum [175]. This is considerably lower than the heat of adsorption of ethylene from the gas phase (which, e.g., on nickel, is about -58 kcal/mole). The low heat of adsorption was attributed to dis-

placement of water molecules from the surface. In support of this
hypothesis, adsorption of ethylene was represented by the following thermodynamic cycle:

$$
\begin{array}{c}
C_2H_{4soln} + 4H_2O_{ads} \xrightarrow{\Delta H_{ads}} C_2H_{4ads} + 4H_2O_{soln} \\
\downarrow{\Delta H_1} \quad \xrightarrow[4\Delta H_2]{} 4H_2O_{gas} \xrightarrow{\Delta H_3} \downarrow{4\Delta H_4} \\
\underline{\qquad C_2H_4 gas \qquad}
\end{array}
\qquad \text{(VIII.27)}
$$

The following numerical values of ΔH were taken for the calcula-
tion: $\Delta H_1 = 4$, $\Delta H_2 = 22.6$, $\Delta H_3 = -58$, and $\Delta H_4 = 9.6$ kcal/mole.
The value of ΔH_2 was calculated as the sum of the dispersion energy
and the image force interaction. The value of -2.0 kcal/mole was
found for ΔH_{ads} from the above cycle, in agreement with the experimental value. For the standard state, with the activity of ethylene equal to unity, Bockris et al. found that the entropy change in
adsorption of ethylene is 16 ± 1 cal $\cdot$ K$^{-1} \cdot M^{-1}$ In contrast, negative values of the entropy are obtained for adsorption from the
gas phase. This difference was also explained on the hypothesis
that adsorption from solution is a substitution process. When a
molecule is adsorbed from the gas phase, it loses degrees of freedom, and this must lead to decrease of entropy. On the other hand,
in adsorption from solutions each ethylene molecule desorbs four
water molecules; this increases the degrees of freedom and the
entropy of the system. Bockris et al. [175] also concluded from
thermodynamic calculations that adsorption of ethylene must be
nondissociative rather than dissociative, and must occur with
double-bond rupture. However, such experimental data on adsorption of ethylene as the occurrence of dehydrogenation and the
reaction with water, discussed above, are not taken into account
in the foregoing views and in scheme (VIII.27). Moreover, it is
not clear whether the experimentally determined dependences of
surface coverage on the concentration or pressure can be treated
as equilibrium adsorption isotherms (see Chapter X).

5. Influence of Adsorption of Inorganic Ions on Adsorption of Organic Substances

Specifically adsorbed anions have a considerable influence on
adsorption of organic substances on metals of the platinum group

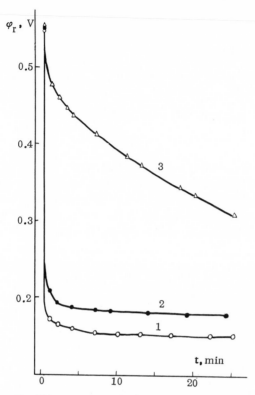

Fig. 108. Potential transients for a Pt/Pt electrode after addition of 0.5 M C_2H_5OH at φ_r = 500 mV to solutions of: 1) 0.1 N H_2SO_4; 2) 0.1 N HCl; 3) 0.09 N KBr + 0.01 N HCl (from data in [179]).

[34, 64, 116, 138, 149, 153, 179-185]. The character of this influence depends both on the nature of the anion and on the nature of the organic substance.

Figures 108 and 109 show that the shape of the curves representing the shift of potential on addition of methanol and of the oxidation curves of the chemisorbed substance changes in presence of halide ions. Chloride ions have little influence on the amount of ethanol chemisorbed, whereas in presence of Br⁻ the degree of ethanol adsorbed on open circuit during the same time is small. The same effects are observed in adsorption of methanol. However, if methanol is adsorbed on the electrode in absence of

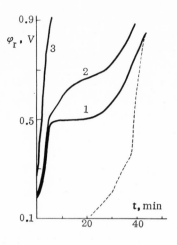

Fig. 109. Anodic charging curves for a Pt/Pt electrode in presence of ethanol chemisorbed on open circuit in solutions of: 1) 0.1 *N* H₂SO₄; 2) 0.1 *N* HCl; 3) 0.09 *N* KBr + 0.01 *N* HCl. The charging curve for 0.1 *N* H₂SO₄ solution is dashed. Current density 1 · 10⁻⁴ A/cm²) (from data in [179]).

specifically adsorbed ions and Br⁻ ions are then introduced into the solution, the amount of previously chemisorbed methanol remains unchanged for several hours [64]. The character of the oxidation curve of the chemisorbed substance undergoes substantial changes, and the influence of anions becomes especially pronounced as oxidation of the chemisorption products proceeds, i.e., with liberation of sites for anion adsorption on the electrode surface.

In presence of specifically adsorbed anions the arrest due to electrochemical oxidation of chemisorbed carbon dioxide on platinum splits into two arrests (Fig. 110) [183, 184].

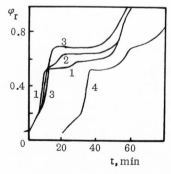

Fig. 110. Influence of Cl⁻ and Br⁻ ion concentrations on charging curves for a moderately aged Pt/Pt electrode in presence of (CO₂)ᵣ in solutions of: 1) 0.01 *N* HCl + 0.99 *N* H₂SO₄; 2) 0.1 *N* HCl + 0.9 *N* H₂SO₄; 3) 1 *N* HCl; 4) 0.01 *N* KBr + 0.1 *N* H₂SO₄. Current density 1 · 10⁻⁴ A/cm² (from data in [184]).

Chemisorption of organic compounds is influenced even by phosphate anions [34, 179-182], the specific adsorption of which on platinum is relatively low [186, 187]. According to Niedrach and Tochner [182], in phosphoric acid solutions the first wave on the potentiodynamic curve for oxidation of adsorbed ethane disappears, and reappears only at higher temperatures than in perchloric or hydrochloric acid solutions. The improvement of the characteristics of hydrocarbon fuel cells by the use of fluoride electrolytes [188] appears to be due to a certain extent to the slight specific adsorbability of F^- ions on platinum at low anodic potentials. Another cause is the higher solubility of hydrocarbons in fluoride electrolytes.

Chemisorption of carbon dioxide on platinum is strongly retarded by Zn^{2+} cations [183]. Halide ions influence this process in the φ_r range in which they are adsorbed on the electrode [79,183].

Catalyst poisons such as mercury and arsenic suppress dehydrogenation of methanol on rhodium; i.e., their influence on this process is analogous to that of halide ions [189]. It was also suggested [189] that the nature of the adsorbed particles changes in presence of mercury and arsenic.

Adsorption of organic substances on platinum metals causes substantial changes in the structure of the electric double layer, as was shown by Kazarinov and Mansurov [190] with the aid of radioactive tracers. Hexyl alcohol suppresses adsorption of phosphate anions at potentials φ_r up to 0.8 V. Adsorption of SO_4^{2-} was also suppressed in $5 \cdot 10^{-3} N$ $Na_2SO_4 + 5 \cdot 10^{-3} N$ H_2SO_4 solution in presence of methyl and hexyl alcohols in the φ_r range from the reversible hydrogen potential to 0.4 and 0.8 V, respectively. Hexyl alcohol added to sodium sulfate solution at the reversible hydrogen potential not only alters the dependence of adsorption of Na^+ cations on the potential; Na^+ cations adsorbed in its presence do not exchange appreciably with cations in the solution volume. Adsorption of naphthalene leads to considerable decrease of the adsorption of Na^+ and SO_4^{2-} ions on a platinized electrode (Fig. 111).

Reshet'ko and Shlygin [191] also concluded that the structure of the double layer on platinum is changed in presence of methanol. Changes in the composition of the solution during formation of the double layer were determined [192].

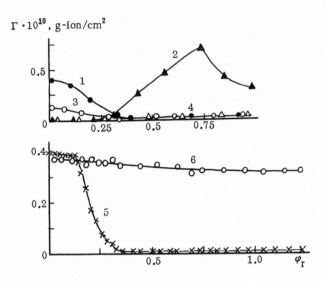

$\Gamma \cdot 10^{10}$, g-ion/cm^2

Fig. 111. Dependence of adsorption of Na$^+$ (1, 3, 5, 6) and
SO$_4^{2-}$ (2, 4) ions on the potential of a Pt/Pt electrode in various
solutions: 1, 2) $5 \cdot 10^{-3}$ N Na$_2$SO$_4$ + $5 \cdot 10^{-3}$ N H$_2$SO$_4$; 3, 4
$5 \cdot 10^{-3}$ N Na$_2$SO$_4$ + $5 \cdot 10^{-3}$ N H$_2$SO$_4$ + $2 \cdot 10^{-4}$ M naphthalene;
5) 10^{-3} N Na$_2$SO$_4$ + 10^{-3} N H$_2$SO$_4$; 6) 10^{-3} N Na$_2$SO$_4$ + 10^{-3} N
H$_2$SO$_4$ + 10^{-3} M hexyl alcohol (from data in [190]).

Gilman [116, 180] observed displacement of Cl$^-$ ions during adsorption of CO and C$_2$H$_6$ on a platinum electrode. In adsorption of
CO, each surface concentration of the adsorbate corresponds to a
definite Cl$^-$ concentration on the surface, which becomes established rapidly. The following empirical relation between the
coverage of the surface with Cl$^-$(θ_{Cl^-}), θ_{CO}, the potential, and the
Cl$^-$ concentration [Cl$^-$]$_0$ near the surface was derived:

$$\theta_{Cl^-}/\varphi = 0.77 + 0.069 \ln [Cl^-]_0 - 0.049 \theta_{CO}.$$

The initial rate of ethane adsorption is lowered by the presence of
Cl$^-$ ions. The adsorption of Cl$^-$ decreases as adsorption of ethane
proceeds, eventually reaching a certain limiting value. This value,
at $\varphi_r = 0.3$-0.4 V, is approximately half the limiting value in absence of Cl$^-$ ions. Adsorption of Cl$^-$ has no influence on the composition of chemisorbed ethane.

6. Joint Adsorption of Hydrogen and Organic Substances

Analysis of the hydrogen region of the charging curve in presence of chemisorbed products leads to some interesting conclusions on the nature of the products formed by adsorption of organic substances and of the surface of platinum.

Adsorption of methanol on platinum on open circuit results in a small preferential decrease of the number of sites corresponding to firmly bonded hydrogen [193]. This distribution of the chemisorbed substance, "uniform" with respect to hydrogen adsorption sites with different $Pt-H_{ads}$ bond energies, may mean that equally probable site distribution predominates on the surface, regardless of the chemical potential μ_H of the adsorbed hydrogen ("microscopic heterogeneity" in the terminology of Chizmadzhev and Markin [194, 195]).

Isotherms of hydrogen adsorption in presence of chemisorbed C_2H_5OH, C_3H_7OH, C_4H_9OH, CH_2OHCH_2OH, and CH_3OH, hydrogenated by cathodic polarization of the electrode [71], virtually coincide with the adsorption isotherm in presence of chemisorbed methanol; this provides additional confirmation of the above conclusions regarding the adsorption relationships of these compounds.

In anodic polarization of the electrode in methanol solutions, when considerable amounts of chemisorption products accumulate on the surface, the decrease of the number of sites corresponding to firmly bonded hydrogen becomes more pronounced [71]. Decrease of the H_{ads} bond energy on the sites free from organic particles may also occur. A similar effect was observed by Breiter in adsorption of CO on a Pt/Pt electrode [113]. It may also be noted that in presence of adsorbed methanol and CO the isotherms of hydrogen adsorption become closer in form to the Temkin isotherm, whereas the isotherm of hydrogen adsorption in sulfuric acid solutions can be only approximately represented by the Temkin isotherm [51]. Thus, the adsorbed products have approximately the same influence as halide anions on the form of the hydrogen adsorption isotherms.

Preferential adsorption at sites of firmly bonded hydrogen has been noted in the case of methane [138] and glycolaldehyde

[76, 77], and at the initial instant of adsorption also in the cases of ethylene, ethane, and propane [156, 171, 172] on platinum. According to Pshenichnikov et al. [156, 171, 172], particle distribution over sites of different bond energies alters in time.

In adsorption of CO on rhodium [122], the numbers of sites having different energies of hydrogen bonding decrease equally. A detailed discussion of the interaction of H_{ads}, CO_{ads}, and $CO_{2\,ads}$ has been published by Brummer and Cahill [195a].

7. Evolution of Molecular Hydrogen during Contact of Organic Substances with Platinum Catalysts

Among the various reactions occurring during adsorption of organic molecules, the cases reported in the literature of evolution of molecular hydrogen in the range $\varphi_r > 0$ during contact of platinum electrodes with solutions of various substances attract special attention.* Some examples of this effect have been given above, and are also examined in Vielstich's monograph [198]. Evolution of hydrogen at $\varphi_r > 0$ has been observed during contact of various saturated and unsaturated hydrocarbons with Raney platinum catalysts [199], and in presence of ethylene and acetylene [163]; however, this contradicts the results reported in [22, 167].

Although under certain conditions molecular hydrogen is evolved during contact of platinum electrodes with methanol [57, 99], this effect generally occurs in solutions of such organic compounds as formaldehyde, formic acid, and ethylene glycol at φ_r close to 0. An especially interesting case of evolution of molecular hydrogen at anodic potentials has been described by Sokol'skii and Beloslyudova [200]. They observed evolution of hydrogen at $\varphi_r \gtrsim 150$ mV in acid solution in presence of phenylacetylene and styrene; cathodic polarization of the platinum electrode in phenylacetylene solutions to a potential of 0.08 V led to turbulent evolution of hydrogen. The nature of the processes of evolution of gaseous hydrogen has not been investigated in detail.

_____

*Evolution of hydrogen has been presumed to occur during contact of platinum with allyl alcohol [5] and in presence of propanol [196]. However, it has been shown by chromatographic analysis [28, 197] that in these cases mainly the reduction products of the organic compounds are evolved.

Daniél'-Bek and co-workers [201] consider hydrogen evolution to be the result of catalytic decomposition of organic compounds into molecular hydrogen and oxidation products, and of coupled reactions of electrochemical oxidation of the organic compounds and cathodic evolution of hydrogen. They assume that the rate of catalytic decomposition of the organic compounds is independent of the potential (however, see p.327 et seq.). They devised a technique of combined gasometric and polarization measurements for separating the catalytic and electrochemical mechanisms of hydrogen evolution on the basis of this assumption.

The concept put forward earlier by Müller et al. [47, 202] in relation to the evolution of hydrogen at considerable anodic polarizations at a number of metals (Cu, Ag, Au, etc.) having low hydrogen exchange currents in alkaline formaldehyde solutions has been used [57, 68, 93, 99] to explain the evolution of gaseous hydrogen. According to this concept, molecular hydrogen is formed as the result of chemical interaction between the conversion products formed on contact of the organic compounds with the platinum electrode. This involves the assumption that the hydrogen adsorption process $H_2 \rightarrow 2H_{ads}$ is slow and that the rate of hydrogen ionization is lowered in presence of organic substances.

A similar mechanism has been suggested [94] for evolution of hydrogen from formic acid on palladium, based on the results of analysis of the isotopic composition of the gas (see p. 345).

It is interesting to note that nonelectrochemical oxidation reactions which are also accompanied by hydrogen evolution are known [203].

References

1. A.I. Shlygin and M.E. Manzhelei, Uch. Zap. Kishinevsk. Univ., 8:13 (1953).
2. T.O. Pavela, Ann. Acad. Sci. Fennicae, Ser. A, II, No. 59 (1954).
3. A.I. Shlygin, Proceedings of the Third Conference on Electrochemistry [in Russian], Izd. Akad. Nauk SSSR (1953), p. 322.
4. G.P. Khomchenko and G.D. Vovchenko, Vestn. Mosk. Gos. Univ., Ser. Estestv. Nauk, No. 8, 91 (1955).
5. Yu.A. Podvyazkin and A.I. Shlygin, Zh. Fiz. Khim., 30:1521 (1956).
6. T.I. Pochekaeva, Zh. Fiz. Khim., 35:1606 (1961).
7. M.W. Breiter and S. Gilman, J. Electrochem. Soc., 109:622 (1962).
8. M.W. Breiter, Electrochim. Acta, 8:477 (1963).
9. D.V. Sokol'skii, Hydrogenation in Solutions [in Russian], Izd. Akad. Nauk Kaz.SSR, Alma-Ata (1962).

10.   G. C. Bond, Catalysis by Metals, Academic Press, New York (1962).
11.   D. O. Hayward and B. M. W. Trapnell, Chemisorption, Plenum Press, New York (1964).
12.   R. W. Roberts, Trans. Faraday Soc., 58, 1159 (1962); Ann. N. Y. Acad. Sci., 101:766 (1963); J. Phys. Chem., 67:2035 (1963); 68:2718 (1964).
13.   C. Kemball, in: Catalysis, Aspects of Catalyst Selectivity and Stereospecificity [Russian translation], Izd. Inostr. Lit., Moscow (1963).
14.   D. W. McKee, J. Am. Chem. Soc., 84:1109, 4427 (1962); J. Phys. Chem., 67:841 (1963).
15.   A. G. Popov and R. E. Mardaleishvili, Vestn. Mosk. Gos. Univ., Ser. Khim., No. 1, 34 (1962).
16.   V. M. Gryaznov, V. I. Shimulis, and T. V. Dilingerova, Vestn. Mosk. Gos. Univ., Ser. Khim., No. 2, 26 (1962).
17.   G. Blyholder and L. D. Neff, J. Catalysis, 2:138 (1963); J. Phys. Chem., 70:1738 (1966); G. Blyholder and W. V. Wyatt, J. Phys. Chem., 70:1745 (1966).
18.   S. L. Kiperman and I. R. Davydova, Zh. Fiz. Khim., 34:262 (1965).
19.   B. A. Morrow and N. Sheppard, J. Phys. Chem., 70:2406 (1966).
20.   R. S. Hansen, J. R. Arthur, Jr., N. Y. Mimeault, and R. R. Rye, J. Phys. Chem., 70:2787 (1966).
21.   B. J. Piersma, Electrosorption (E. Gileadi, editor), Plenum Press, New York (1967), p. 19.
22.   L. W. Niedrach, J. Electrochem. Soc., 111:1309 (1964).
23.   E. Gileadi and B. Piersma, Modern Aspects of Electrochemistry, Vol. 4 (J. O'M. Bockris, editor), Plenum Press, New York (1966), p. 47.
24.   J. Giner, Electrochim. Acta, 8:857 (1963).
25.   J. Giner, Electrochim. Acta, 9:63 (1964).
26.   J. Giner, 15th CITCE Meeting, London (1964).
27.   A. N. Frumkin and B. I. Podlovchenko, Dokl. Akad. Nauk SSSR, 150:349 (1963).
28.   B. I. Podlovchenko, O. A. Petrii, and A. N. Frumkin, Dokl. Akad. Nauk SSSR, 153:379 (1963).
29.   B. I. Podlovchenko and E. P. Gorgonova, Dokl. Akad. Nauk SSSR, 156:673 (1964).
30.   M. J. Schlatter, Fuel Cells (G. J. Young, editor), Reinhold Publishing Co., New York (1963), p. 190; 145th National Meeting, American Chemical Society, Division of Fuel Chemistry, 7:(4), New York (1963).
31.   W. T. Grubb, Nature, 198:883 (1963); J. Electrochem. Soc., 111:1086 (1964).
32.   G. Bianchi and P. Longhi, Chim. Ind. (Milan), 46:501 (1964).
33.   S. Gilman, Hydrocarbon Fuel Cell Technology (S. Baker, editor), Academic Press, New York—London (1965), p. 349.
34.   B. I. Podlovchenko, O. A. Petrii, and E. P. Gorgonova, Élektrokhimiya, 1:182 (1965).
35.   O. A. Petrii, B. I. Podlovchenko, A. N. Frumkin, and Hira Lal, J. Electroanal. Chem., 10:253 (1965).
36.   B. I. Podlovchenko and O. A. Petrii, Methods for Investigation of Catalysts and Catalytic Reactions [in Russian], Vol. 1, Izd. Sib. Otd. Akad. Nauk SSSR, Novosibirsk (1965), p. 266.

37. J. E. Oxley, G. K. Johnson, and B. T. Buzalski, Electrochim. Acta, 9:897 (1964).
38. D. M. Dražić and V. Dražić, Electrochim. Acta, 11:1235 (1966).
39. N. K. Khristoforova and A. I. Shlygin, Uch. Zap. Dal'nevost. Gos. Univ., 8:69 (1966).
40. A. D. Obrucheva, Zh. Fiz. Khim., 32:2155 (1958); Dokl. Akad. Nauk SSSR, 141:1413 (1961).
41. Hira Lal, O. A. Petrii, and B. I. Podlovchenko, Élektrokhimiya, 1:316 (1965).
42. A. I. Shlygin and G. A. Bogdanovskii, Proceedings of the Fourth Conference on Electrochemistry, 1956 [in Russian], Izd. Akad. Nauk SSSR (1959), p. 282; Zh. Fiz. Khim., 31:2428 (1957); 33:1769 (1959); 34:57 (1960); G. A. Martinyuk and A. I. Shlygin, Zh. Fiz. Khim., 32:164 (1958); A. K. Korolev and A. I. Shlygin, Zh. Fiz. Khim., 36:314 (1962).
43. V. S. Bagotskii and Yu. B. Vasil'ev, Electrochim. Acta, 9:869 (1964).
44. S. S. Beskorovainaya, Yu. B. Vasil'ev, and V. S. Bagotskii, Élektrokhimiya, 2:167 (1966).
45. V. S. Bagotskii and Yu. B. Vasil'ev, Electrochim. Acta, 11:1439 (1966).
46. V. S. Bagotskii and Yu. B. Vasil'ev, Advances in the Electrochemistry of Organic Compounds [in Russian], Izd. "Nauka," Moscow (1966), p. 38.
47. E. Müller, Z. Eletrochem., 29:264 (1923); E. Müller and S. Takegami, Z. Elektrochem., 34:704 (1928); S. Tanaka, Z. Elektrochem., 35:38 (1929).
48. E. Justi and W. Winsel, Kalte Verbrennung (1962), p. 242.
49. R. Buck and L. Griffith, J. Electrochem. Soc., 109:1005 (1962).
50. A. Kutschker and W. Vielstich, Electrochim. Acta, 8:985 (1963).
51. A. N. Frumkin, Advances in Electrochemistry and Electrochemical Engineering, Vol. 3 (P. Delahay, editor), Interscience (1963), p. 287.
52. M. I. Temkin, Zh. Fiz. Khim., 15:296 (1941).
53. S. Z. Roginskii, Adsorption and Catalysis on Heterogeneous Surfaces [in Russian], Izd. Akad. Nauk SSSR (1948).
54. S. L. Kiperman, Introduction to Kinetics of Heterogeneous Catalytic Reactions [in Russian], Izd. Akad. Nauk SSSR (1964).
55. S. Gilman, Electroanalytical Chemistry (A. J. Bard, editor), Vol. 2, Marcel Dekker, Inc., New York (1967), p. 111.
56. R. E. Smith, H. B. Urbach, and N. L. Hatfield, J. Phys. Chem., 71:4121 (1967).
57. V. F. Stenin and B. I. Podlovchenko, Élektrokhimiya, 3:481 (1967).
58. T. Biegler and D. F. A. Koch, J. Electrochem. Soc., 114:904 (1967).
59. T. Biegler, J. Phys. Chem., 72:1571 (1968).
60. M. W. Breiter, J. Electroanal. Chem., 14:407 (1967).
61. M. W. Breiter, J. Electroanal. Chem., 15:221 (1967).
62. B. I. Podlovchenko, A. N. Frumkin, and V. F. Stenin, Élektrokhimiya, 4:339 (1968).
63. G. Bianchi and P. Longhi, Chim. Ind., 46:1286 (1964).
64. B. I. Podlovchenko, V. F. Stenin, and V. E. Kazarinov, The Double Layer and Adsorption on Solid Electrodes [in Russian], Izd. Tartusk. Gos. Univ. Tartu (1968), p. 121; Élektrokhimiya (in press).

65.   V. S. Éntina, O. A. Petrii, and V. T. Rysikova, Élektrokhimiya, 3:758 (1967).

66.   O. A. Petrii and N. Lokhanyai, Élektrokhimiya, 4:514, 656 (1968).

67.   B. I. Podlovchenko, Élektrokhimiya, 1:101 (1965).

68.   B. I. Podlovchenko, O. A. Petrii, A. N. Frumkin, and Hira Lal, J. Electroanal. Chem., 11:12 (1966).

69.   T. A. Geissman, Principles of Organic Chemistry [Russian translation], Vol. 2, Izd. Inostr. Lit., Moscow (1950), p. 106.

70.   F. Nagy and G. Horanyi, Acta Chim. Acad. Sci. Hung., 49:243 (1966).

71.   B. I. Podlovchenko and V. F. Stenin, Élektrokhimiya, 3:649 (1967).

72.   Ya. Veber, Yu. B. Vasil'ev, and V. S. Bagotskii, Élektrokhimiya, 2:515, 522 (1966).

73.   M. J. Schlatter, Fuel Cells, Vol. 2 (G. J. Young, editor), Reinhold Publishing Co., New York (1963), p. 200; H. Binder, A. Köhling, and G. Sandstede, Fuel Cell Systems, Advances in Chemistry Series, No. 47, Washington, D. C. (1965), p. 283.

74.   P. R. Johnson and A. T. Kuhn, J. Electrochem. Soc., 112:599 (1965).

75.   D. F. A. Koch, Proceedings of the First Australian Conference on Electrochemistry, 1963, Pergamon Press (1965), p. 657.

76.   L. Formaro and S. Trassatti, Chim. Ind. (Milan), 48:706 (1966).

77.   S. Trassatti and L. Formaro, J. Electroanal. Chem., 17:343 (1968).

78.   A. G. Polyak, Yu. B. Vasil'ev, and V. S. Bagotskii, Élektrokhimiya, 1:968 (1965).

79.   V. N. Kamath and Hira Lal, J. Electroanal. Chem., 19:137 (1968).

80.   R. A. Munson, J. Electrochem. Soc., 112:572 (1964).

81.   V. S. Éntina, O. A. Petrii, and I. V. Shelepin, Élektrokhimiya, 2:457 (1966).

82.   M. H. Gottlieb, J. Electrochem. Soc., 111:465 (1964).

83.   C. W. Fleischmann, G. K. Johnson, and A. T. Kuhn, J. Electrochem. Soc., 111:602 (1964).

84.   W. Vielstich and U. Vogel, Ber. Bunsenges, Phys. Chem., 68:688 (1964).

85.   M. W. Breiter, J. Electrochem. Soc., 111:1298 (1964).

86.   S. B. Brummer and A. C. Makrides, J. Phys. Chem., 68:1448 (1964).

87.   D. R. Rhodes and E. F. Steigelman, J. Electrochem. Soc., 112:16 (1965).

88.   M. W. Breiter, Electrochim. Acta, 10:503 (1965).

89.   S. B. Brummer, J. Phys. Chem., 69:562 (1965).

90.   S. B. Brummer, J. Phys. Chem., 69:1363 (1965).

91.   M. W. Breiter, J. Electrochem. Soc., 112:1244 (1965).

92.   A. B. Brummer, J. Electrochem. Soc., 113:1041 (1966).

93.   N. Minakshisundaram, Yu. B. Vasil'ev, and V. S. Bagotskii, Élektrokhimiya, 3:193, 283 (1967).

94.   A. G. Polyak, Yu. B. Vasil'ev, V. S. Bagotskii, and R. M. Smirnova, Élektrokhimiya, 3:1076 (1967).

95.   C. Liang and T. C. Franklin, Electrochim. Acta, 9:517 (1964).

96.   H. Binder, A. Köhling, and G. Sandstede, J. Electroanal. Chem., 17:111 (1968).

97.   T. Takamura and F. Mochimaru, Extended Abstracts CITCE General Meeting, Tokyo (1966), p. 103.

98.   K. Sasaki, M. Joi, and S. Nagaura, Electrochim. Acta, 11:1776 (1966).

99. V. S. Éntina, O. A. Petrii, and Yu. N. Zhitnev, Élektrokhimiya, 3:344 (1967).
100. H. Wieland, Ber., 45:679 (1912).
101. K. A. Hofmann, Ber., 51:1526 (1918); 52:1185 (1919); 53:914 (1920).
102. B. Bruns and A. Frumkin, Z. Phys. Chem. (A), 147:125 (1930); Zh. Fiz. Khim., 1:219 (1930); B. Bruns and O. Zarubina, Zh. Fiz. Khim., 2:68 (1931).
103. R. A. Munson, J. Electroanal. Chem., 5:292 (1963).
104. A. B. Fasman, G. N. Padyukova, and D. V. Sokol'skii, Dokl. Akad. Nauk SSSR, 150:856 (1963).
105. T. B. Warner and S. Schuldiner, J. Electrochem. Soc., 111:992 (1964).
106. S. B. Brummer and J. I. Ford, J. Phys. Chem., 69:1355 (1965).
107. G. N. Padyukova, A. B. Fasman, and D. V. Sokol'skii, Élektrokhimiya, 2:885 (1966).
108. A. B. Fasman, Z. N. Novikova, and D. V. Sokol'skii, Zh. Fiz. Khim., 40:556 (1966).
109. Z. N. Novikova, A. B. Fasman, and D. V. Sokol'skii, Élektrokhimiya, 2:1015 (1966).
110. G. N. Padyukova, A. B. Fasman, and I. V. Khizhnyak, Élektrokhimiya, 4:191 (1968).
111. Z. N. Novikova and A. B. Fasman, Élektrokhimiya, 4:307 (1968).
112. M. W. Breiter, Electrochim. Acta, 12:1213 (1967).
113. M. W. Breiter, J. Phys. Chem., 72:1305 (1968).
114. S. Gilman, J. Phys. Chem., 66:2657 (1962); 67:78, 1898 (1963).
115. S. Gilman, J. Phys. Chem., 68:70 (1964).
116. S. Gilman, J. Phys. Chem., 70:2880 (1966).
117. V. I. Veselovskii, N. B. Miller, and O. G. Tyurikova, Élektrokhimiya (in press).
118. H. Pichler, Advances in Catalysis, Vol. 4, Academic Press, New York–London (1952), p. 271.
119. L. W. Niedrach, D. M. McKee, J. Paynter, and I. F. Danzig, Electrochem. Technol., 5:318 (1967).
120. D. W. McKee, L. W. Niedrach, J. Paynter, and I. F. Danzig, Electrochem. Technol., 5:419 (1967).
121. D. W. McKee and A. J. Scarpellino, Electrochem. Technol., 6:101 (1968).
122. S. Gilman, J. Phys. Chem., 71:4330, 4339 (1967).
123. T. B. Warner and S. Schuldiner, J. Phys. Chem., 69:4048 (1965).·
124. S. B. Brummer, J. Phys. Chem., 69:4049 (1965).
125. R. P. Eischens and W. A. Pliskin, Advances in Catalysis, Vol. 10, Academic Press, New York–London (1958), p. 1.
126. N. N. Kavtaradze and N. P. Sokolova, Dokl. Akad. Nauk SSSR, 112:847 (1965); 168:140 (1966); 172:386 (1967); Methods for Investigation of Catalysts and Catalytic Reactions, Vol. 1 [in Russian], Izd. Sib. Otd. Akad. Nauk SSSR, Novosibirsk (1965), pp. 95, 109.
127. D. W. McKee, J. Catalysis, 8:240 (1967).
128. B. J. Piersma, T. B. Warner, and S. Schuldiner, J. Electrochem. Soc., 113:841 (1966).
129. B. I. Podlovchenko, V. F. Stenin, and A. A. Ekibaeva, Élektrokhimiya, 4:1004 (1968).

384 CHAPTER 8

130. M. W. Breiter, J. Electroanal. Chem., 19:131 (1968); Extended Abstracts, 19th Meeting of CITCE, Detroit, Mich. (1968), p. 135.
131. S. B. Brummer and M. J. Turner, J. Phys. Chem., 71:3902 (1967).
132. R. P. Eischens and W. A. Pliskin, Actes du Deuxième Congrès International de Catalyse, Paris, 1960 (1961), p. 789; K. Hirota, K. Kuwatta, T. Otaki and S. Asia, ibid., p. 809; W. M. H. Sachtler and J. Fahrenfort, ibid., p. 831.
133. J. Chatt, B. L. Shaw, and A. E. Field, J. Chem. Soc., 3466 (1964).
134. C. E. Heath and C. H. Worsham, Fuel Cells (G. J. Young, editor), Vol. 2, Reinhold Publishing Co., New York (1963), p. 182.
135. J. O'M. Bockris, Proceedings of the First Australian Conference on Electrochemistry, 1963, Pergamon Press (1965), p. 691.
136. L. W. Niedrach, S. Gilman, and J. Weinstock, J. Electrochem. Soc., 112:1161 (1965).
137. L. W. Niedrach, J. Electrochem. Soc., 113:645 (1956).
138. R. V. Marvet and O. A. Petrii, Élektrokhimiya, 3:153 (1967).
139. A. H. Taylor and S. B. Brummer, J. Phys. Chem., 72:2856 (1968).
140. J. A. Shropshire and H. H. Horowitz, J. Electrochem. Soc., 113:490 (1966).
141. H. J. Barger and M. L. Savitz, J. Electrochem. Soc., 115:686 (1968).
142. R. Kh. Burshtein, A. G. Pshenichnikov, and V. S. Tyurin, Dokl. Akad. Nauk SSSR, 160:629 (1965).
143. R. Kh. Burshtein, A. G. Pshenichnikov, V. S. Tyurin, and A. A. Michri, Catalytic Reactions in the Liquid Phase, Proceedings of the Second All-Union Conference [in Russian], Izd. "Nauka," Kaz.SSR, Alma-Ata (1967), p. 383.
144. A. M. Bograchev, A. G. Pshenichnikov, and R. Kh. Burshtein, Élektrokhimiya, 4:358 (1968).
145. R. Kh. Burshtein, A. G. Pshenichnikov, V. S. Tyurin, and L. L. Knots, Élektrokhimiya, 1:1268 (1965).
146. R. J. Flannery and D. C. Walker, Hydrocarbon Fuel Cell Technology (S. Baker, editor), Academic Press, New York–London (1965), p. 335.
147. S. Gilman, Trans. Faraday Soc., 61:2546, 2561 (1965).
148. S. Gilman, J. Electrochem. Soc., 113:1036 (1966).
149. L. W. Niedrach and M. Tochner, J. Electrochem. Soc., 114:17 (1967).
150. J. Giner, Extended Abstracts, 19th CITCE Meeting, Detroit, Mich. (1968), p. 135.
151. W. T. Grubb and M. E. Lazarus, Extended Abstracts, Battery Division, Electrochemical Society Meeting, Philadelphia, Vol. 11 (1966), p. 14; J. Electrochem. Soc., 114:360 (1967).
152. S. B. Brummer, J. I. Ford, and M. J. Turner, J. Phys. Chem., 69:3424 (1965).
153. S. B. Brummer and M. J. Turner, Hydrocarbon Fuel Cell Technology (S. Baker, editor), Academic Press, New York–London (1965), p. 409.
154. S. B. Brummer and M. J. Turner, Extended Abstracts, Battery Division, Electrochemical Society Meeting, Philadelphia, Vol. 11 (1966), p. 13; J. Phys. Chem., 71:2825 (1967).
155. J. Giner, J. Electrochem. Soc., 111:376 (1964).
156. A. G. Pshenichnikov, A. M. Bograchev, and R. Kh. Burshtein, Élektrokhimiya (in press).

157. H. J. Barger and A. J. Coleman, J. Phys. Chem., 72:2285 (1968).
158. E. J. Cairns and A. M. Breitenstein, J. Electrochem. Soc., 114:764 (1967).
159. E. J. Cairns, A. M. Breitenstein, and A. J. Scarpellino, J. Electrochem. Soc., 115:569 (1968).
160. M. J. Turner and S. B. Brummer, Extended Abstracts, Battery Division, Electrochemical Society Meeting, Philadelphia (1966), p. 6; J. Phys. Chem., 71:3494 (1967).
161. B. I. Podlovchenko, E. A. Kolyadko, and R. P. Petukhova, The Double Layer and Adsorption on Solid Electrodes [in Russian], Izd. Tartusk. Gos. Univ., Tartu (1968), p. 118.
162. V. S. Tyurin, A. G. Pshenichnikov, and R. Kh. Burshtein, Élektrokhimiya, 2:948 (1966).
163. T. M. Beloslyudova and D. V. Sokol'skii, Élektrokhimiya, 1:1182 (1965).
164. A. T. Kuhn, Electrochim. Acta, 13:477 (1968).
165. G. Bianchi, L. de Carlo, and G. Faita, Chim. Ind. (Milan), 47:830 (1965); Extended Abstracts, 17th CITCE Meeting, Tokyo (1966), p. 117.
166. G. Bianchi, L. de Carlo, and G. Faita, Chim. Ind. (Milan), 49:16 (1967).
167. Zh. A. Kravchenko, O. V. Altshuller, O. M. Vinogradova, N. A. Shurmovskaya, and R. Kh. Burshtein, Élektrokhimiya (in press).
168. S. Gilman, Trans. Faraday Soc., 62:466, 481 (1966).
169. A. G. Pshenichnikov (Pschenichnikov), R. Kh. Burshtein, V. S. Tyurin (Tiurin), and A. A. Michri, Deuxième Journeés Internationales Étude de Combustion, Bruxelles (1967), p. 181.
170. A. A. Michri, A. G. Pshenichenikov, and R. Kh. Burshtein, Élektrokhimiya, 4:508 (1968).
171. A. G. Pshenichnikov, V. S. Tyurin, and R. Kh. Burshtein, The Double Layer and Adsorption on Solid Electrodes [in Russian], Izd. Tartusk. Gos. Univ., Tartu (1968), p. 131.
172. V. S. Tyurin, A. G. Pshenichnikov, and R. Kh. Burshtein, Élektrokhimiya (in press).
173. J. O'M. Bockris and D. A. J. Swinkels, J. Electrochem. Soc., 111:736 (1964).
174. J. O'M. Bockris, M. Green, and D. A. J. Swinkels, J. Electrochem. Soc., 111:742 (1964).
175. E. Gileadi, B. T. Rubin, and J. O'M. Bockris, J. Phys. Chem., 69:3335 (1965).
176. W. Heiland, E. Gileadi, and J. O'M. Bockris, J. Phys. Chem., 70:1207 (1966).
177. E. Gileadi, Comptes Rendus, Deuxième Symposium Européen sur les Inhibiteurs de Corrosion, Ferrara, 1965, Vol. 2 (1966), p. 543.
178. E. Gileadi, J. Electroanal. Chem., 11:137 (1966).
179. B. I. Podlovchenko and Z. A. Iofa, Zh. Fiz. Khim., 38:211 (1964).
180. S. Gilman, Extended Abstracts, Battery Division, Electrochemical Society Meeting, Philadelphia (1966), p. 9; J. Phys. Chem., 71:2424 (1967).
181. O. A. Petrii, Zh. N. Malysheva, and R. V. Marvet, Élektrokhimiya, 3:962, 1141 (1967).
182. L. W. Niedrach and M. Tochner, J. Electrochem. Soc., 114:233 (1967).
183. B. I. Podlovchenko, A. A. Ekibaeva, and V. F. Stenin, Advances in the Electrochemistry of Organic Compounds [in Russian], Izd. "Nauka" (1968), p. 3; Élektrokhimiya, 4:1374 (1968).

184.   V. N. Kamath and Hira Lal, J. Electroanal. Chem., 19:249 (1968).
184a. M. W. Breiter, Discussions Faraday Soc., No. 45, 79 (1968).
185.   Ya. Veber, Yu. B. Vasil'ev, and V. S. Bagotskii, Élektrokhimiya (in press).
186.   S. Gilman, J. Phys. Chem., 68:2098, 2112 (1964).
187.   V. E. Kazarinov and N. A. Balashova, Collection Czech. Chem. Commun., 30:4184 (1965).
188.   E. J. Cairns, Hydrocarbon Fuel Cell Technology (S. Baker, editor), Academic Press, New York—London (1965), p. 465; Nature, 210:161 (1966); J. Electrochem. Soc., 113:1200 (1966).
189.   Zh. I. Bobanova, G. A. Bogdanovskii, and G. D. Vovchenko, Élektrokhimiya, 4:798 (1968).
190.   V. E. Kazarinov and G. N. Mansurov, Élektrokhimiya, 2:1388 (1966).
191.   P. K. Reshet'ko and A. I. Shlygin, Uch. Zap. Dal'nevost. Gos. Univ., 8:79 (1966).
192.   A. Shlygin, A. Frumkin, and V. Medvedovskii, Acta Physicochim. URSS, 4:911 (1936).
193.   B. I. Podlovchenko, Hira Lal, and O. A. Petrii, Élektrokhimiya, 1:744 (1965).
194.   Yu. A. Chizmadzhev and V. S. Markin, Élektrokhimiya, 3:127 (1967).
195.   V. S. Markin and Yu. A. Chizmadzhev, Élektrokhimiya, 4:123 (1968).
195a. S. B. Brummer and K. Cahill, Discussions Faraday Soc., No. 45 (1968).
196.   T. I. Pochekaeva, Zh. Fiz. Khim., 36:2751 (1962).
197.   M. E. Manzhelei and A. F. Sholin, Dokl. Akad. Nauk SSSR, 141:897 (1961).
198.   W. Vielstich, Brennstoffelemente, Verlag Chemie, Weinheim (1965).
199.   H. Binder, A. Köhling, and G. Sandstede, Advanced Energy Conversion, 6:135 (1966).
200.   T. M. Beloslyudova and D. V. Sokol'skii, Dokl. Akad. Nauk SSSR, 162:1297 (1965).
201.   V. S. Daniél'-Bek and G. V. Vitvitskaya, Zh. Prikl. Khim., 37:1724 (1964); Élektrokhimiya, 1:354, 494, 759 (1965); T. N. Glazatova and V. S. Daniél'-Bek, Élektrokhimiya, 2:1042 (1966).
202.   E. Müller and E. Hochstetter, Z. Elektrochem., 20:367 (1914).
203.   A. N. Bakh, Collected Works on Chemistry and Biochemistry [in Russian], Izd. Akad. Nauk SSSR, Moscow (1950), p. 300.

Chapter 9

Thermodynamics of Surface Phenomena on Electrodes Adsorbing Hydrogen and Oxygen

It was shown in the preceding chapter that adsorption of organic compounds on platinum metals, accompanied by far-reaching chemical changes, differs substantially from adsorption phenomena on a mercury electrode. However, even if molecular breakdown does not occur and the adsorption process is fully reversible, the ability of metals of the platinum group (and to various degrees also of other solid electrodes) to adsorb hydrogen and oxygen should lead to a number of peculiarities in the adsorption of organic compounds. The thermodynamic approach to adsorption of organic substances accompanied by adsorption of hydrogen and oxygen on electrode surfaces was used by Frumkin [1].

Before analyzing the thermodynamic relationships characterizing joint adsorption of organic substances with hydrogen or oxygen, it is appropriate to examine the thermodynamics of surface phenomena on metals of the platinum group in electrolyte solutions in absence of organic compounds. Knowledge of the state of the surface under these conditions is a theoretical prerequisite for understanding processes of chemisorption and oxidation of organic substances on the platinum metals. This problem is also of independent scientific interest, because the thermodynamics of surface phenomena has been hitherto developed predominantly in relation to the mercury electrode and to amalgam electrodes [1a]. In distinction from the mercury electrode, the platinum electrode is not ideally polarized, as the following reaction may occur at

387

its interface with the solution:

$$H_{ads} + H_2O \rightleftharpoons H_3O^+ + e \qquad\qquad (IX.1)$$

or, in the oxygen region

$$O_{ads} + H_2O + 2e \rightarrow 2OH^- \qquad\qquad (IX.2)$$

However, in a certain range of potentials the platinum electrode is "perfectly polarized" or "completely polarized" (vollkomen polarisierbar) in Planck's terminology [2-4]; i.e., its state in a given solution is determined by the quantity of electricity imparted to it starting at a certain initial potential. For ideal (perfect) polarizability to be achieved the change of potential must be slow enough to allow establishment of equilibrium in reactions (IX.1) and (IX.2). This is attained without difficulty for reaction (IX.1) in the hydrogen region and, as will be shown below, also for reaction (IX.2) in a certain potential range of the oxygen region.

Another necessary condition for ideal polarizability of the elec trode is that the amount of dissolved molecular hydrogen interacting with the electrode surface is negligible in comparison with the amount of adsorbed hydrogen. This condition can be realized by the use of Pt/Pt electrode of true surface area 5 m^2, with 20 cm^3 of solution at $\varphi_r \gtrsim 30$ mV.

As Planck showed nearly 80 years ago, the Lippmann equation is applicable to an ideally polarized electrode, although in this case the physical meaning of the charge density differs from that given to ε in the case of an ideally polarized mercury electrode [2, 4] (see below).

A platinum and a mercury electrode in an electrolyte solution also differ in the following respect. The state of the mercury electrode depends on only one variable (apart from the concentration of the salt CA), the electrode potential; the state of the platinum electrode also depends on the solution pH.

The thermodynamic theory of the platinum hydrogen electrode was put forward by Frumkin and Shlygin [5] in 1936. Various thermodynamic relationships were subsequently examined by Frumkin [6, 7] and by Frumkin, Balashova, and Kazarinov [8]. The thermodynamic theory of electrodes which adsorb hydrogen and oxygen was developed further and verified experimentally in a series of investigations by Frumkin, Petrii, and co-workers [3, 9-26].

1. Dependence of the Potential of a Platinum Electrode on the Solution pH under Isoelectric Conditions

We assume that the platinum electrode may be regarded as reversible with respect to ionization of adsorbed hydrogen and to discharge of hydrogen ions in a solution containing a neutral salt CA and an acid HA. We suppose further that the amount of hydrogen dissolved in the bulk solution and in the metal is negligible in comparison with the amount of hydrogen on the electrode surface.

The state of the system under consideration can be defined by the chemical potential μ_H of atomic hydrogen and the chemical potentials μ_{HA} and μ_{CA} of the acid and salt. Hydrogen atoms, acid and salt ions, and water molecules are chosen as the components forming the boundary layer.* We denote the surface densities of the components in the Gibbs sense, i.e., the amounts of the substances which must be introduced into the system so that the composition of the volume phase remains constant when the interfacial area is increased by unity, by Γ_H, Γ_{HA}, and Γ_{CA}. The position of the interface is determined by the condition $\Gamma_{H_2O} = 0$. The values of Γ and μ are expressed in electrical units.

The value of Γ_H includes both the amount of hydrogen adsorbed in atomic form on the platinum surface and the amount expended as the result of ionization in formation of unit interfacial area. This is precisely the value which would be found experimentally by direct determination of the adsorption of hydrogen on an electrode immersed in the solution at given μ_H and μ_{H^+}.

The value of $-\Gamma_H$ represents the quantity of electricity stored per unit area of the electrode (the sign is negative because ionization of a hydrogen atom leads to the appearance of a negative charge). In the same way the total negative charge stored on the negative plate of a lead storage cell is determined by the amount of lead deposited, with a correction for the negative charge of the metallic part of the double layer. In this sense $-\Gamma_H$ can be called the total charge per unit electrode area. If no electrochemically active substances are supplied to the electrode, Γ_H remains con-

---

*The system could also be composed of ions and electrons (instead of atoms and ions). The ultimate results are the same [3, 20].

stant on open circuit. Measurement conditions under which Γ_H remains constant will be called isoelectric, in accordance with the terminology proposed by Frumkin et al. [7, 27].

By Faraday's law, the consumption of hydrogen represented by Γ_H is equivalent to the quantity of electricity Q which must be supplied to the electrode if the same change of the surface state is produced by current supplied from an external source:

$$-\Gamma_H = Q \tag{IX.3}$$

This relation is derived in greater detail in [3].

We denote the surface free energy density by σ, and the electrode potential measured against a hydrogen electrode in the same solution in equilibrium with molecular hydrogen under atmospheric pressure by φ_r. Several relationships exist between the above values:

$$(d\varphi_r)_{\mu_{HA},\mu_{CA}} = -(d\mu_H)_{\mu_{HA},\mu_{CA}} ; \tag{IX.4}$$

$$\Gamma_{HA} = \Gamma_{H^+}; \ \Gamma_{CA} = \Gamma_{C^+}; \ \Gamma_{A^-} = \Gamma_{HA} + \Gamma_{CA}; \ \Gamma_{H^+} = \Gamma_{A^-} - \Gamma_{C^+}; \tag{IX.5}$$

$$d\sigma = -\Gamma_H d\mu_H - \Gamma_{HA} d\mu_{HA} - \Gamma_{CA} d\mu_{CA} . \tag{IX.6}$$

Equation (IX.4) is the condition for equilibrium between the electrode and the solution, and Eq. (IX.5) is the condition for electroneutrality in the adsorption process. Equation (IX.6) is the Gibbs expression for the total differential of the surface free energy in the case under discussion.

From Eqs. (IX.6), (IX.4), and (IX.3) it follows that

$$\Gamma_H = -Q = \left(\frac{\partial\sigma}{\partial\varphi_r}\right)_{\mu_{HA},\mu_{CA}} . \tag{IX.7}$$

Thus, the charging curve of the platinum electrode, i.e., the dependence of Q on φ_r, is the equilibrium dependence of the surface tension of the electrode on the potential. Integration of the charging curve gives the electrocapillary curve accurate to the integration constant. The potential corresponding to the condition $\Gamma_H = 0$ (or Q = 0) is the potential of the maximum of the electrocapillary curve. It will be shown below how this potential can be found from experimental data.

The following four relations can be obtained from Eqs. (IX.6) and (IX.4) by the usual operations with the partial derivatives of the total differential:

$$\left(\frac{\partial \Gamma_H}{\partial \mu_{HA}}\right)_{\varphi_2, \mu_{CA}} = -\left(\frac{\partial \Gamma_{HA}}{\partial \varphi_r}\right)_{\mu_{HA}, \mu_{CA}} ; \qquad \text{(IX.8)}$$

$$\left(\frac{\partial \varphi_r}{\partial \mu_{HA}}\right)_{\Gamma_H, \mu_{CA}} = \left(\frac{\partial \Gamma_{HA}}{\partial \Gamma_H}\right)_{\mu_{HA}, \mu_{CA}} ; \qquad \text{(IX.9)}$$

$$\left(\frac{\partial \varphi_r}{\partial \mu_{HA}}\right)_{\Gamma_{HA}, \mu_{CA}} = \left(\frac{\partial \Gamma_{HA}}{\partial \Gamma_H}\right)_{\varphi_r, \mu_{CA}} ; \qquad \text{(IX.10)}$$

$$\left(\frac{\partial \Gamma_{HA}}{\partial \varphi_r}\right)_{\Gamma_H, \mu_{CA}} = -\left(\frac{\partial \Gamma_H}{\partial \mu_{HA}}\right)_{\Gamma_{HA}, \mu_{CA}} . \qquad \text{(IX.11)}$$

Equation (IX.9) will be examined more fully in this section. In the light of the foregoing, it can also be written in the following form:

$$\left(\frac{\partial \varphi_r}{\partial \mu_{HA}}\right)_{Q, \mu_{CA}} = -\left(\frac{\partial \Gamma_{HA}}{\partial \varphi_r}\right)_{\mu_{HA}, \mu_{CA}} \bigg/ \left(\frac{\partial Q}{\partial \varphi_r}\right)_{\mu_{HA}, \mu_{CA}}$$

$$= -\left(\frac{\partial \Gamma_{H^+}}{\partial \varphi_r}\right)_{\mu_{HA}, \mu_{CA}} \bigg/ \left(\frac{\partial Q}{\partial \varphi_r}\right)_{\mu_{HA}, \mu_{CA}} . \qquad \text{(IX.12)}$$

It is significant for practical use of Eq. (IX.12) that the influence of dissolved hydrogen on μ_{HA} and μ_{CA} can be neglected; therefore μ_{HA} and μ_{CA} remain constant when [HA] and [CA] are kept constant while φ_r and therefore μ_H varies.

Two particular cases of Eq. (IX.12) are of special interest.

1. Solution of the pure acid, [CA] = 0.

In this case Eq. (IX.12) is simplified:

$$\left(\frac{\partial \varphi_r}{\partial \mu_{HA}^{\pm}}\right)_Q = -2 \left(\frac{\partial \Gamma_{H^+}}{\partial \varphi_r}\right)_{\mu_{HA}^{\pm}} \bigg/ \left(\frac{\partial Q}{\partial \varphi_r}\right)_{\mu_{HA}^{\pm}} , \qquad \text{(IX.13)}$$

where $\mu_{HA}^{\pm}$ is the average chemical potential of HA ions.

2. An acidified solution of the neutral salt, where [CA] $\gg$ [HA]. In this case μ_{A^-} remains practically constant with variation of

[HA] if [CA] = const, and therefore

$$d\mu_{HA} = d\mu_{H^+} + d\mu_{A^-} = d\mu_{H^+},$$ (IX.14)

where $d\mu_{H^+}$ can be found from the change of potential of the hydrogen electrode if [CA] remains constant. In this case we have from Eq. (IX.12)

$$\left(\frac{\partial\varphi_r}{\partial\mu_{H^+}}\right)_{Q,\mu_{CA}} = -\left(\frac{\partial\Gamma_{H^+}}{\partial\varphi_r}\right)_{\mu_{H^+},\mu_{CA}} \Big/ \left(\frac{\partial Q}{\partial\varphi_r}\right)_{\mu_{H^+},\mu_{CA}},$$ (IX.15)

Equation (IX.15) remains valid for acids of the type H_2A.

In a number of cases it is more convenient to refer the measured potential to a constant reference electrode, the potential of which does not vary with μ_{HA}, rather than to a hydrogen electrode in the same solution. We denote the potential measured against such an electrode by φ. With the condition [CA] ≫ [HA] and at constant μ_{CA}, it is evident that

$$d\varphi = d\varphi_r + d\mu_{H^+}.$$ (IX.16)

From Eqs. (IX.15) and (IX.16) it follows that

$$\left(\frac{\partial\varphi}{\partial\mu_{H^+}}\right)_{Q,\mu_{CA}} = 1 - \left(\frac{\partial\Gamma_{H^+}}{\partial\varphi}\right)_{\mu_{H^+},\mu_{CA}} \Big/ \left(\frac{\partial Q}{\partial\varphi}\right)_{\mu_{H^+},\mu_{CA}}.$$ (IX.17)

In HA solutions of variable concentration, extending beyond the limits of applicability of the laws of dilute solutions, direct reference of the potential to a constant electrode is a problem outside the framework of thermodynamic treatment. However, as was shown by Frumkin [28], if the potential is referred to an imaginary reference electrode the potential of which differs from that of an electrode reversible with respect to the cation (the H^+ ion in this instance) by $-\mu_{HA}^{\pm}$, the thermodynamic relations of the electrocapillary theory are of the same form for concentrated as for dilute solutions. Therefore such an electrode can be tentatively considered as a constant reference electrode; in the case of dilute solutions this definition becomes the generally accepted one. Accordingly, for pure acid solutions of variable concentration we write

$$d\varphi = d\varphi_r + d\mu_{HA}^{\pm}.$$ (IX.18)

From Eqs. (IX.18) and (IX.13) it follows that

$$\left(\frac{\delta\varphi}{\delta\mu_{HA}^{\pm}}\right)_Q = 1 - 2\left(\frac{\delta\Gamma_{H^+}}{\delta\varphi}\right)_{\mu_{HA}^{\pm}} \Big/ \left(\frac{\delta Q}{\delta\varphi}\right)_{\mu_{HA}^{\pm}}. \tag{IX.19}$$

The above treatment can be extended to alkali solutions, COH, and to alkaline solutions of the neutral salt, CA + COH (assuming, as before, that [CA] ≫ [COH]). In alkaline solutions, evidently

$$\Gamma_{COH} = \Gamma_{OH^-}; \; \Gamma_{C^+} = \Gamma_{CA} + \Gamma_{COH}; \; \Gamma_{A^-} = \Gamma_{CA}; \; \Gamma_{OH^-} = \Gamma_{C^+} - \Gamma_{A^-}. \tag{IX.20}$$

The stoichiometry of charging of a hydrogen electrode in alkaline solutions is represented by the reaction

$$H_{ads} + OH^- \rightleftarrows H_2O + \bar{e}. \tag{IX.21}$$

For solutions of the pure alkali, instead of Eqs. (IX.13) and (IX.19) we have

$$\left(\frac{\delta\varphi_r}{\delta\mu_{COH}^{\pm}}\right)_Q = -2\left(\frac{\delta\Gamma_{HO^-}}{\delta\varphi_r}\right)_{\mu_{COH}^{\pm}} \Big/ \left(\frac{\delta Q}{\delta\varphi_r}\right)_{\mu_{COH}^{\pm}}, \tag{IX.22}$$

$$\left(\frac{\delta\varphi}{\delta\mu_{COH}^{\pm}}\right)_Q = -1 - 2\left(\frac{\delta\Gamma_{OH^-}}{\delta\varphi}\right)_{\mu_{COH}^{\pm}} \Big/ \left(\frac{\delta Q}{\delta\varphi}\right)_{\mu_{COH}^{\pm}}, \tag{IX.23}$$

where $\mu_{COH}^{\pm}$ is the average chemical potential of C^+ and OH^- ions, and for alkaline solutions of the neutral salt we have instead of Eqs. (IX.15) and (IX.17)

$$\left(\frac{\delta\varphi_r}{\delta\mu_{OH^-}}\right)_{Q,\mu_{CA}} = -\left(\frac{\delta\Gamma_{OH^-}}{\delta\varphi_r}\right)_{\mu_{OH^-},\mu_{CA}} \Big/ \left(\frac{\delta Q}{\delta\varphi_r}\right)_{\mu_{OH^-},\mu_{CA}}, \tag{IX.24}$$

$$\left(\frac{\delta\varphi}{\delta\mu_{OH^-}}\right)_{Q,\mu_{CA}} = -1 - \left(\frac{\delta\Gamma_{OH^-}}{\delta\varphi}\right)_{\mu_{OH^-},\mu_{CA}} \Big/ \left(\frac{\delta Q}{\delta\varphi}\right)_{\mu_{OH^-},\mu_{CA}}. \tag{IX.25}$$

In alkaline solutions the potential φ is referred to a reference electrode which differs from an electrode reversible with respect to H_2 under atmospheric pressure by μ_{OH^-} in alkaline solutions of the neutral salt and by $\mu_{COH}^{\pm}$ in the pure alkali. It is evident that with the condition $\Gamma_{H_2O} = 0$, Eqs. (IX.13) and (IX.15) are valid over the entire pH range.

The applicability region of the above relationships can be extended to potentials at which adsorbed hydrogen on the electrode surface is replaced by adsorbed oxygen, i.e., to the oxygen region of the charging curve. Assuming that under these conditions the system can again be regarded as reversible (the limits in which this is permissible are evident from experimental data), Eqs. (IX.4), (IX.6), and (IX.3) must be replaced by the following:

$$d\mu_o = d\varphi_r \qquad \text{(IX.26)}$$

$$d\sigma = -\Gamma_o d\mu_o - \Gamma_{HA} d\mu_{HA} - \Gamma_{CA} d\mu_{CA} \qquad \text{(IX.27)}$$

$$\Delta\Gamma_o = \Delta Q \qquad \text{(IX.28)}$$

where μ_0 and Γ_0 are also expressed in electrical units. As the changes of sign of $d\varphi_r$ in Eq. (IX.26) as compared with (IX.4), and of ΔQ in Eq. (IX.28) as compared with (IX.3), compensate each other, Eq. (IX.11) and all its corollaries remain valid; the same applies to Eqs. (IX.22)-(IX.25), derived for alkaline solutions. Thus, as long as the assumption that the ionization of the adsorbed hydrogen is a reversible reaction is valid, the experimental results can be treated regardless of whether adsorbed hydrogen or oxygen is present on the electrode surface. This result is understandable even from the fact that under the condition $\Gamma_{H_2O} = 0$ adsorption of oxygen may be regarded as negative adsorption of hydrogen. The case of simultaneous presence of H_{ads} and O_{ads} does not require separate examination, as their presence in equivalent amounts is thermodynamically indistinguishable from chemisorption of water, and as a result of the condition $\Gamma_{H_2O} = 0$, must be disregarded. Only the amount of Γ_H or Γ_O in excess of the equivalent of the other enters the calculation.

The quantities $\left(\dfrac{\partial Q}{\partial \varphi_r}\right)_{\mu_{HA}, \mu_{CA}}$ and $\left(\dfrac{\partial \varphi_r}{\partial \mu_{HA}}\right)_{Q, \mu_{CA}}$ in Eq. (IX.12) can be found experimentally; the former from the equilibrium charging curves and the latter from the change of potential of the isolated electrode with the solution pH, i.e., by the method of isoelectric shifts of potential. It is further possible to calculate the dependence of Γ_{H^+} on φ_r and to compare it with the dependence found experimentally from the change of solution pH in contact with an electrode having an extensive surface (the adsorption curve method [29]). Thus, Eq. (IX.12) can be used for experimental verification of the applicability of the thermodynamic theory to electrodes consisting of metals of the platinum group.

The validity of the thermodynamic relationships derived above is independent of the interpretation of the physical meaning of the quantities they contain. However, if we wish to correlate these quantities with model concepts of the structure of the interface, we must inevitably go beyond the framework of purely thermodynamic concepts.

Let us first consider the quantity $\Gamma_{H^+} = \Gamma_{A^-} - \Gamma_{C^+}$. Changes of the H^+ ion concentration on contact of the electrode with the solution may occur both by transfer of H^+ ions from the solution to the surface, with formation of adsorbed atoms and appearance of positive charges on the surface, and by transfer of H^+ ions into the ionic part of the double layer (and the corresponding reverse processes). The adsorbed H^+ ions may be attracted by the negative surface charges, or may be adsorbed "specifically," retaining their ionic character. Independently of this, we will not consider the possible appearance of species intermediate between H_{ads} and electrically attracted H^+ ions in the double layer, and if the surface layer nevertheless contains specifically adsorbed H^+ ions we will regard them as adsorbed atoms and include their charge with the surface charge of the metal.

If we assume further that the H^+ ion concentration is low in comparison with the concentration of other cations in the solution, we can disregard electrostatic adsorption of H^+ ions in comparison with Γ_{C^+}. We assume that all the other ions present in the solution retain their charges on adsorption; in other words, we do not include the part of their charge which passes to the metal as the result of adsorption with the surface charge of the metal. The calculation method is conventional in this respect and does not correspond to the real physical picture of the process, because the bonds between platinum and ions of pronounced specific adsorbability, such as I^- or Tl^+, approximate to the covalent type [8, 30]. However, determinations carried out by Lorenz [31] indicate that partial charge transfer also occurs in specific adsorption of ions on mercury, and yet in this case the transferred charge is also not included in the surface charge of the metal when the structure of the double layer is discussed.

With the above assumptions, Γ_{H^+} becomes identical to ε, the charge of the part of the double layer facing the metal at the electrode-solution interface:

$$\Gamma_{H^+} = \Gamma_{A^-} - \Gamma_{C^+} = \varepsilon . \qquad (IX.29)$$

In distinction from Γ_H, Γ_{H^+} should be called the free charge of the surface. With these assumptions Γ_H can be subdivided into two terms, one of which is determined by adsorption of hydrogen in the form of H atoms, and the other by transfer of hydrogen in the form of H^+ ions into the bulk solution:

$$\Gamma_H = A_H - \Gamma_{H^+} = A_H - \varepsilon , \qquad (IX.30)$$

where A_H is the amount of atomic hydrogen per cm^2 of surface, expressed in electrical units. Increase of Γ_{H^+} evidently corresponds to decrease of Γ_H.

In the case of adsorption from alkaline solutions we assume similarly that OH^- ions do not enter the ionic part of the double layer, being displaced by other anions, and that the change of OH^- ion concentration is determined by the change of the surface charge and therefore by ionization of adsorbed hydrogen.

The validity of this model of the double layer structure is confirmed by experiments on displacement of H^+ ions by alkali – metal cations from the surface layer into the bulk solution [8].

From Eqs. (IX.15) and (IX.29) it follows that

$$\left(\frac{\partial\varepsilon}{\partial\varphi_r}\right)_{\mu_{A^+},\mu_{CA}} = -\left(\frac{\partial\varphi_r}{\partial\mu_{H^+}}\right)_{Q,\mu_{CA}}\left(\frac{\partial Q}{\partial\varphi_r}\right)_{\mu_{H^+},\mu_{CA}} . \qquad (IX.31)$$

The left-hand side of Eq. (IX.31) represents the equilibrium value of the differential capacity of the double layer at the electrode – solution interface. It might seem that in principle this value could also be found by differentiation of experimental Γ_{H^+} vs φ_r curves. However, this cannot be achieved in practice owing to the insufficient accuracy in determination of Γ_{H^+}. Another possible way of determining $\partial\varepsilon/\partial\varphi_r$ could be ac measurement of the differential electrode capacity at high frequencies, at which the pseudocapacity of the reaction

$$H^+ + e \rightleftharpoons H_{ads}$$

could be disregarded. Apart from the experimental difficulties which must be overcome in this case (e.g., see [8]), there is no certainty that the value of $\partial\varepsilon/\partial\varphi_r$ found in this way is the equilibrium value of the differential capacity of the double layer, in view of the possible slowness of the processes leading to establishment of equilibrium during adsorption of ions on metals of the platinum group [8]. Therefore the use of Eq. (IX.31) is probably the only method at present available for determination of equilibrium values of the differential capacities of electrodes of this type.

In solutions which do not contain excess of the foreign cation C^+ the value of Γ_{H^+} cannot be equated to ε, because some of the H^+ ions, the surface density of which is denoted by $\Gamma_{H^+}^i$, are involved in formation of the ionic part of the double layer. Evidently,

$$\Gamma_{H^+} = \varepsilon + \Gamma_{H^+}^i . \qquad \text{(IX.32)}$$

From Eqs. (IX.13) and (IX.31) it follows that

$$\left(\frac{\delta \varphi_r}{\delta \mu_{HA}^{\pm}}\right)_Q = -2\left(\frac{\delta(\varepsilon + \Gamma_{H^+}^i)}{\delta \varphi_r}\right)_{\mu_{HA}^{\pm}} \Big/ \left(\frac{\delta Q}{\delta \varphi_r}\right)_{\mu_{HA}^{\pm}} . \qquad \text{(IX.33)}$$

From the thermodynamic theory of electrocapillarity [28] it follows that at low surface charges, i.e., near the point of zero charge, in absence of specific adsorption the surface excesses of the cation and the anion (opposite in sign) are equal in absolute magnitude to half the charge density. Assuming that this conclusion is applicable to the present case, we have

$$\Gamma_{H^+} = -\Gamma_{A^-} = -\tfrac{1}{2}\varepsilon . \qquad \text{(IX.34)}$$

From Eqs. (IX.33) and (IX.34) we have

$$\left(\frac{\delta \varepsilon}{\delta \varphi_r}\right)_{\mu_{HA}^{\pm}} = -\left(\frac{\delta \varphi_r}{\delta \mu_{HA}^{\pm}}\right)_Q \left(\frac{\delta Q}{\delta \varphi_r}\right)_{\mu_{HA}^{\pm}} . \qquad \text{(IX.35)}$$

Comparison of Eqs. (IX.35) and (IX.31) shows that near the point of zero charge the same change of potential with variation of acidity under isoelectric conditions is to be expected in solutions of pure acids as in presence of excess neutral salt.

An expression for determining the Esin−Markov coefficient for electrodes of the platinum group can be easily derived from Eq. (IX.33). For this purpose we replace $-Q$ by $A_H - \varepsilon$. Then, taking into account that in acid solution

$$\varepsilon = \Gamma_{A^-} - \Gamma_{H^+}^i , \qquad \text{(IX.36)}$$

we obtain

$$dQ = d\varepsilon - dA_H = d\Gamma_{A^-} - d\Gamma_{H^+}^i - dA_H . \qquad \text{(IX.37)}$$

From Eqs. (IX.16), (IX.33), (IX.36), and (IX.37) we have

$$\left(\frac{\partial\varphi}{\partial\mu_{HA}^{\pm}}\right)_Q = 1 + 2\left(\frac{\partial\Gamma_{A^-}}{\partial\varphi}\right)_{\mu_{HA}^{\pm}} \Big/ \left[\frac{\partial(A_H\text{-}\Gamma_{A^-} + \Gamma_{H^+}^i)}{\partial\varphi}\right]_{\mu_{HA}^{\pm}} =$$

$$= \left[\frac{\partial(A_H + \Gamma_{A^-} + \Gamma_{H^+}^i)}{\partial(A_H - \Gamma_{A^-} + \Gamma_{H^+}^i)}\right]_{\mu_{HA}^{\pm}} . \qquad (IX.38)$$

When $A_H = 0$, the condition $Q = const$ becomes $\varepsilon = const$, and Eq. (IX.38) assumes the form

$$\left(\frac{\partial\varphi}{\partial\mu_{HA}^{\pm}}\right)_\varepsilon = -\left(\frac{\partial\Gamma_{H^+}^i}{\partial\varepsilon}\right)_{\mu_{HA}^{\pm}} - \left(\frac{\partial\Gamma_{A^-}}{\partial\varepsilon}\right)_{\mu_{HA}^{\pm}} . \qquad (IX.39)$$

Equation (IX.39) coincides with the thermodynamic expression for the Esin–Markov effect, known from the electrocapillary theory [28, 32]. However, in distinction from mercury, it is inapplicable when $\varepsilon = 0$ because the condition $A_H = 0$, as will be shown later, is satisfied only in hydrogen halide solutions in a certain range of positive ε.

The range of potentials where $A_H = 0$ can be easily found by determination of curves representing isoelectric shifts of potential. From Eqs. (IX.15) and (IX.30) it follows that

$$\left(\frac{\partial\varphi_r}{\partial\mu_{H^+}}\right)_{Q,\mu_{CA}} = \frac{1}{(\partial A_H/\partial\Gamma_{H^+})_{\mu_{H^+},\mu_{CA}} - 1} \qquad (IX.40)$$

or

$$\left(\frac{\partial\varphi}{\partial\mu_{H^+}}\right)_{Q,\mu_{CA}} = \frac{(\partial A_H/\partial\Gamma_{H^+})_{\mu_{H^+},\mu_{CA}}}{(\partial A_H/\partial\Gamma_{H^+})_{\mu_{H^+},\mu_{CA}} - 1} . \qquad (IX.41)$$

The quantity $\left(\frac{\partial A_H}{\partial\Gamma_{H^+}}\right)_{\mu_{H^+},\mu_{CA}} = 0$ only when the derivative $\left(\frac{\partial\varphi}{\partial\mu_{H^+}}\right)_{Q,\mu_{CA}}$ is equal to zero. This may be true, in particular, when $A_H = 0$.

If we consider the behavior of a hydrogen electrode in presence of hydrogen dissolved in the electrolyte or in the metal, assuming equilibrium between dissolved and adsorbed hydrogen, we can obtain the relation

$$\left(\frac{\partial\varphi_r}{\partial\mu_{HA}}\right)_{\Pi_H,\mu_{CA}} = -\left(\frac{\partial\Gamma_{HA}}{\partial\varphi_r}\right)_{\mu_{HA},\mu_{CA}} \cdot \left(\frac{\partial Q}{\partial\varphi_r}\right)_{\mu_{HA},\mu_{CA}} . \qquad (IX.42)$$

The only difference between this equation and (IX.12), derived with the presence of dissolved hydrogen disregarded, is that the dependence of φ_r on μ_{HA} is considered at constant Π_H and not at constant Γ_H, where Π is defined by

$$\Pi_H = \Gamma_H + v_s C_H , \qquad (IX.43)$$

where C_H is the volume concentration of dissolved hydrogen in electrical units, and v_S is the electrode volume (or solution volume) per unit surface area. Thus, the presence of dissolved hydrogen in the metal or in the electrolyte does not influence the form of the thermodynamic expressions if it is assumed that equilibrium between dissolved and adsorbed hydrogen is maintained in all changes of state of the system.

As an example, let us examine the shift of potential with variation of the solution acidity in the case of a Pt/Pt electrode in $0.01\,N$ HCl $+ 1\,N$ KCl. The vessel for the working electrode, and the vessel for the reference electrode (a reversible hydrogen electrode), separated by two efficient ground-glass stopcocks, are first filled with a solution of lower HCl concentration than indicated above; namely, $0.001\,N$ HCl $+ 1.009\,N$ KCl. When the working electrode has reached an equilibrium with this solution (through which an inert gas is blown), at the chosen φ_r the solution is replaced by a carefully degassed solution of correspondingly higher HCl concentration, namely, $0.1\,N$ HCl $+ 0.91\,N$ KCl. A similar change is made in the reference electrode, and φ_r of the new electrode is then determined in the new solution. The compositions of the solutions are chosen so that the Cl$^-$ ion activity remains strictly constant.

Equation (IX.13)* was verified experimentally for a Pt/Pt electrode in the following solutions: $0.01\,N$ H$_2$SO$_4$; $0.01\,N$ HCl; $0.01\,N$ HBr; $0.01\,N$ KOH. Equation (IX.15) was applied in studies of the following systems: Pt/Pt electrode in $0.01\,N$ H$_2$SO$_4$ $+ 1\,N$ Na$_2$SO$_4$, $0.01\,N$ HCl $+ 1\,N$ KCl, $0.01\,N$ HBr $+ 1\,N$ KBr, $0.002\,N$ HBr $+ 0.05\,N$ KBr, $0.01\,N$ KOH $+ 1\,N$ KCl, $0.01\,N$ KOH $+ 1\,N$ KBr, and $0.01\,N$ KOH $+ 1\,N$ KI; platinum coated with rhodium black, in

---

*For an acid of the H$_2$A type this equation is of the form [24]

$$\left(\frac{\delta\varphi_r}{\delta\mu^{\pm}_{H_2A}}\right)_Q = -\,3\left(\frac{\delta\Gamma_{H_2A}}{\delta\varphi_r}\right)_{\mu^{\pm}_{H_2A}} \Big/ \left(\frac{\delta Q}{\delta\varphi_r}\right)_{\mu^{\pm}_{H_2A}} .$$

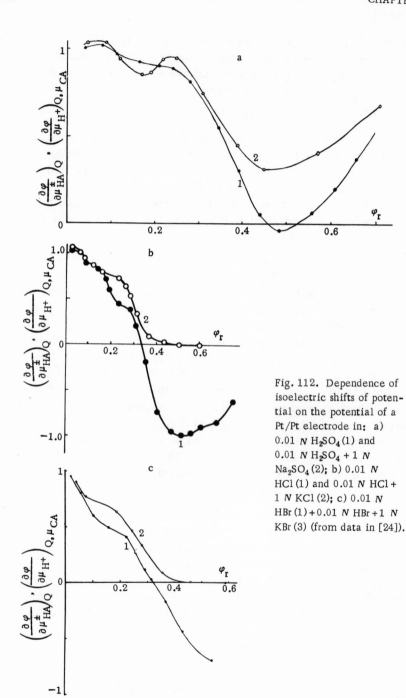

Fig. 112. Dependence of isoelectric shifts of potential on the potential of a Pt/Pt electrode in: a) 0.01 N H_2SO_4 (1) and 0.01 N H_2SO_4 + 1 N Na_2SO_4 (2); b) 0.01 N HCl (1) and 0.01 N HCl + 1 N KCl (2); c) 0.01 N HBr (1) + 0.01 N HBr + 1 N KBr (3) (from data in [24]).

0.01 N H$_2$SO$_4$ + 1 N Na$_2$SO$_4$, 0.01 N HCl + 1 N HCl, and 0.01 N KOH + 1 N KCl; platinum coated with iridium black, in 0.01 N H$_2$SO$_4$ + 1 N Na$_2$SO$_4$, 0.01 N HCl + 1 N KCl, 0.01 N HBr + 1 N KBr, 0.01 N KOH + 1 N KCl, and 0.01 N KOH + 1 N KBr; platinum coated with ruthenium black, in 0.01 N HCl + 1 N KCl and in 0.01 N KOH + 1 N KCl; platinum$-$ruthenium (10% ruthenium by wt.) electrode in 0.01 N HCl + 1 N KCl. Characteristics of the electrodes and experimental details are given in the literature [9, 11, 13, 23].

In illustration, Fig. 112 shows plots of $\left(\dfrac{\partial \varphi}{\partial \mu_{HA}^{\pm}}\right)_Q$ vs φ_r for a Pt/Pt electrode in 0.01 N H$_2$SO$_4$, HCl, and HBr solutions, compared with corresponding curves for solutions with an excess of the indifferent electrolyte. At low φ_r the values of $\left(\dfrac{\partial \varphi}{\partial \mu_{HA}^{\pm}}\right)_Q$ and $\left(\dfrac{\partial \varphi}{\partial \mu_{H^+}}\right)_{Q,\mu_{CA}}$ for different solutions are close to unity. It is significant that at potentials approximately corresponding to the points of zero charge of platinum in the solutions investigated (see Section 4), the isoelectric shifts in acid solutions with and without the indifferent electrolyte are fairly similar. The probable explanation is that Eq. (IX.35) is only approximately valid.

With increase of the anodic value of the potential the isoelectric shifts in hydrohalic acid solutions diminish, but in accordance with different laws. For example at sufficiently anodic φ_r the value of $\left(\dfrac{\partial \varphi}{\partial \mu_{H^+}}\right)_{Q,\mu_{CA}}$ for 0.01 N HCl + 1 N KCl and 0.01 N HBr + 1 N KBr solutions becomes zero, because A$_H$ becomes zero [see Eq. (IX.41)]. In solutions of the pure acids, in the φ_r range where A$_H$ = 0, $\left(\dfrac{\partial \varphi}{\partial \mu_{HA}^{\pm}}\right)_Q$ reaches a minimum value, close to -1. Thus, in the stated φ_r range the Pt/Pt electrode behaves in acid solutions in absence of neutral salts approximately like a reversible chlorine or a reversible bromine electrode. This result directly indicates that the Esin$-$Markov effect, due to the discrete structure of the double layer [33] does not occur with Cl$^-$ and Br$^-$ ions on the Pt/Pt electrode, in distinction from the mercury electrode. It may be supposed that the absence of the discreteness effect is due to strong chemisorptive interaction between the adsorbed anions and the platinum surface, as the result of which the negative anion charges are collectivized to a considerable extent with the electron

Γ_{H^+}, $\mu C / cm^2$

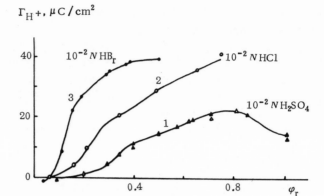

Fig. 113. Adsorption curves for a Pt/Pt electrode in 0.01 N H_2SO_4 (1), 0.01 N HCl (2), and 0.01 N HBr (3). The points represent experimental data; the continuous curves are calculated (from data in [20, 24]).

gas of the metal. Absence of the Esin−Markov effect also follows from the absence of any appreciable superequivalent adsorption of anions and cations on platinum, demonstrated by Balashova and Kazarinov [30, 34] by the radioactive tracer method.

In presence of the SO_4^{2-} anion, the specific adsorbability of which on platinum is low [30, 34, 34a] the isoelectric shifts of potential at sufficiently anodic potentials are only slightly less than zero in pure acid solutions and greater than zero in presence of an excess of the indifferent electrolyte. The cause of this is overlapping of the regions of hydrogen and oxygen adsorption on platinum in presence of SO_4^{2-} ions, as the result of which a true "double-layer" region is absent [35].

In Fig. 113, Γ_{H^+} vs φ_r curves calculated from Eq. (IX.12) are compared with experimental curves found from variations of the hydrogen ion concentration. The calculated and experimental curves are in qualitative agreement over almost the entire range of potentials investigated. It follows that the thermodynamic theory is applicable to the systems studied. These data indicate that the specific adsorbability of the anions increases in the series $SO_4^{2-} < Cl^- < Br^-$.

Curves representing isoelectric shifts of potential at platinum, rhodium, iridium, and ruthenium electrodes in various solutions

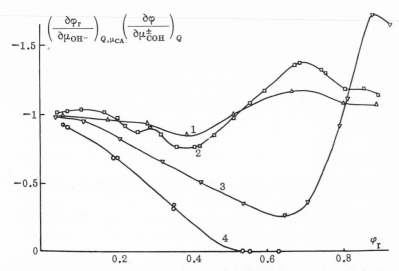

Fig. 114. Dependence of isoelectric shifts of potential on the potential of a Pt/Pt electrode in: 0.01 N KOH (1), 0.01 N KOH + 1 N KCl (2), 0.01 N KOH + 1 N KBr (3), and 0.01 N KOH + 1 N KI (4) (from data in [15]).

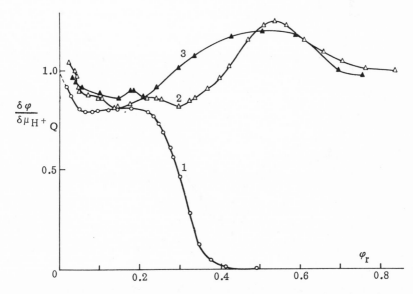

Fig. 115. Dependence of isoelectric shifts of potential on the potential of a rhodium electrode in: 0.01 N HCl + 1 N KCl (1); 0.01 N H_2SO_4 + 1 N Na_2SO_4 (2); 0.01 N KOH + 1 N KCl (3) (from data in [10, 14]).

containing indifferent electrolytes in excess are shown in Figs.
114-117. The common feature of the curves is that the isoelectric shifts of potential at low φ_r are close to unity. Thus, the behavior of platinum-metal electrodes at low φ_r tends to the behavior of the electrodes in equilibrium with gaseous hydrogen at
constant pressure. In accordance with Eq. (IX.12), in view of the
fact that $\left(\dfrac{\partial Q}{\partial\varphi_r}\right)_{\mu_{H^+},\mu_{CA}} \neq 0$, it may be concluded that the derivative
$\left(\dfrac{\partial\Gamma_{H^+}}{\partial\varphi_r}\right)_{\mu_{H^+},\mu_{CA}}$ is close to zero at low φ_r.

At more anodic potentials the curves for the isoelectric shifts
of potential are strongly dependent both on the nature of the electrodes and the composition of the solution. The higher the specific
adsorbability of the anion, the more does $\left(\dfrac{\partial\varphi}{\partial\mu_{H^+}}\right)_{Q,\mu_{CA}}$ deviate
from unity; this is particularly evident from data for the platinum
electrode in alkaline solutions. It follows that conclusions with
regard to specific adsorbability of ions on electrodes can be drawn

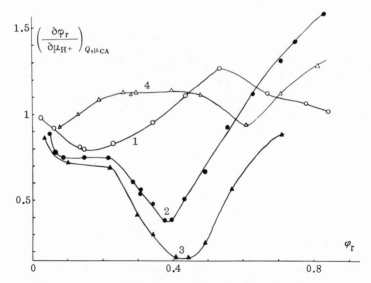

Fig. 116. Dependence of isoelectric shifts of potential on the potential
of an iridium electrode in: 0.01 N H_2SO_4 + 1 N Na_2SO_4 (1); 0.01 N HCl +
1 N KCl (2); 0.01 N HBr + 1 N KBr (3); 0.01 N KOH + 1 N KCl (4) (from
data in [23]).

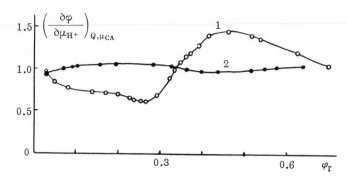

Fig. 117. Dependence of isoelectric shifts of a potential on the potential of a ruthenium electrode in: 0.01 N HCl + 1 N KCl (1) and 0.01 N KOH + 1 N KCl (2) (from data in [13]).

directly from measurements of isoelectric shifts of potential. At a platinum electrode in acid solutions containing halide ions (Cl⁻, Br⁻) and in alkaline solutions in presence of I⁻ anions, and at a rhodium electrode in acidified potassium chloride solution in a certain range of potentials, $\left(\dfrac{\partial \varphi}{\partial \mu_{H^+}}\right)_{Q,\mu_{CA}}$ is zero, i.e., a true double-layer region is observed. In other cases this region can-

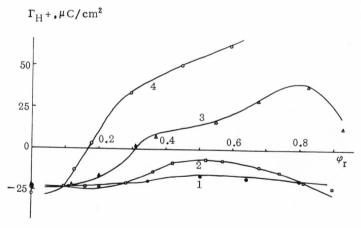

Fig. 118. Adsorption curves for a Pt/Pt electrode in: 0.01 N KOH (1); 0.01 N KOH + 1 N KCl (2); 0.01 N KOH + 1 N KBr (3); 0.01 N KOH + 1 N KI (4). The points represent experimental data; the continuous curves are calculated (from data in [15]).

not be detected owing to overlapping of the hydrogen and oxygen adsorption regions. The overlapping is especially pronounced in the case of the ruthenium electrode and all metals in alkaline solutions.

Intersection of the curves for the isoelectric potential shifts with the horizontal straight line corresponding to $\left(\dfrac{\partial\varphi}{\partial\mu_{H^+}}\right)_{Q,\mu_{CA}} = 1$ gives the positions of the extrema on the Γ_{H^+} vs φ_r curves. Thus, the shape of the adsorption curves can be established with great precision by determinations of $\left(\dfrac{\partial\varphi}{\partial\mu_{H^+}}\right)_{Q,\mu_{CA}}$.

Experimental Γ_{H^+} vs φ_r curves are compared in Figs. 118-121 with curves calculated theoretically from Eq. (IX.15).

The following point is significant for the calculations. Determination of slow charging curves by the usual method (current density of the order of 10^{-7} A per cm^2 of true area) at the potentials of hydrogen adsorption gives coinciding curves in the anodic and cathodic directions. However, if the measurements are taken up to the potentials of oxygen adsorption (to $\varphi_r \sim 0.9$ V) hysteresis is observed between the anodic and cathodic directions of the curves, introducing uncertainty into determination of the $\partial Q / \partial \varphi_r$ values

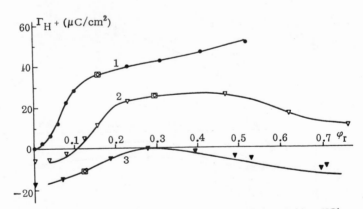

Fig. 119. Adsorption curves for a rhodium electrode in: 0.01 N HCl + 1 N KCl (1); 0.01 N H$_2$SO$_4$ + 1 N Na$_2$SO$_4$ (2); 0.01 N KOH + 1 N KCl (3). The points represent experimental data; the continuous curves are calculated (from data in [10, 14]).

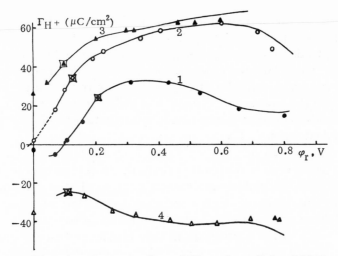

Fig. 120. Adsorption curves for an iridium electrode in: 0.01 N
HCl + 1 N KCl (1); 0.01 N H$_2$SO$_4$ + 1 N Na$_2$SO$_4$ (2); 0.01 N HBr +
1 N KBr (3); 0.01 N KOH + 1 N KCl (4). The points represent
experimental data; the continuous curves are calculated (from
data in [23]).

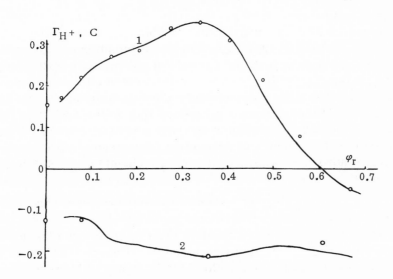

Fig. 121. Adsorption curves for a ruthenium electrode in: 0.01 N HCl +
1 N KCl (1); 0.01 N KOH + 1 N KCl (2). The points represent experimental
data; the continuous curves are calculated (from data in [13]).

required for the calculations. The hysteresis is due to slow establishment of adsorption equilibrium in adsorption of oxygen on the electrode (see [36-40] for a more detailed discussion of this effect). Therefore the charging curves were determined as follows. A certain quantity of electricity was imparted to the electrode and the polarizing circuit was then broken. After a constant potential had become established, polarization of the electrode was repeated, etc. The potentials established on open circuit were plotted against the quantity of electricity passed. Such charging curves, in distinction from those determined by the usual method, may be called equilibrium curves. The anodic and cathodic equilibrium charging curves coincide completely at the potentials of hydrogen adsorption, while at potentials of oxygen adsorption (up to ~ 0.9 V) the hysteresis is only slight, so that the choice of the direction in which the potential is changed has little influence on calculation of the Γ_{H^+} vs φ_r curves. Charging curves recorded in the anodic direction were used for the calculations.

Figures 118-121 show that the theoretically calculated and experimental curves in most cases coincide quantitatively over the potential range investigated. This justifies the treatment of surfaces of electrodes consisting of platinum metals in various solutions as equilibrium systems the state of which at $\mu_{CA} = \text{const}$ is determined by the independent variables μ_H and μ_{H^+} or μ_{OH^-}. A fact of particular interest is that the assumption of reversibility of these systems is justified not only at potentials of hydrogen adsorption but also within certain limits at potentials of oxygen adsorption. Even when the calculated and experimental curves are not in complete quantitative agreement they follow a similar course. The deviations between theory and experiment are due to difficulties in establishment of adsorption equilibrium in presence either of adsorbed oxygen or of strongly chemisorbed ions (e.g., in acid KBr solutions).

The Γ_{H^+} vs φ_r curves indicate that appearance of adsorbed oxygen on the surface of the platinum metals decreases anion adsorption; on iridium and ruthenium even Cl^- ions are displaced by adsorbing oxygen. Similar relationships, although less pronounced, are observed between adsorption of cations and of hydrogen. As φ_r approaches zero, i.e., with increasing coverage of the surface by hydrogen, cation adsorption becomes virtually independent of the potential or even diminishes.

In alkaline solutions not containing strongly adsorbed anions the surface charge on all the electrodes studied remains positive (cations are adsorbed) and depends little on the potential, so that the change of potential is apparently determined in the main by variation of the dipole moment of the bonds between the metal surface and chemisorbed hydrogen and oxygen atoms. All these data indicate that adsorption phenomena on metals of the platinum group are strongly dependent on the presence of adsorbed hydrogen and oxygen on the surface, and that the structure of the double layer on these metals differs greatly from that on mercury [41a].

Equilibrium differential capacity curves, which can be calculated from Eq. (IX.31), also reveal a number of interesting peculiarities (Fig. 122). These curves have a well-defined maximum

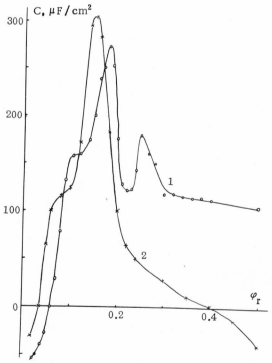

Fig. 122. Dependence of the equilibrium differential capacity of the electric double layer·on the potential of a Pt/Pt electrode in 0.01 N HCl + 1 N KCl (1), and of a rhodium electrode in 0.01 N H$_2$SO$_4$ + 1 N Na$_2$SO$_4$ (2) (from data in [10]).

at the potentials of hydrogen adsorption, the capacity at the maximum reaching values of the order of 300-400 $\mu F/cm^2$. The capacity maximum with decreasing φ_r probably arises as the result of displacement of adsorbed anions by adsorbed hydrogen which forms dipoles with their negative ends facing the solution (see below), which lower the capacity of the double layer. The relationship between equilibrium differential capacity curves and differential capacity curves determined by the ac method was examined by Damaskin and Petrii [41], who made an attempt at a model interpretation of the dependence of the surface charge on the potential of an electrode adsorbing hydrogen and oxygen.

2. Dependence of the Potential of a Platinum Electrode on the Composition of the Solution under Isoelectric Conditions

Thermodynamic relationships of the kind discussed above can be derived for a platinum electrode in solutions containing specifically adsorbed ions. Equations which are valid over the entire concentration range of these ions are given in a paper by Frumkin, Petrii, et al. [10]. Relationships for cases which have been investigated experimentally [17, 20, 25, 26] are presented below.

Suppose that the concentration of a specifically adsorbable ion is so low that it can be varied while the chemical potentials of the other ions in the solution remain virtually constant. Under these conditions the state of the system is determined by μ_H and by the chemical potential μ_i of the specifically adsorbed ion.

From the relation

$$d\sigma = -\Gamma_H d\mu_H - \Gamma_i d\mu_i ,$$ (IX.44)

where Γ_i is the surface density of the surface-active ion, it follows that

$$\left(\frac{\partial\Gamma_H}{\partial\mu_i}\right)_\varphi = -\left(\frac{\partial\Gamma_i}{\partial\varphi_r}\right)_{\mu_i} ,$$ (IX.45)

$$\left(\frac{\partial\varphi}{\partial\mu_i}\right)_{\Gamma_H} = \left(\frac{\partial\Gamma_i}{\partial\Gamma_H}\right)_{\mu_i} ,$$ (IX.46)

$$\left(\frac{\partial\varphi}{\partial\mu_i}\right)_{\Gamma_i} = \left(\frac{\partial\Gamma_i}{\partial\Gamma_H}\right)_{\varphi}, \tag{IX.47}$$

$$\left(\frac{\partial\Gamma_i}{\partial\varphi}\right)_{\Gamma_H} = -\left(\frac{\partial\Gamma_H}{\partial\mu_i}\right)_{\Gamma_i}. \tag{IX.48}$$

Equation (IX.46) can also be written in the form

$$\left(\frac{\partial\varphi}{\partial\mu_i}\right)_Q = -\left(\frac{\partial\Gamma_i}{\partial\varphi}\right)_{\mu_i} \bigg/ \left(\frac{\partial Q}{\partial\varphi}\right)_{\mu_i}. \tag{IX.49}$$

This expression makes it possible to determine the dependence of adsorption of the ion on the electrode potential from the slope of the charging curve and from the potential shift $\left(\frac{\partial\varphi}{\partial\mu_i}\right)_Q$ with variation of the concentration of the specifically adsorbed ions.

The method for investigating ion adsorption on platinum by determination of the shift of potential of an isolated electrode on addition of specifically adsorbed ions to the solution was proposed by Obrucheva [42-44], and is known as the method of adsorption potentials. It was used for studying adsorption of halide ions, and revealed specific adsorption of Tl^+, Cd^{2+}, Zn^{2+}, and Pb^{2+} cations on platinum [43, 45, 46]. Obrucheva [47] also showed that shifts of potential due to adsorption on smooth platinum can be measured, although the experimental technique with smooth electrodes is unusually difficult and involves very thorough purification of the solutions. Results were obtained with smooth platinum which were almost identical with results for platinized platinum.

Equation (IX.49) is the basis for using adsorption shifts of potential for quantitative investigations of ion adsorption on platinum metals.

Adsorption of Br^- ions on platinum-, rhodium-, and iridium-coated electrodes from $0.01\,N$ HBr $+ 1\,N$ H_2SO_4 was determined with the aid of Eq. (IX.49) from the results of measurements of adsorption shifts of potential [17, 20, 25].

In Fig. 123 the derivative $\left(\frac{\partial\varphi}{\partial\mu_{Br^-}}\right)_Q$ is plotted as a function φ_r. At low φ_r the derivative $\left(\frac{\partial\varphi}{\partial\mu_{Br^-}}\right)_Q$ tends to zero. According-

ing to Eq. (IX.49), this means that $\left(\frac{\partial\Gamma_{Br^-}}{\partial\varphi}\right)_{\mu_i}$ tends to zero, evi-

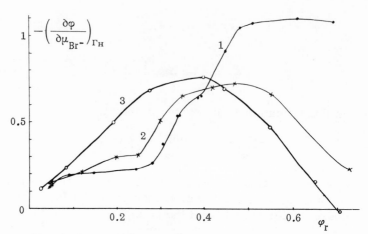

Fig. 123. Dependence of the shift of potential due to adsorption on the potential of Pt/Pt (1), rhodium (2), and iridium (3) electrodes in 0.01 N HBr + 1 N H_2SO_4 (from data in [25]).

dently owing to the decrease of adsorption Γ_{Br^-} when $\varphi_r \to 0$. In the potential range 500-700 mV on platinum $\left(\dfrac{\partial\varphi}{\partial\mu_{Br^-}}\right)_Q \sim -1$; in other words, the electrode potential changes by 58 mV with a tenfold change of the Br^- ion concentration, i.e., the Pt/Pt electrode behaves like a reversible bromine electrode. This result confirms that the Esin–Markov effect does not occur on platinum in solutions containing Br^-. Indeed, Eq. (IX.49) can be transformed as follows:

$$\left(\frac{\partial\varphi}{\partial\mu_i}\right)_Q = \frac{1}{\left(\dfrac{\partial A_H}{\partial\Gamma_i}\right)_{\mu_i} - \left(\dfrac{\partial\varepsilon}{\partial\Gamma_i}\right)_{\mu_i}} \, . \qquad (IX.50)$$

We now suppose that the adsorbability of the ion i is so high that it displaces other ions of the same sign from the ionic part of the double layer. Then in the case of a surface-active anion

$$\varepsilon = \Gamma_i - \Gamma_{C^+} \qquad (IX.51)$$

and

$$\left(\frac{\partial\varphi}{\partial\mu_i}\right)_Q = 1 \Big/ \left[\left(\frac{\partial A_H}{\partial\Gamma_i}\right)_{\mu_i} - 1 + \left(\frac{\partial\Gamma_{C^+}}{\partial\Gamma_i}\right)_{\mu_i}\right] . \qquad (IX.52)$$

At $\varphi_r \simeq 500$-700 mV in presence of Br^-, $A_H = 0$, and consequently

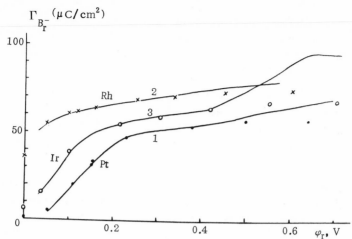

Fig. 124. Dependence of the adsorption of Br⁻ ions on the potentials of Pt/Pt (1), rhodium (2), and iridium (3) electrodes in 0.01 N HBr + 1 N H_2SO_4. Points represent experimental data; the continuous curves are calculated (from data in [25]).

$\left(\dfrac{\partial A_H}{\partial \Gamma_{Br^-}}\right)_{\mu_{Br^-}} = 0$ and $\Gamma_H = -\varepsilon$. Thus, the equation $\left(\dfrac{\partial \varphi}{\partial \mu_{Br^-}}\right)_Q = -1$ means that $\left(\dfrac{\partial \Gamma_{C^+}}{\partial \Gamma_i}\right)_{\mu_i} = 0$; i.e., that the Esin–Markov effect is absent.*

In the case of rhodium and iridium the shift of φ due to adsorption passes through a maximum without reaching -1. This is due to overlapping of the hydrogen and oxygen adsorption regions on these metals.

In Fig. 124 Γ_{Br^-} vs φ_r curves calculated from Eq. (IX.49) are compared with experimental data obtained by titration of the solutions by the Fajans method. Quantitative agreement between

---

*It should be noted that the data presented here differ from those reported by Obrucheva [42, 43], who found $\left(\dfrac{\partial \varphi}{\partial \mu_{Br^-}}\right)_Q \sim -2$. Frumkin et al. [8] pointed out that this result contradicts Eq. (IX.52) and is inconsistent with data on the low superequivalent adsorption of ions on platinum. It was suggested [8] that high values of $\left(\dfrac{\partial \varphi}{\partial \mu_{Br^-}}\right)_Q$ might have been obtained because adsorption equilibrium was not attained in the more dilute solutions. This suggestion was confirmed by Petrii and Kotlov [17].

calculations and experiment is observed up to ~0.4–0.5 V. Some discrepancies between the experimental and calculated Γ_{Br^-} values may possibly be due to the fact that equilibrium is not attained under these conditions.

It can be concluded from the results presented in Fig. 124 that the thermodynamic theory of the shifts of potential due to adsorption is valid and that the method of adsorption potentials can be used for quantitative investigation of reversible specific adsorption of ions on platinum.

If a change in the concentration of specifically adsorbed ions results in changes of the chemical potentials of these ions and also of oppositely charged ions, the following relationships [3] are valid:

$$\left(\frac{\partial \Gamma_H}{\partial \mu_{CA}}\right)_{\mu_{H^+},\varphi_r} = -\left(\frac{\partial \Gamma_{C^+}}{\partial \varphi_r}\right)_{\mu_{H^+},\mu_{CA}} - \left(\frac{\partial \Gamma_{A^-}}{\partial \varphi_r}\right)_{\mu_{H^+},\mu_{CA}} ; \qquad \text{(IX.53)}$$

$$\left(\frac{\partial \varphi_r}{\partial \mu_{CA}}\right)_{\Gamma_H,\mu_{H^+}} = -\left(\frac{\partial \Gamma_H}{\partial \mu_{CA}}\right)_{\mu_{H^+},\varphi_r} \Big/ \left(\frac{\partial \Gamma_H}{\partial \varphi_r}\right)_{\mu_{H^+},\mu_{CA}} ; \qquad \text{(IX.54)}$$

and

$$\left(\frac{\partial \varphi}{\partial \mu_{CA}}\right)_{\Gamma_H,\mu_{H^+}} = \left(\frac{\partial \Gamma_{C^+}}{\partial \Gamma_H}\right)_{\mu_{CA},\mu_{H^+}} + \left(\frac{\partial \Gamma_{A^-}}{\partial \Gamma_H}\right)_{\mu_{CA},\mu_{H^+}} . \qquad \text{(IX.55)}$$

On the other hand, Eq. (IX.15) can be reduced to the form

$$\left(\frac{\partial \varphi_r}{\partial \mu_{H^+}}\right)_{\Gamma_H,\mu_{CA}} = \left(\frac{\partial \Gamma_{A^-}}{\partial \Gamma_H}\right)_{\mu_{H^+},\mu_{CA}} - \left(\frac{\partial \Gamma_{C^+}}{\partial \Gamma_H}\right)_{\mu_{H^+},\mu_{CA}} . \qquad \text{(IX.56)}$$

It is evident that comparison of results of determinations of $\left(\frac{\partial \varphi_r}{\partial \mu_{H^+}}\right)_{\Gamma_{HA},\mu_{CA}}$ and $\left(\frac{\partial \varphi_r}{\partial \mu_{CA}}\right)_{\Gamma_H,\mu_{H^+}}$, with Eqs. (IX.55) and (IX.56) taken into account, can be used for determining the dependence of Γ_{C^+} and Γ_{A^-} on φ_r, i.e., for separate determinations of anion and cation adsorption on the electrode surface. The results of such determinations and calculations for a Pt/Pt electrode in 0.009 N HBr + $5 \cdot 10^{-2} N$ KBr are shown in Fig. 125. They are in quantitative agreement with direct adsorption measurements [26].

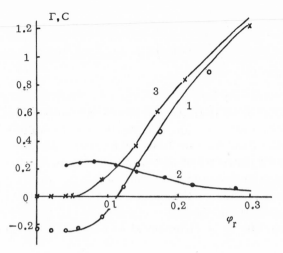

Fig. 125. Dependence of the surface charge (1) and of adsorption of K^+ cations (2) and Br^- anions (3) on the potential of a Pt/Pt electrode in $2 \cdot 10^{-3}$ N HBr + 5 $\cdot$ 10^{-2} N KBr. Points represent experimental data; the continuous curves are calculated (from data in [26]).

Equation (IX.55) is analogous to the equation

$$\left(\frac{\partial\varphi}{\partial\mu_{CA}}\right)_{\varepsilon} = -\left(\frac{\partial\Gamma_{C^+}}{\partial\varepsilon}\right)_{\mu_{CA}} - \left(\frac{\partial\Gamma_{A^-}}{\partial\varepsilon}\right)_{\mu_{CA}}, \qquad \text{(IX.57)}$$

known from the thermodynamic theory of electrocapillary [28, 32]. However, while (IX.57) refers to variation of φ with μ_{CA} at constant ε, in Eq. (IX.55) Γ_H should be taken as constant. Accordingly, in the case of a Pt-hydrogen electrode Γ_H plays the same part in the thermodynamic theory of electrocapillarity as ε in the theory of electrocapillarity for an ideally polarized mercury electrode. With this taken into account, Eq. (IX.46) may be regarded as the analog of the Dutkiewicz – Parsons equation [48],

$$\left(\frac{\partial\varphi}{\partial\mu_i}\right)_{\varepsilon} = -\left(\frac{\partial\Gamma_i}{\partial\varepsilon}\right)_{\mu_i}, \qquad \text{(IX.58)}$$

derived for the case of replacement of surface-inactive by surface-active ions while the over-all concentration remains constant.

3. Influence of Adsorption of Hydrogen and Oxygen on Adsorption of Organic Compounds

Let us now examine the behavior of a platinum hydrogen electrode in a solution of definite pH and definite electrolyte concentration in presence of an adsorbable organic substance. In this case the state of the system is defined by μ_H and μ_{org}, where μ_{org} is the chemical potential of the organic substance. The concentration of the organic substance is assumed to be so low that its variation during adsorption on the electrode does not influence the chemical potentials of the other components in the solution. The expression for the total differential of the free surface energy then becomes

$$d\sigma = -\Gamma_H \, d\mu_H - \Gamma_{org} \, d\mu_{org} , \qquad \text{(IX.59)}$$

where Γ_{org} is the Gibbs adsorption of the organic substance.

From Eq. (IX.59) it follows that

$$\left(\frac{\partial \Gamma_H}{\partial \mu_{org}} \right)_\varphi = -\left(\frac{\partial \Gamma_{org}}{\partial \varphi} \right)_{\mu_{org}} , \qquad \text{(IX.60)}$$

$$\left(\frac{\partial \varphi}{\partial \mu_{org}} \right)_{\Gamma_H} = \left(\frac{\partial \Gamma_{org}}{\partial \Gamma_H} \right)_{\mu_{org}} , \qquad \text{(IX.61)}$$

$$\left(\frac{\partial \varphi}{\partial \mu_{org}} \right)_{\Gamma_{org}} = \left(\frac{\partial \Gamma_{org}}{\partial \Gamma_H} \right)_\varphi , \qquad \text{(IX.62)}$$

$$\left(\frac{\partial \Gamma_{org}}{\partial \varphi} \right)_{\Gamma_H} = -\left(\frac{\partial \Gamma_H}{\partial \mu_{org}} \right)_{\Gamma_{org}} . \qquad \text{(IX.63)}$$

The dependence of the adsorption of the organic substance on the potential can be obtained from Eq. (IX.62). With the aid of (IX.30), we write Eq. (IX.62) in the following form:

$$\left(\frac{\partial \mu_{org}}{\partial \varphi} \right)_{\Gamma_{org}} = -\left(\frac{\partial \mu_{org}}{\partial \Gamma_{org}} \right)_\varphi \left(\frac{\partial \Gamma_{org}}{\partial \varphi} \right)_{\mu_{org}} = \left(\frac{\partial A_H}{\partial \Gamma_{org}} \right)_\varphi - \left(\frac{\partial \varepsilon}{\partial \Gamma_{org}} \right)_\varphi . \qquad \text{(IX.64)}$$

Since

$$\left(\frac{\partial \mu_{\text{org}}}{\partial \varphi}\right)_{\Gamma_{\text{org}}} = \left(\frac{\partial \Delta G_{\text{org}}}{\partial \varphi}\right)_{\Gamma_{\text{org}}},$$

where ΔG_{org} is the standard free energy of adsorption of the organic substance, Eq. (IX.64) may be written in the form

$$\left(\frac{\partial \Delta G_{\text{org}}}{\partial \varphi}\right)_{\Gamma_{\text{org}}} = -\left(\frac{\partial \varepsilon}{\partial \Gamma_{\text{org}}}\right)_{\varphi} + \left(\frac{\partial A_{\text{H}}}{\partial \Gamma_{\text{org}}}\right)_{\varphi}. \qquad \text{(IX.65)}$$

The variation of adsorption of organic substances on a mercury electrode is determined by the first term in the right-hand side of Eq. (IX.65). To compare the values of the first and second terms we can assume, in the usual way (see Part I), that

$$\varepsilon = \varepsilon_1 \left(1 - \frac{\Gamma_{\text{org}}}{\Gamma_{\infty}}\right) + \varepsilon_2 \frac{\Gamma_{\text{org}}}{\Gamma_{\infty}}, \qquad \text{(IX.66)}$$

where Γ_{∞} is the limiting adsorption of the organic substance, and ε_1 and ε_2 are the charges per unit area of the uncovered and covered parts of the surface. Similarly we can put [8]

$$A_{\text{H}} = (A_{\text{H}})_1 \left[1 - \frac{\Gamma_{\text{org}}}{\Gamma_{\infty}}\right] + (A_{\text{H}})_2 \frac{\Gamma_{\text{org}}}{\Gamma_{\infty}}, \qquad \text{(IX.67)}$$

where $(A_{\text{H}})_1$ and $(A_{\text{H}})_2$ are the amounts of adsorbed hydrogen per cm^2 on the free metal surface and on the surface with limiting coverage by the organic substance. From Eqs. (IX.66) and (IX.67) it follows that

$$\frac{\partial \varepsilon}{\partial \Gamma_{\text{org}}} = \frac{\varepsilon_2 - \varepsilon_1}{\Gamma_{\infty}}, \qquad \text{(IX.68)}$$

$$\frac{\partial A_{\text{H}}}{\partial \Gamma_{\text{org}}} = \frac{(A_{\text{H}})_2 - (A_{\text{H}})_1}{\Gamma_{\infty}}. \qquad \text{(IX.69)}$$

In the case of the mercury electrode $|\varepsilon_1|$ is several times greater than $|\varepsilon_2|$ for many organic compounds over a wide range of potential, and this determines the desorbing influence of the electric field. As far as can be judged from the results obtained by the cathodic pulse method, a similar relation holds between

$(A_H)_2$ and $(A_H)_1$. Thus, the relative significance of the first and second terms in the right-hand side of Eq. (IX.65) as factors determining the magnitude of $(\partial \Delta G/\partial \varphi)_{r_{org}}$, depends on the relation between A_H and ε. Since the maximum value of A_H in the hydrogen region of the platinum electrode is $\sim 2 \cdot 10^{-4}$ C/cm^2 while the maximum value of ε, e.g., in acid sodium sulfate solution, is of the order of 10^{-5} C/cm^2, it is evident that the second term in the right-hand side of Eq. (IX.65) has the decisive influence. Even at the positive boundary of the hydrogen region, and within the "double layer" region, where (in the case of solutions which do not contain surface-active anions) the amount of adsorbed gases, although small, is not zero [5, 29, 35], ε and A_H may be of the same order of magnitude and the term $(\partial A_H/\partial \Gamma_{org})_{\varphi}$ must be taken into account in determination of the dependence of G_{org} on φ. It follows that the relationships derived for the mercury electrode cannot be used in the case of metals which adsorb hydrogen for determining the position of maximum adsorption of neutral molecules relative to the point of zero charge. Owing to the desorbing action of hydrogen, the potential of maximum adsorption of organic substances on metals of the platinum group must be shifted into the double-layer region.

The magnitude of this shift is limited by the appearance of adsorbed hydrogen in addition to positive charges on the surface with increase of φ. An analogous approach to adsorption on a metal which adsorbs oxygen reversibly leads to the following expression [49]:

$$\left(\frac{\partial \Delta G_{org}}{\partial \varphi} \right)_{\Gamma_{org}} = -\left(\frac{\partial \varepsilon}{\partial \Gamma_{org}} \right)_{\varphi} - \left(\frac{\partial A_0}{\partial \Gamma_{org}} \right)_{\varphi}, \qquad (IX.70)$$

where A_0 is the amount of oxygen adsorbed per cm^2 and not entering the volume in the form of OH^- ions. In reality, equilibrium adsorption of oxygen is observed only in the first part of the oxygen region in acid solution at low surface coverages, which restricts the applicability range of Eq. (IX.70). Despite this, there is no doubt that the presence of oxygen must also diminish adsorption of organic compounds.

In the case of adsorption of neutral organic compounds, it may be assumed in the first approximation that the adsorption maximum must be near the potential corresponding to $\Gamma_H = 0$, which is the potential at the maximum of the equilibrium electrocapillary curve for electrodes of the platinum type.

Green and Dahms [50] proposed a method for determining the position of the point of zero charge of solid electrodes from the potential at which the adsorption of an organic compound is independent of the electrolyte concentration. In the light of the above theory, this method can be regarded as valid only if it is shown that A_H does not vary with the electrolyte concentration [8].

Bockris et al. [51-59] interpret the dependence of reversible adsorption of organic substances on metals of the platinum group on the potential in terms of competition between organic molecules and water molecules for sites on the electrode surface, with the assumption that the standard free energy of adsorption of water depends on the electric field at the electrode−solution interface. Calculation of the dependence of adsorption of organic compounds on the potential is based on the theoretical concepts developed by Bockris, Devanathan, and Müller [60]. As was shown in Part I, this theory leads to a relation between the energy of adsorption of the organic compound and the potential to which there are a number of objections. In certain publications from Bockris's laboratory (e.g., [59]) the original form of the theory was modified somewhat. However, the change in the energy of the system as the result of a change of the surface charge is disregarded in the calculations in [59]. Moreover, it is assumed in [59] that the potential drop across the compact layer is constant, whereas in reality the complete potential difference in the double layer is constant in adsorption.

4. Potentials of Zero Charge of the Platinum Metals and Their Dependence on the Solution pH

For verification of various views on adsorption of organic substances on metals of the platinum group it is important to know the potential at the point of zero charge, φ_n. The following methods have been described in the literature for determination of φ_n of the platinum metals: direct determination of the surface charge from the change of hydrogen ion concentration, and comparison of experimental and calculated Γ_{H^+} vs φ_r curves (I); deter-

TABLE 24. Potentials of Zero Charge of the Platinum
Metals and Gold

No.	Solution	φ_n, V (N.H.E.)	Method	Source
	Platinum			
1	$2 \cdot 10^{-5}$ N H_2SO_4	0.16	III	[61]
2	1 N $Na_2SO_4 + 10^{-2}$ N H_2SO_4	0.11	I	[29]
3	10^{-3} N $Na_2SO_4 + 10^{-3}$ N H_2SO_4	0.18	II	[30]
4	10^{-2} N $Cs_2SO_4 + 10^{-2}$ N H_2SO_4	0.19	II	[30]
5	$2 \cdot 10^{-5}$ N HCl	0.19	III	[61]
6	1 N NaCl $+ 10^{-2}$ N HCl	0.06	I	[29]
7	10^{-3} N NaCl $+ 10^{-3}$ N HCl	0.10	II	[30]
8	10^{-3} N NaBr $+ 10^{-3}$ N H_2SO_4	0.04	II	[30]
9	1 N NaBr $+ 10^{-2}$ N HBr	−0.02	I	[29]
10	10^{-2} N $CdSO_4 + 10^{-2}$ N H_2SO_4	0.65	II	[30]
11	1 N NaBr $+ 5 \cdot 10^{-2}$ N NaOH	−0.26	I	[29]
12	10^{-2} N CsI $+ 10^{-2}$ N CsOH	−0.58	II	[30]
13	10^{-2} N KCl and 10^{-3} N KCl	0.20	IV	[62]
14	1 N $Na_2SO_4 + 10^{-2}$ N H_2SO_4	0.27	V	[63]
15	1 N $Na_2SO_4 + 10^{-2}$ N H_2SO_4	0.16	I	[21]
16	1 N KCl $+ 10^{-2}$ N HCl	0.06	I	[3]
17	1 N KBr $+ 10^{-2}$ N HBr	−0.03	I	[21]
18	1 N KBr $+ 10^{-2}$ N KOH	−0.39	I	[15]
19	1 N KI $+ 10^{-2}$ N KOH	−0.52	I	[15]
	Palladium			
	10^{-2} N $Na_2SO_4 + 10^{-3}$ N H_2SO_4	0.11	II	[64]
	Rhodium			
1	1 N $Na_2SO_4 + 10^{-2}$ N H_2SO_4	−0.04	I	[14]
2	$2 \cdot 10^{-3}$ $Na_2SO_4 + 10^{-3}$ N H_2SO_4	−0.03	II	[65]
3	$2 \cdot 10^{-2}$ N $Na_2SO_4 + 10^{-2}$ N H_2SO_4	−0.03	II	[65]
4	1 N KCl $+ 10^{-2}$ N HCl	−0.12	I	[11]
	Iridium			
1	1 N $Na_2SO_4 + 10^{-2}$ N H_2SO_4	−0.06	I	[23]
2	1 N KCl $+ 10^{-2}$ N HCl	−0.12 (∼ −0.14)	I	[23]

Table 24 (continued)

No.	Solution	φ_n, V (N.H.E.)	Method	Source
	Ruthenium			
1	1 N KCl + 10^{-2} N HCl	< -0.12 (~ -0.3) 0.5	I	[13]
	Pt/Ru alloy (10% Ru by wt.)			
1	1 N KCl + 10^{-2} N HCl	-0.01	I	[13]
	Gold			
1	10^{-3} N KCl	0.05	IV hardness maximum	[66] [67]
2	1 N KCl	0.15	minimum creep rate	[68]
3	0.02 N H$_2$SO$_4$	0.23	capacity minimum	[69]
4	0.015 M K$_2$SO$_4$	0.09	capacity minimum	[70]
5	0.05 N KNO$_3$	0.17	capacity minimum	[71]

mination of the potential at which the adsorptions of anions and cations, expressed in electrical units, become equal (II); determination of the deflection of a platinum filament in an electric field (III); determination of the work expended in mutual approach of two crossed filaments before contact on the potential (IV); determination of the potential corresponding to the maximum contact angle (V). Method (V) is very rough.

Certain results obtained by these methods are given in Table 24. It is seen that the results given by the different methods are in satisfactory agreement.

It must be specially stressed that the far-reaching agreement between the results of determinations of adsorption curves on the one hand, and charging curves and isoelectric shifts of potential on the other (see Section 1) leaves no doubt about the reliability of

CHAPTER 9

the methods used for determining the points of zero charge of the platinum metals and the accuracy of the φ_n values obtained.

The potentials of zero charge become progressively more negative in the series Pt > Pd > Rh > Ir > Ru. A point of zero charge cannot be observed on reduced ruthenium, because it lies at $\varphi_r < 0$. Therefore the value found by rough extrapolation of the adsorption curve is given in the table. A potential of 0.5 V may be regarded as the point of zero charge of oxidized ruthenium.

The potentials of zero charge of the platinum metals are shifted in the negative direction upon adsorption of Cl^- and Br^- anions, and in the positive direction upon adsorption of Cd^{2+}.*

In alkaline solutions in absence of strongly adsorbed ions a point of zero charge, in the usual meaning of the term, is not observed on platinum metals.

In addition to the methods enumerated above, several attempts have been made to determine φ_n of platinum, rhodium, and iridium [69, 72-76] from the position of the minimum of the differential capacity vs potential curve.[†] It was assumed that the minimum corresponds to maximum diffuseness of the double layer. This method gives good results for metals which do not adsorb hydrogen, and possibly for gold; however, as was noted in Chapter V, the use of this method involves considerable difficulties in the case of the platinum metals with adsorbed hydrogen and oxygen on the surface, as the hydrogen and oxygen ionization pseudo-capacity is superposed on the capacity of the double layer. The slow establishment of adsorption equilibrium during formation of a double layer on platinum gives rise to additional difficulties [8]. Although in principle determinations in dilute solutions at fairly high frequencies should give correct values of φ_n the difficulties which arise in such determinations with highly active electrodes

---

*The physical meaning of φ_n in presence of strongly chemisorbed ions has already been discussed in Chapter VII.

[†]Eyring et al. [83-85] developed a method for determination of φ_n of solid metals based on determination of the potential of a rapidly cleaned surface on open circuit. The values of φ_n obtained for gold by this method are appreciably more negative than those given in Table 24. The φ_n values for platinum in acid solutions were in satisfactory agreement with the data in Table 24. However, in contrast to the results in [29], φ_n were also detected in alkaline solutions. Therefore, the φ_n values reported in [83-85] should be regarded with caution.

have not yet been overcome.* The φ_n values obtained by different investigators from measurements of the capacity of Pt electrodes [72, 73] differ substantially. Moreover, it was shown by Kheifets and Krasikov [72] that the values obtained greatly depend on the solution pH; this was confirmed by Gileadi, Argade, and Bockris [74], who derived the following relationship for $HClO_4$ + $NaClO_4$ and $NaOH$ + $NaClO_4$ solutions:

$$\varphi_n = 0.56 - 2.3 \frac{RT}{F} \, pH \, . \tag{IX.71}$$

However, Balashova et al. [77], using the tracer atom method, could not detect dependence of φ_n of platinized platinum on pH in Na_2SO_4 + H_2SO_4 solutions of total concentration $10^{-2} - 10^{-3} \, N$ in the pH range from 2 to 5.

We will examine the dependence of φ_n of the platinum metals on pH in more detail. This dependence was first demonstrated experimentally by Frumkin, Shlygin, and Medvedovskii [29] in determinations of adsorption curves in acid and alkaline KBr solutions. It was pointed out by Frumkin and Shlygin [5] that this dependence is the consequence of the dipolar character of the bonds between adsorbed hydrogen and the electrode surface.

In the light of the thermodynamic theory discussed above, the dependence of φ_n on pH is evidently represented by Eq. (IX.10) which, in the case of excess indifferent electrolyte, can be written in the following form:

$$\left(\frac{\partial \varphi_r}{\partial \mu_{H^+}} \right)_{\varepsilon, \mu_{CA}} = - \left(\frac{\partial \varepsilon}{\partial Q} \right)_{\varphi_r, \mu_{CA}} \tag{IX.72}$$

or

$$\left(\frac{\partial \varphi}{\partial \mu_{H^+}} \right)_{\varepsilon, \mu_{CA}} = 1 + \frac{\left(\dfrac{\partial \varepsilon}{\partial \mu_{H^+}} \right)_{\varphi_r, \mu_{CA}}}{\left(\dfrac{\partial \varphi_r}{\partial \mu_{H^+}} \right)_{Q, \mu_{CA}} \left(\dfrac{\partial Q}{\partial \varphi_r} \right)_{\mu_{H^+}, \mu_{CA}}} = \frac{\left(\dfrac{\partial A_H}{\partial \Gamma_i} \right)_{\varphi_r, \mu_{CA}}}{\left(\dfrac{\partial A_H}{\partial \varepsilon} \right)_{\varphi_r, \mu_{CA}} - 1} . \tag{IX.73}$$

---

*The results of Shevchenko, Pshenichnikov, and Burshtein [8, 75] show that in the case of electrodes of moderate activity a minimum can be observed on the differential capacity curves near the potential of zero charge of platinum in $HClO_4$ + $NaClO_4$ solutions.

The values of $\left(\dfrac{\partial\varphi}{\partial\mu_{H^+}}\right)_{\varepsilon,\mu_{CA}}$ can be found from experimental Γ_{H^+} vs φ_r curves corresponding to different pH. On the other hand, they can be calculated with the aid of Eq. (IX.73) from experimental values of Γ_{H^+} at different pH and results of independent determinations of isoelectric shifts of potential and charging curves. Agreement between the experimental and calculated values of $\left(\dfrac{\partial\varphi}{\partial\mu_{H^+}}\right)_{\varepsilon,\mu_{CA}}$ is evidently a criterion of correct experimental determination of the electrode surface charge in solutions of different pH. It is significant that the comparison is made in this case for different charges; it is therefore possible to demonstrate the reliability of adsorption curves determined at different pH over a wide range of potentials.

Determinations of the dependence of the point of zero charge on the solution pH by a combination of the adsorption curve method and isoelectric shifts of potential were carried out with the following systems [21]: a Pt/Pt electrode in H_2SO_4 + $1\,N$ Na_2SO_4 (pH

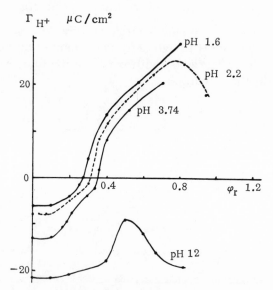

Fig. 126. Adsorption curves for a Pt/Pt electrode in
$1\,N$ Na_2SO_4 + H_2SO_4 and $1\,N$ Na_2SO_4 + NaOH solutions
of different pH (from data in [21]).

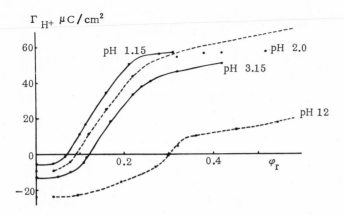

Fig. 127. Adsorption curves for a Pt/Pt electrode in 1 N KBr + HBr and 1 N KBr + KOH solutions of different pH (from data in [21]).

1.60–3.74), in HCl + 1 N KCl (pH 1.20–3.24), in HBr + 1 N KBr (pH 1.15–3.15), and in KOH + 1 N KBr (pH 10.3–12.3), and a rhodium electrode in H_2SO_4 + 1 N Na_2SO_4 (pH 1.67–3.55).

The surface charge is plotted against the potential for these systems in Figs. 126 and 127. Plots of $\left(\dfrac{\partial\varphi}{\partial\mu_{H^+}}\right)_\varepsilon$ vs ε, calculated from Eq. (IX.73) and found experimentally, are compared in Fig.

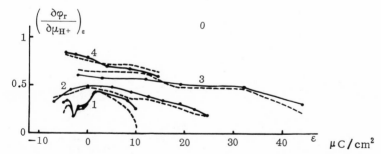

Fig. 128. Dependence of the potential of constant charge on the surface charge for various systems: 1) Pt/Pt electrode in H_2SO_4 + 1 N Na_2SO_4; 2) Pt/Pt electrode in HCl + 1 N KCl; 3) Pt/Pt electrode in HBr + 1 N KBr; 4) rhodium electrode in H_2SO_4 + 1 N Na_2SO_4. The points and continuous curves represent experimental data; the dash curves are calculated (from data in [21]).

128. The satisfactory agreement of the curves indicates that the experimental data conform to the thermodynamic relationships.

The dependence of φ_n on pH for the same system was calculated from these data (Fig. 129). According to Eq. (IX.73), φ_n would be independent of pH if $\left(\dfrac{\partial A_H}{\partial \varepsilon}\right)_{\varphi_r, \mu_{CA}}$ was zero and, in particular, if $A_H = 0$, as is the case with a mercury electrode [78]. In the case of a Pt electrode in acidifed solutions of the weakly adsorbed SO_4^{2-} anion, φ_n depends but slightly on pH; the change of φ_n per unit pH is only 18 mV. It is possible that it is even somewhat less in $10^{-2}-10^{-3} N$ solutions, i.e., at lower concentrations than were used by Petrii et al. [21] (see [77]). The dependence of φ_n on pH becomes more pronounced in passing from sulfates to acidified chlorides and especially to bromides ($\partial \varphi_H / \partial pH = -35$ mV) and from Pt to Rh ($\partial \varphi_H / \partial pH = -45$ mV), but it is still weaker than the dependence found by Gileadi, Argade, and Bockris ($\partial \varphi_H / \partial pH = -58$ mV) from determinations of the position of the potential of the minimum ac capacity of the electrode in dilute

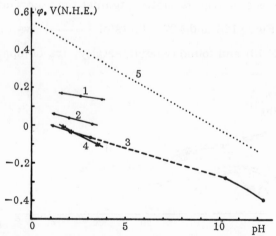

Fig. 129. Dependence of the point of zero charge on the solution pH for a Pt/Pt electrode in $H_2SO_4 + 1\ N\ Na_2SO_4$ (1), HCl + 1 N KCl (2), and HBr (or KOH) + 1 N KBr (3), and for a rhodium electrode in $H_2SO_4 + 1\ N\ Na_2SO_4$ (4) (from data in [21]); 5) dependence of the point of zero charge of platinum on pH, found by Gileadi et al. [74].

solutions. The dependence found by these workers is indicated by a dashed curve in Fig. 129. The cause of the deviation between the results obtained by the two methods* has not been finally elucidated, but agreement between them cannot be expected, because the purpose of the preliminary treatment of the electrode in the experiments of Gileadi, Argade, and Bockris was to avoid establishment of equilibrium in the ionization of adsorbed hydrogen (1), whereas in [21] the object was to attain this equilibrium, since it is the prerequisite of the theory put forward.

The most probable explanation, in our view, is that in the experiments of Gileadi et al. a considerable amount of oxygen (or OH groups) was present on the surface of the Pt electrodes. A point of zero free charge on oxidized surfaces of platinum metals has already been detected by a number of methods. In the case of a ruthenium electrode it is obtained in equilibrium determinations of Γ_{H^+} in acidifed KCl as shown above; in the case of a platinum electrode, it is found in determinations of the electrokinetic potential in very dilute solutions of acids [61], and of $\partial \sigma / \partial \varphi$ at higher frequencies by the Gokhshtein method [80].

The potential of zero charge is usually taken to mean the value of φ at which the charge ε of the metal part of the double layer becomes zero. As was stated earlier, in the case of solutions containing foreign cations in excess this corresponds to the condition $\Gamma_{H^+} = \Gamma_{A^-} - \Gamma_{C^+} = 0$. However, as explained on p. 388 this definition is somewhat conventional. The potential at which the total electrode charge becomes zero, $\Gamma_H = 0$, i.e., the potential at which the surface layer can be formed from ions of the solution, without involving electrons or adsorbed H and O atoms, is of greater fundamental significance.

No methods are available for direct experimental determinations of A_H, but from the fact that in chloride and bromide solutions in a certain range of potentials $\Delta Q = \Delta_{H^+}$ it can be concluded that in this range (known as the double-layer region) $A_H = 0$ and therefore $Q = \Gamma_{H^+}$. If Q for a certain value of φ is known, it is easy to find $\varphi_{Q=0}$ with the aid of a charging curve. This calculation is represented graphically in Fig. 130. After $\varphi_{Q=0}$ has been

---

*The difficulties in explaining the dependence on pH found in [74] are noted by the authors themselves [79].

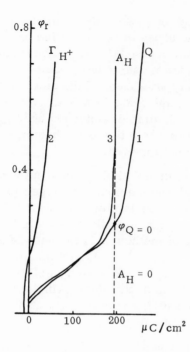

Fig. 130. Charging curve (1), adsorption curve (2), and dependence of the amount of hydrogen adsorbed on φ_r (3) for a Pt/Pt electrode in 0.01 N HCl + 1 N KCl (from data in [9]).

determined, e.g., for an acidified chloride solution, it can easily be found for any other solution which gives a charging curve without a double-layer region, effecting the transition to it from the acidified chloride solution by change of composition under isoelectric conditions. The method is applicable to Pt and Rh; however, in the case of Ir and Ru, solutions giving charging curves with a double-layer region are yet to be found. Values of $\varphi_{Q=0}$ and φ_n for several systems [21] are compared in Table 25. Values of $\varphi_{Q=0}$ can also be determined for solutions such as Na_2SO_4 + NaOH, where Γ_{H^+} does not become zero at any φ_r and therefore the φ_n concept becomes meaningless.

Adsorbed atoms and ions are both involved in giving rise to the potential difference at the interface between the platinum metal and the electrolyte solution. Their relative contributions are quantitatively expressed by the quantities $X = \left(\dfrac{\partial \varphi}{\partial A_H}\right)_{\Gamma_{H^+}}$ and $Y = \left(\dfrac{\partial \varphi}{\partial \mu_{H^+}}\right)_{A_H}$. The connection between X and Y and measurable quantities is given by the thermodynamic relations [3]:

TABLE 25

Electrode	Solution	$\varphi_{Q=0}$, V (N.H.E.)	φ_n, V (N.H.E.)
Pt/Pt	1 N Na$_2$SO$_4$ + 0.01 N H$_2$SO$_4$	0.23	0.16
	1 N Na$_2$SO$_4$ + 0.01 N NaOH	−0.25	−
	1 N KCl + 0.01 N KOH	−0.30	−
	1 N KCl + 0.01 N HCl	0.14	0.04
	1 N KBr + 0.01 N HBr	0.06	−0.03
	1 N KBr + 0.01 N KOH	−0.33	−0.39
Rh	1 N KCl + 0.01 N HCl	0	−0.12

$$X = - \left[\left(\frac{\partial \Gamma_{H^+}}{\partial \mu_{H^+}} \right)_{\varphi_r} - \left(\frac{\partial \Gamma_{H^+}}{\partial \varphi_r} \right)_{\mu_{H^+}} \right] / Z \ , \qquad \text{(IX.74)}$$

$$X + Y = \left[\left(\frac{\partial Q}{\partial \varphi_r} \right)_{\mu_{H^+}} - \left(\frac{\partial \Gamma_{H^+}}{\partial \varphi_r} \right)_{\mu_{H^+}} \right] / Z \ , \qquad \text{(IX.75)}$$

$$Z = \left(\frac{\partial Q}{\partial \varphi_r} \right)_{\mu_{H^+}} \left(\frac{\partial \Gamma_{H^+}}{\partial \mu_{H^+}} \right)_{\varphi_r} - \left(\frac{\partial \Gamma_{H^+}}{\partial \varphi_r} \right)_{\mu_{H^+}}^2 \ . \qquad \text{(IX.76)}$$

As was shown above, $\left(\frac{\partial \Gamma_{H^+}}{\partial \mu_{H^+}} \right)_{\varphi_r}$ can be found experimentally and by calculation. This makes it possible to determine X and Y with the aid of Eqs. (IX.74)-(IX.76) [22]. The results of determinations of X and Y are shown in Figs. 131 and 132.

On platinum X is positive at low and moderate coverages of the surface by adsorbed hydrogen, and becomes negative at high coverages. On rhodium X > 0 at all φ_r. A positive value of X means that the adsorbed hydrogen dipoles are oriented with their negative ends toward the solution. Therefore the H$_{ads}$ dipoles on Rh have the same orientation at all φ_r; this confirms the conclusion drawn from an analysis of the influence of pH on adsorption of hydrogen on rhodium [8, 81].

The change in the sign of X on platinum may be attributed to appearance of dipoles of the reverse orientation (positive ends toward the solution). As has already been noted [6, 8], this is consistent with results of determinations of the electronic work function on Pt [82].

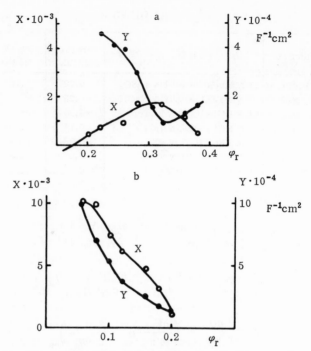

Fig. 131. Dependence of the contributions of hydrogen atoms (X) and ions of the double layer (Y) to the potential difference on the potential at a platinum (a) and rhodium (b) electrode in 0.01 N $H_2SO_4 + 1$ N Na_2SO_4 (from data in [22]).

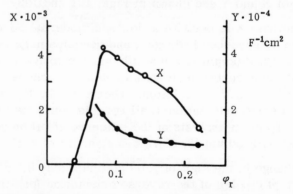

Fig. 132. Dependence of the contributions of hydrogen atoms (X) and ions of the double layer (Y) to the potential difference on the potential at a platinum electrode in 0.01 N HBr $+ 1$ N KBr (from data in [22]).

In the cases under consideration the contribution of H_{ads} to the potential difference is considerably less than that of ions in the double layer, with equal amounts of the adsorbed species. Thus, at certain φ_r in sulfate solutions the difference between the two is almost two orders of magnitude. However, by virtue of the relation between the maximum values of A_H and Γ_{H^+}, $A_H \gg \Gamma_{H^+}$, the total contributions of adsorbed atoms and ions are comparable in magnitude.

The contribution of atoms increases with increasing specific adsorbability of the electrolyte anions. This provides an explanation of the dependence of $\partial\varphi_n/\partial pH$ on the adsorbability of the anions. It follows from Eqs. (IX.73), (IX.74), and (IX.75) that

$$\left(\frac{\partial\varphi_r}{\partial\mu_{H^+}}\right)_{\Gamma_H} = -\frac{X}{X+Y}\left[1 \pm \left(\frac{\partial\mu_{H^+}}{\partial\varphi_r}\right)_Q\right]. \qquad (IX.77)$$

The coefficient $X/(X + Y)$ in Eq. (IX.77) increases with increasing ratio X:Y. If the quantity in the square brackets does not change substantially, this should lead to increase of the absolute magnitude of $\left(\dfrac{\partial\varphi}{\partial\mu_{H^+}}\right)_{\Gamma_{H^+}}$. However, an unambiguous relation between $\left(\dfrac{\partial\varphi}{\partial\mu_{H^+}}\right)_{\Gamma_{H^+}}$ and X and Y has as yet not been derived. The physical explanation of the increased contribution of atoms in presence of specifically adsorbed anions may be that the anions are adsorbed predominantly on the same sites as H_{ads} dipoles oriented with their positive ends toward the solution and displace them, whereas dipoles of the opposite orientation are adsorbed on other sites. However, this explanation appears to be inconsistent with the influence of anions on the $Pt-H_{ads}$ bond energy [6].

The variation of φ_n with pH is greater for rhodium than for platinum. The ratio X:Y is also greater than the corresponding ratio for Pt in sulfate solutions.

The quantity Y^{-1} is the differential capacity of the electrode at constant A_H. The value of Y^{-1} is 50 $\mu F/cm^2$ in presence of SO_4^{2-}, 60 $\mu F/cm^2$ in presence of Cl^-, and 100 $\mu F/cm^2$ in presence of Br^- (for platinum); i.e., it increases in the series $SO_4^{2-} < Cl^- < Br^-$. On decrease of φ_r to zero, i.e., with increasing surface coverage by H_{ads}, Y^{-1} increases by a factor of 2–5 on Pt and almost tenfold on Rh. According to Frumkin et al. [8], this can be attributed both to replacement of anions by cations in the double layer and to increasing surface coverage by H_{ads}.

The cause of the sharp decrease of X at low surface coverages by H_{ads}, observed both for Pt and for Rh (Figs. 131 and 132), is still obscure.

The values of X and Y represent the differential effects of introduction of atoms and ions into the surface layer. Tentative estimates of the absolute contribution of H_{ads} to the potential difference at the Pt−solution interface [22] show that it corresponds to tenths of 1 V (from 0.1 to 0.45) and increases in the series sulfates−chlorides−bromides. The over-all potential difference at the metal−solution interface due to adsorbed hydrogen dipoles is always positive. However, these estimates are crude approximations as yet.

References

1.   A. N. Frumkin, Dokl. Akad. Nauk SSSR, 154:1432 (1964).
1a.   P. Delahay, Double Layer and Electrode Kinetics, Interscience, New York−London−Sydney (1965).
2.   M. Planck, Ann. Phys., 44:385 (1891).
3.   A. N. Frumkin, O. A. Petrii (Petry), and R. V. Marvet, J. Electroanal. Chem., 12:504 (1966).
4.   A. N. Frumkin, J. Electroanal. Chem., 18:328 (1968).
5.   A. N. Frumkin and A. I. Shlygin, Izv. Akad. Nauk SSSR, Ser. Khim., 773 (1936); Acta Physicochim. URSS, 5:819 (1936).
6.   A. N. Frumkin, Advances in Electrochemistry and Electrochemical Engineering (P. Delahay, editor), Vol. 3, Interscience, New York (1963), p. 287.
7.   A. N. Frumkin, Élektrokhimiya, 2:387 (1966).
8.   A. N. Frumkin, N. A. Balashova, and V. E. Kazarinov, J. Electrochem. Soc., 113:1011 (1966).
9.   O. A. Petrii, R. V. Marvet, and A. N. Frumkin, Élektrokhimiya, 3:117 (1967); Proceedings of the Symposium on Electrode Processes, Jodhpur (India) (1967).
10.   A. N. Frumkin, O. A. Petrii (Petry), A. M. Kossaya, V. S. Éntina, and V. V. Topolev, J. Electroanal. Chem., 16:175 (1968).
10a. A. N. Frumkin and O. A. Petrii (Petry), Abhandl. Sächsische Akad. Wiss., Lepizig, Math.-Naturwiss. Kl., 49:17 (1968).
11.   O. A. Petrii, A. M. Kossaya, and Yu. M. Tyurin, Élektrokhimiya, 3:617 (1967).
12.   A. N. Frumkin, O. A. Petrii, and R. V. Marvet, Élektrokhimiya, 3:1311 (1967).
13.   V. S. Éntina and O. A. Petrii, Élektrokhimiya, 4:457 (1968).
14.   A. N. Frumkin, O. A. Petrii, and A. M. Kossaya, Élektrokhimiya, 4:475 (1968).
15.   O. A. Petrii and Yu. G. Kotlov, Élektrokhimiya, 4:774 (1968).
16.   O. A. Petrii, A. N. Frumkin, and Yu. G. Kotlov, J. Res. Hokkaido Univ. (1968).

17. O. A. Petrii and Yu. G. Kotlov, Élektrokhimiya, 4:1256 (1968).
18. A. N. Frumkin and O. A. Petrii (Petry), Extended Abstracts, 19th CITCE Meeting, Detroit, Mich. (1968), p. 33.
19. O. A. Petrii, The Double Layer and Adsorption on Solid Electrodes [in Russian], Izd. Tartusk. Gos. Univ., Tartu (1968), p. 19.
20. A. N. Frumkin and O. A. Petrii (Petry), Electrochim. Acta (in press).
21. O. A. Petrii, A. N. Frumkin, and Yu. G. Kotlov, Élektrokhimiya (in press).
22. O. A. Petrii, A. N. Frumkin, and Yu. G. Kotlov, Élektrokhimiya (in press).
23. O. A. Petrii and Nguyen Van Tue, Élektrokhimiya (in press).
24. A. N. Frumkin, O. A. Petrii, and V. V. Topolev, Élektrokhimiya (in press).
25. O. A. Petrii, Nguyen Van Tue, and Yu. G. Kotlov, Élektrokhimiya (in press).
26. O. A. Petrii and Yu. G. Kotlov, Élektrokhimiya (in press).
27. E. Ponomarenko, A. N. Frumkin, and R. Burshtein, Izv. Akad. Nauk SSSR, Ser. Khim., 1549 (1963).
28. A. N. Frumkin, Zh. Fiz. Khim., 30:2066 (1956).
29. A. I. Shlygin, A. N. Frumkin, and V. Medvedovskii, Acta Physicochim. URSS, 4:911 (1936).
30. N. A. Balashova and V. E. Kazarinov, Usp. Khim., 34:1721 (1965).
31. W. Lorenz, Z. Phys. Chem., 224:145 (1963).
32. R. Parsons, Proceedings of the Second Congress of Surface Activity, Electrical Phenomena, 1957 (J. H. Schulman, editor), Plenum Press, New York (1957), p. 38.
33. R. Parsons, Modern Aspects of Electrochemistry, Vol. 3 (J. O'M. Bockris, editor), Chapt. 3, Plenum Press, New York (1964).
34. N. A. Balashova and V. E. Kazarinov, Electroanalytical Chemistry (A. J. Bard, editor), Vol. 3, M. Dekker, New York (in press).
34a. G. Horanyi, J. Solt, and F. Nagy, Magy. Kem. Folyoirat, 73:414, 561 (1967).
35. R. V. Marvet and O. A. Petrii, Élektrokhimiya, 3:901 (1967).
36. A. N. Frumkin, Usp. Khim., 18:9 (1949).
37. V. I. Nesterova and A. N. Frumkin, Zh. Fiz. Khim., 26:1178 (1952).
38. A. D. Obrucheva, Zh. Fiz. Khim., 26:1448 (1952).
39. S. Gilman, Electroanalytical Chemistry (A. J. Bard, editor), Vol. 2, M. Dekker, New York (1967), p. 111.
40. D. Gilroy and B. E. Conway, Can. J. Chem., 46:875 (1968).
41. B. B. Damaskin and O. A. Petrii, Elektrokhimiya, 4:598 (1968).
41a. A. N. Frumkin, J. Res. Inst. Catalysis Hokkaido Univ., 15:61 (1967).
42. A. D. Obrucheva, Zh. Fiz. Khim., 32:2155 (1958).
43. A. D. Obrucheva, Dokl. Akad. Nauk SSSR, 120:1072 (1958).
44. A. D. Obrucheva, Dokl. Akad. Nauk SSSR, 141:1413 (1961).
45. A. N. Frumkin, Electrochim. Acta, 5:266 (1961).
46. A. N. Frumkin, Acta Univ. Debrecen, Phys. et Chim., 37 (1966).
47. A. D. Obrucheva, Dokl. Akad. Nauk SSSR, 142:859 (1962).
48. E. Dutkiewicz and R. Parsons, J. Electroanal. Chem., 11:100 (1966).
49. A. N. Frumkin, Élektrokhimiya, 1:394 (1965).
50. M. Green and H. Dahms, J. Electrochem. Soc., 110:466 (1963).
51. J. O'M. Bockris and D. A. J. Swinkels, J. Electrochem. Soc., 111:736 (1964).

52.   J. O'M. Bockris, M. Green, and D. A. J. Swinkels, J. Electrochem. Soc.,
       111:742 (1964).

53.   E. Gileadi, B. T. Rubin, and J. O'M. Bockris, J. Phys. Chem., 69:3335 (1965).

54.   W. Heiland, E. Gileadi, and J. O'M. Bockris, J. Phys. Chem., 70:1207 (1966).

55.   E. Gileadi, Comptes Rendus, Deuxième Symposium Européen sur les Inhibiteurs
       de Corrosion, Ferrara, Italia, 1965, Vol. 2 (1966), p. 543.

56.   E. Gileadi, J. Electroanal. Chem., 11:137 (1966).

57.   B. Piersma and E. Gileadi, Modern Aspects of Electrochemistry, Vol. 4
       (J. O'M. Bockris, editor), Plenum Press, New York (1966), p. 47.

57a.   E. Gileadi (editor), Electrosorption, Plenum Press, New York (1967).

58.   H. Dahms and M. Green, J. Electrochem. Soc., 110:1075 (1963).

59.   H. Wroblowa and M. Green, Electrochim. Acta, 8:679 (1963).

60.   J. O'M. Bockris, M. A. V. Devanathan, and K. Müller, Proc. Roy. Soc.
       (London), A274:55 (1963).

61.   N. A. Balashova and A. N. Frumkin, Dokl. Akad. Nauk SSSR, 20:449 (1938).

62.   T. N. Voropaeva, B. V. Deryagin, and B. N. Kabanov, Dokl. Akad. Nauk SSSR,
       128:881 (1959); Kolloidn. Zh., 24:396 (1962).

63.   A. Gorodetskaya and B. Kabanov, Phys. Z. Sovjetunion, 5:418 (1934).

64.   N. A. Balashova, N. T. Gorokhova, and M. I. Kulezneva, The Double Layer
       and Adsorption on Solid Electrodes [in Russian], Izd. Tartusk. Gos. Univ.,
       Tartu (1968), p. 28.

65.   N. A. Balashova, A. M. Kossaya, and N. T. Gorokhova, Élektrokhimiya, 3:656
       (1967).

66.   T. N. Voropaeva, B. V. Deryagin, and B. N. Kabanov, Izv. Akad. Nauk SSSR,
       Otd. Khim. Nauk, No. 2, 257 (1963).

67.   A. Pfützenreuter and G. Mazing, Z. Metallkunde, 42:361 (1951).

68.   E. K. Venstrem, V. I. Likhtman, and P. A. Rebinder, Dokl. Akad. Nauk SSSR,
       107:105 (1956).

69.   B. S. Krasikov, Zh. Prikl. Khim., 37:2420 (1964).

70.   M. Petit and J. Clavilier, Compt. Rend., C265:145 (1967).

71.   J. Clavilier and Van Hoong Nguyen, Compt. Rend., C267:207 (1968).

72.   V. M. Kheifets and B. S. Krasikov, Zh. Fiz. Khim., 31:1992 (1952).

73.   T. N. Birintseva and B. N. Kabanov, Zh. Fiz. Khim., 37:2000 (1963).

74.   E. Gileadi, S. D. Argade, and J. O'M. Bockris, J. Phys. Chem., 70:2044 (1966).

75.   L. A. Shevchenko, A. G. Pshenichnikov, and R. Kh. Burshtein, Élektrokhimiya
       (in press).

76.   B. S. Krasikov and V. S. Krivonos, Zh. Prikl. Khim., 39:1332 (1966).

77.   N. A. Balashova, N. T. Gorokhova, and M. I. Kulezneva, Élektrokhimiya,
       4:871 (1968).

78.   G. Gouy, Ann. Chim. Phys., 29(7):145 (1903); D. C. Grahame, E. U. Coffin,
       J. I. Cummings, and M. A. Poth, J. Am. Chem. Soc., 74:1207 (1952).

79.   S. D. Argade and E. Gileadi, Electrosorption (E. Gileadi, editor), Plenum Press,
       New York (1967), p. 87.

80.   A. Ya. Gokhshtein, Élektrokhimiya, 2:1318 (1966); 4:665 (1968); Electrochim.
       Acta (in press).

81.   Yu. M. Tyurin and A. M. Kossaya, Dokl. Akad. Nauk SSSR, 159:1140 (1964).

82.   J. Mignolet, J. Chim. Phys., 54:19 (1957).

83.   T. N. Andersen, R. S. Perkins, and H. Eyring, J. Am. Chem. Soc., 86:4496 (1964).

84.   R. S. Perkins, R. C. Livingston, T. N. Andersen, and H. Eyring, J. Phys. Chem., 69:3329 (1965).

85.   D. D. Bode, T. N. Andersen, and H. Eyring, J. Electrochem. Soc., 114:72 (1967).

Chapter 10

Relationships in Adsorption of Organic Compounds on Metals of the Platinum Group

1. Dependence of Adsorption of Organic Compounds on the Electrode Potential and Solution pH

The dependence of adsorption of numerous compounds on the potential has been investigated.

The concepts developed by Frumkin [1], which were discussed in the preceding chapter, apparently have a direct bearing on interpretation of some of the results obtained by Bockris et al. [2-5] in studies of the dependence of adsorption of naphthalene and benzene on platinum on the potential. These compounds are relatively stable chemically, and undergo only slight chemical changes during adsorption at normal temperatures.* According to Bockris et al. [2], maximum adsorption of naphthalene on platinum in acid solutions corresponds to 0.1-0.4 V (N.H.E.), and in alkaline solutions to −0.4 V (N.H.E.). The existence of an adsorption maximum in alkaline solutions is a particularly important argument in support of the view that adsorption of hydrogen and oxygen influences

_____

*It has been shown [6-8] that benzene is hydrogenated and oxidized (dehydrogenated [9]) on the surface of a Pt/Pt electrode even at room temperature, although at a low rate. A stably bound product which undergoes cathodic reduction accumulates on the electrode during prolonged contact with benzene solutions [6].

adsorption of organic compounds because, as was shown earlier, in alkaline solutions of weakly adsorbed ions a point of zero charge cannot be observed on platinum.

The dependence of adsorption of benzene on the potential is also represented by a bell-shaped curve, the maximum of which lies near $\varphi_{Q=0}$ in H_2SO_4 + Na_2SO_4 and NaOH + Na_2SO_4 solutions in the pH range from 0.8 to 11 [3-5] (Fig. 133).

Thus, the potentials of maximum adsorption of naphthalene and benzene in both acid and alkaline solutions correspond to the φ_r region where adsorption of hydrogen and oxygen on platinum is minimal, and do not lie close to φ_n of platinum.

Interpretation of the relation between adsorption of other organic compounds and the electrode potential is more complicated.

The dependence of θ and φ in the case of methanol is determined by the experimental conditions.

The results obtained in a study of the dependence of adsorption of methanol on the potential of a smooth platinum electrode [10-12] are shown in Fig. 134. For determination of these curves the electrode potential was lowered sharply to the chosen value from 1.2 V, at which methanol is not adsorbed appreciably. The surface coverage was determined by the cathodic and anodic pulse

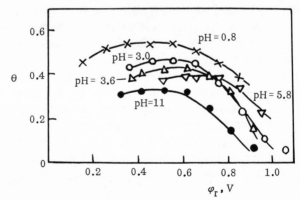

Fig. 133. Dependence of adsorption of benzene on the potential at a platinum electrode in H_2SO_4 + Na_2SO_4 and NaOH + Na_2SO_4 solutions of various pH values (from data in [3]).

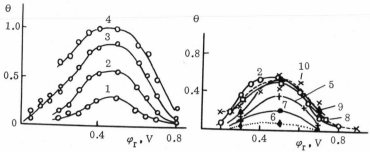

Fig. 134. Dependence of the surface coverage of smooth platinum by organic particles in 1 N H_2SO_4 solutions of various methanol concentrations (M): 1) 10^{-3}; 2) 10^{-2}; 3) 10^{-1}; 4) 1, and at 10^{-2} methanol concentration in buffer solutions of various pH values: 5) 0.32; 6) 1.4; 7) 2.8; 8) 6.2; 9) 10.1; 10) 12.2; 11) 13.6 (from data in [11, 12]).

method after adsorption equilibrium had been attained. According to these data, the maximum of methanol adsorption at various concentrations corresponds to $\varphi_r \sim 0.35$–0.55 V. The position of the maximum is close to $\varphi_{Q=0}$. This conclusion is confirmed by curves determined at various pH. The degrees of adsorption themselves depend on the solution pH; this may be due to changes in the surface state of platinum with variation of pH in the region of maximum adsorption of the alcohol. The rising branch of the θ vs φ_r curve is shifted by ~100 mV in the cathodic direction upon a tenfold increase of methanol concentration. The corresponding shift of the descending branch is ~60 mV.

Determinations by the pulse method with the use of rhodium and iridium electrodes treated under similar conditions showed that maximum adsorption of methanol occurs at $\varphi_r = 0.2$–0.3 V on rhodium and at $\varphi_r = 0.3$–0.4 V on iridium; i.e., as on the platinum electrode, it corresponds in general to the potential of minimum adsorption of hydrogen and oxygen on the metal surfaces [13–16]. A θ vs φ_r curve of similar form was also obtained for glycolaldehyde on platinum [17, 18].

However, a dependence of θ on φ_r of a different kind is obtained under other experimental conditions.* For example, if the

---

*A dependence of θ on φ_r different from that described above was also obtained by the radioactive tracer method by Smith et al. [19], although their aim was to reproduce the conditions used by Bagotskii et al. [10-12]. The reasons for the deviations from the results obtained by pulse measurements are not clear.

organic substance is adsorbed at potentials of maximum adsorption, e.g., at 0.5 V, and the electrode potential is then decreased, no appreciable decrease of surface coverage is observed for a considerable time [20-23]. It follows that the substance previously adsorbed on the electrode is not displaced by hydrogen. The same experimental conditions were also used by Breiter and Gilman [24] and by Hira Lal, Petrii, and Podlovchenko [25]. These data indicate that adsorption of methanol on platinum is irreversible. On the basis of these results the decrease of methanol adsorption with decrease of φ_r in the experiments described above can be attributed to a fall in the rate of methanol adsorption with increasing surface coverage by adsorbed hydrogen. In consequence, steady-state values of surface coverage at low φ_r should be reached only after very long time intervals.

On the other hand, although the decrease of methanol adsorption at anodic potentials coincides with the start of oxygen adsorption [24], it is apparently due to oxidation of the adsorbed particles to CO_2. In fact, as will be shown below, in the range of potentials where θ decreases with φ_r the rate of steady-state oxidation of methanol becomes comparable with the rate of methanol adsorption. This hypothesis is also supported by experiments on adsorption of methanol on a platinum—ruthenium electrode [26]. In this case the descending branch of the θ vs φ_r curve is shifted by ~150 mV in the cathodic direction by comparison with the descending branch of the θ vs φ_r curve for the platinum electrode. The overpotential of steady-state electrochemical oxidation of methanol at a platinum—ruthenium electrode is also lower by ~150 mV than the overpotential at a platinum electrode.

For interpretation of the dependence of adsorption of higher (e.g., amyl [27]) alcohols on the potential it is necessary to take into account both oxidation processes and hydrogenation processes, which were discussed in Chapter VIII.

Irreversible adsorption of organic compounds due to the irreversibility of processes of oxygen adsorption and ionization on platinum during polarization of the electrode to $\varphi_r \gtrsim 1.5$ V is noted in a number of publications [24, 28-30]. Kazarinov [30] demonstrated the influence of the surface state of platinum on adsorption of n-hexyl alcohol. For example, previous oxidation of the electrode lowers its ability to adsorb hexyl alcohol substantially over the entire range of potentials. The results also depend on the direction in which the potential is varied during the adsorption determinations. All this indicates difficulties in establishment of adsorption equilibrium at a platinum electrode.

The dependence of adsorption of HCOOH on platinum on the potential has been determined in several investigations [29, 31–33], the results of which differ greatly. This problem was discussed by Bagotskii et al. [34], who found that adsorption of formic acid occurs at φ_r of the hydrogen region (0.1–0.35 V), and considerable adsorption of HCOOH persists even at $\varphi_r < 0$ in concentrated solutions. This result is apparently due to possible interaction of HCOOH with adsorbed hydrogen, with simultaneous formation of a chemisorption product, e.g., by the reaction

$$H_{ads} + HCOOH \rightarrow H_2 + COOH_{ads}. \qquad (X.1)$$

The decrease of formic acid adsorption at $\varphi_r > 0.35$ V can only be attributed to desorption as the result of oxidation of the adsorbed particles, as the coverage of the platinum surface with oxygen at these potentials is still too low.

Adsorption of CO on platinum is maximal at $\varphi_r = 0.4$–0.9 V and drops sharply to zero at $\varphi_r \simeq 0.91$ V as the result of oxidation of the adsorbed layer [28]. The course of the θ vs φ_r curve at potentials where adsorption diminishes depends on the experimental conditions [35]. In the φ_r range from -0.2 to $+0.5$ V a monolayer of adsorbed carbon monoxide is formed on rhodium at CO partial pressures from 0.01 to 1 atm [36]. At $\varphi_r > 0.5$ V the coverage θ decreases owing to oxidation of the adsorbate.

Adsorption of CO_2 on platinum from $1 N$ H_2SO_4 at 20°C, which is the result of interaction of CO_2 with H_{ads}, is highest at $\varphi_r \sim 0.06$ V and falls to zero at $\varphi_r \geqslant 0.35$ V [37, 38]. According to Brummer and Turner [39], adsorption of CO_2 at high temperatures is also possible at more anodic φ_r. However, it is not certain that this result is not the consequence of adsorption of impurities from the solution.

According to Niedrach [40–42], maximum adsorption of methane and ethane on platinum occurs in the region $\varphi_r \sim 0.3$ V. Below 0.1 V both substances are adsorbed very slightly. Therefore adsorption of methane and ethane is suppressed by adsorption of hydrogen. At $\varphi_r > 0.5$ V adsorption of methane falls to zero owing to rapid oxidation of the adsorption products. Marvet and Petrii

[43] and Taylor and Brummer [44] obtained similar results. Coverage of the platinum surface by adsorbed ethane falls to zero at $\varphi_r > 0.9$ V, and even after adsorption of ethane at 0.6 V the first oxidation wave of the adsorbed ethane, corresponding to the more easily oxidized species, disappears.

Similarly to these results, Flannery and Walker [45] showed by the radioactive tracer method that adsorption of butane on a platinum electrode reaches the maximum value at 0.3 V, regardless of the butane concentration, and diminishes at higher potentials as the result of oxidation of the adsorbate. They consider that butane is partially displaced by adsorbed hydrogen in the range from 0.3 to 0.1 V. A similar result was obtained for ethane.

Adsorption of propane reaches a maximum at φ_r between 0.1 and 0.7 V [46-50] (Figs. 104 and 105), and of n-hexane [51] and n-octane [52] in the region of 0.2-0.3 V.

The dependence of adsorption of ethylene on platinum on the potential has been investigated by the radioactive tracer method [53, 54]. Maximum adsorption was observed at $\varphi_r = 0.4-0.45$ V in $1 N$ H_2SO_4; the maximum was shifted slightly in the negative direction with increase of the ethylene partial pressure.

Electrochemical methods have been used for studying the dependence of adsorption of ethylene and acetylene on platinum [55-58] and of ethylene on rhodium [59] on the potential. According to Gilman [55-57], the quantity of electricity Q_{org} required for oxidation of adsorbed ethylene and acetylene is highest if the adsorption is performed at $\varphi_r = 0.2-0.4$ V. The decrease of Q_{org} at low potentials was attributed to electrochemical hydrogenation of ethylene and acetylene [60]. The decrease of Q_{org} at high φ_r may be due to a number of causes: changes in the structure of the adsorbed layer resulting from further dehydrogenation of the hydrocarbons, appearance of adsorbed oxygen which competes with the organic substance for surface sites, decrease of the ethylene and acetylene adsorption rates, and oxidation of the adsorbed particles. Thus, a large number of different factors must be taken into consideration in interpretation of the dependence of θ on φ_r on metals of the platinum group.

Data reported by Matsuda and Tamura [61] indicate that methylamines and pyridine are adsorbed on platinum at the potentials of the hydrogen and double-layer regions.

The adsorption maximum of mono- and dimethylamine in 0.5 M Na_2CO_3 corresponds to $\varphi_r \sim 0.8$ V; this was attributed by the authors to competitive adsorption of the amines, water, ions of the double layer, and hydrogen, and to oxidation of the amines at anodic potentials [61a].

The influence of aromatic amines and phenols on evolution of hydrogen on platinum has been noted [62]; this is probably due to adsorption of these substances at fairly high cathodic potentials.

Gold is of considerable interest as a material for adsorption studies, because it appears that on gold catalytic conversions of certain organic compounds either do not occur or proceed at low rates, but oxygen and (to a small extent) hydrogen are adsorbed, so that the influence of gas adsorption of organic compounds can be investigated. However, there have been few quantitative studies of adsorption on gold. Green and Dahms [63, 64] used the radioactive tracer method for investigating adsorption of certain organic compounds on gold. It was shown that adsorption of aromatic compounds (benzene, naphthalene, and phenanthrene) is considerably greater than that of aliphatic compounds (cyclohexane, fatty acids); adsorption of the latter could not be detected by the method used. The higher adsorption of organic compounds was attributed to π-electron interaction.

In a study of the adsorption of benzene, naphthalene, and phenanthrene on gold Dahms and Green found that the adsorption maximum in 0.5 M H_2SO_4 corresponds to $\varphi_r = 0.5$ V (Fig. 135). Assuming that φ_n of gold is about 0.3 V (however, see Chapter IX,

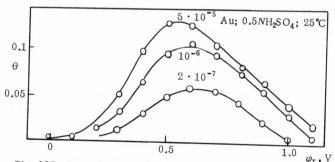

Fig. 135. Dependence of the adsorption of naphthalene on gold on the potential in 0.5 N H_2SO_4 with additions of naphthalene: 1) $2 \cdot 10^{-7}$; 2) 10^{-6}; 3) $5 \cdot 10^{-5}$ M (from data in [64]).

Section 4), they attributed the decrease of adsorption on either side of 0.5 V to the desorbing influence of the electric double layer. However, it follows from Frumkin's theory [1] that the influence of adsorbed hydrogen and oxygen on adsorption cannot be ignored even when the amounts of these adsorbed gases are small. Therefore the conclusion that the electric field of the double layer has a decisive influence on adsorption of organic compounds on gold requires further investigation.

The adsorption of pyridine on gold was investigated in a series of studies [65-67]. The most interesting conclusion drawn from these studies is that pyridine desorption peaks are present on the capacity curves at cathodic potentials. However, desorption of pyridine does not occur at anodic potentials, possibly as the result of π-electron interaction.

2. Dependence of Adsorption of Organic Compounds on Concentration.

Adsorption Isotherms

Breiter and Gilman [24] showed that the surface coverage of platinum increases linearly with the logarithm of the methanol concentration by volume. However, the surface coverages were determined on open circuit at the "rest potential," which is strongly dependent on the methanol concentration. A logarithmic adsorption isotherm was also obtained by Breiter for formic acid at $\varphi_r = 0.2$ V [68].

The dependence of adsorption of various organic compounds on concentration was investigated in detail by Bagotskii, Vasil'ev, et al. [10-16, 34, 69, 70]. Figure 136 shows isotherms of methanol adsorption in $1 N$ H_2SO_4 at various potentials. At medium values of θ (from 0.1 to 0.85), a linear relationship is observed between θ and log c over a wide range (about four orders of magnitude) of methanol concentrations. Isotherms corresponding to different potentials are parallel. Therefore adsorption of methanol is represented by the equation for the Temkin logarithmic isotherm:

$$\theta = a + \frac{1}{f}\ln c, \qquad\qquad (X.2)$$

where f is the heterogeneity factor of the platinum surface. The slope of the linear part of the isotherms gives $f \simeq 10$, which is

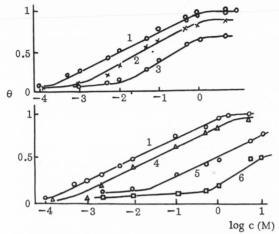

Fig. 136. Isotherms of methanol adsorption on smooth platinum in 1 N H_2SO_4 at various potentials: 1) 0.4; 2) 0.3; 3) 0.2; 4) 0.6; 5) 0.7; 6) 0.8 V (from data in [11, 12]).

of the same order of magnitude as f calculated from isotherms of hydrogen adsorption on platinum, which may be regarded in the first approximation as logarithmic ($f \sim 14$). The heterogeneity factor f represents the difference between the maximum and minimum values of the adsorption energy, G_{max} and G_{min}:

$$f = \frac{G_{max} - G_{min}}{RT}. \qquad (X.3)$$

Similar relationships between θ and log c were found at various temperatures and pH values for methanol, ethylene glycol, and formic acid. The slopes of the logarithmic isotherms for different organic compounds are similar to the slope of the methanol adsorption isotherm.

Logarithmic adsorption isotherms were also obtained for glyoxal and glycolaldehyde [17, 18] and f was close to the value obtained from the hydrogen adsorption isotherms.

Bagotskii, Vasil'ev, et al. [13-16] studied the dependence of adsorption of methanol on rhodium and iridium on the concentration. Here again the adsorption isotherms were of the same form as the isotherms of hydrogen adsorption on the same electrodes. Thus, in the case of rhodium the dependence of adsorption of meth-

anol and of hydrogen on the concentration and pressure may be
represented by two Temkin isotherms with different heterogeneity
factors ($f \simeq 25$ at $\theta < 0.25$ and $f = 6$ at $\theta > 0.25$). In the case
of iridium the isotherms of hydrogen and methanol adsorption are
linear in log θ vs log c coordinates; i.e., in this case we have a
Freundlich power isotherm with an index of 5.5-6.5.

Comparison of the adsorption relationships of organic compounds and hydrogen on platinum, rhodium, and iridium led to the
conclusion that the determining factor in adsorption is the nature
of the heterogeneity of the electrode surface rather than the nature
of the organic substance.

Isotherms of the logarithmic type were obtained for adsorption of n-hexane on platinum [51].

A logarithmic dependence of the adsorption on concentration
also persists in the range of potentials where oxidation of adsorbed methanol occurs. Coverage of the electrode surface with
methanol during passage of an electric current was examined by
Bagotskii and Vasil'ev [70] for the steady-state conditions:

$$CH_3OH_{soln} \rightleftarrows CH_3OH_{ads} \rightarrow \text{reaction products} . \qquad (X.4)$$

Exponential dependence of the rates of adsorption, desorption, and
oxidation of methanol on the surface coverage was assumed. The
relation between the steady-state coverage and concentration under
these conditions is again represented by an equation of the type
(X.2), but the value of the constant a depends on the electrode
potential.

Heiland, Gileadi, and Bockris [3] used the radioactive tracer
method for determining isotherms of benzene adsorption on platinum in $1 N$ H_2SO_4 and H_3PO_4. Partial surface coverages were calculated on the assumption that $\Gamma_\infty = 2.5 \cdot 10^{-10}$ moles /cm^2. This
value of Γ_∞ means that one benzene molecule occupies 10 surface
sites. The electrode area was found from the capacity of the double layer, on the assumption that the capacity per cm^2 of true surface area is 18 μF. The form of the isotherms indicates that in
the concentration range studied surface saturation is attained only
at 70°, and corresponds to ~0.5 in this case. The isotherms were
normalized for quantitative analysis, with maximum coverage
taken as unity. The fact that surface coverage in adsorption of

benzene was not complete was attributed to existence of surface defects inaccessible to large molecules, and of sites having high activation energy or low heats of adsorption. It is possible, however, that the conclusion of incomplete coverage may be the consequence of inaccuracies in calculations of the true surface area. Indeed, it follows from literature data [71, 72] that the double-layer capacity of a Pt electrode in sulfuric acid is ~40–50 $\mu F/cm^2$. The normalized adsorption isotherms are linear in θ vs log c coordinates and therefore conform to the Temkin isotherm. However, application of the Temkin isotherm leads to an anomalously large difference f RT between the maximum and minimum heats of adsorption. Therefore the authors prefer to use the more general Frumkin isotherm, written in the form

$$\log \; \frac{\theta}{1-\theta} + \frac{f\theta}{2.3} = \log Bc. \tag{X.5}$$

for representing adsorption of benzene.

The dependence of adsorption on concentration has also been studied for ethylene [54, 55–57] and acetylene [55–57]. In the case of ethylene there is a considerable divergence between results obtained by the tracer atom method [54] and by the potentiodynamic pulse method [55–57]. By the former method it was found that saturation of the surface with ethylene is reached at $2 \cdot 10^{-5} M$ ethylene concentration in the solution, giving a surface coverage of ~0.4 at $\varphi_r = 0.4$ V. According to Gilman's data [55–57], maximum coverage of the platinum surface with ethylene is reached at considerably lower ethylene concentrations and has a value of ~0.78. According to Gileadi, Rubin, and Bockris [54], adsorption of ethylene on platinum conforms to the Langmuir isotherm. Gilman [55–57] found that in the region of maximum adsorption (0.2–0.4 V) adsorption of ethylene is independent of its concentration. Weak dependence of adsorption on the concentration is observed outside the region of maximum adsorption; this was attributed to hydrogenation of ethylene and oxidation of the adsorbed particles. Similar relationships between adsorption and concentration were found for acetylene [55–57]. The differences between the experimental data reported in [54, 55–57] are apparently due to the different conditions of electrode preparation and to different estimates of the true electrode areas (see [73]).

Dahms and Green [64] analyzed the isotherm of naphthalene adsorption on a gold electrode. Their experimental results conform to the relation

$$Bc = \frac{\theta}{(1 - \theta)^5} \qquad (X.6)$$

derived for adsorption on a uniform surface with the assumption
that a naphthalene molecule in flat orientation displaces five water
molecules.

It follows from the above data that adsorbed organic molecules occupy more than one site on the electrode surface. Chizmadzhev and Markin [74-76] examined theoretically the adsorption isotherm of molecules occupying two area elements on a
heterogeneous surface. Calculation of the adsorption isotherm
for multisite adsorption requires knowledge not only of the distribution of the adsorption sites by heats of adsorption but also of
the relation between the adsorption energies of neighboring adsorption sites. Adsorption isotherm equations were derived for
two limiting cases: "domain" heterogeneity, when sites of the same
adsorption energy form fairly large continuous regions or domains,
and "microscopic" heterogeneity, when the energies of neighboring
regions are unrelated. In both cases the dependence of surface
coverage on the pressure or concentration of the adsorbate in the
region of medium coverages is represented by a logarithmic isotherm of type (X.2). In the case of domain heterogeneity, however,
the coefficient f is determined by the distribution of energies
referred to the whole adsorbed molecule. The slope of the isotherm in the region of medium surface coverages in the case of
microscopic heterogeneity is determined by the distribution of
adsorption energies referred to one site (or bond). Comparison
of this theory with experimental data [11, 12] shows that the microscopic heterogeneity concept is in better agreement with the adsorption properties of platinum electrode surfaces. It was also
shown that the form of the isotherm is insensitive to the distribution
of adsorption sites by adsorption energies. An isotherm close to
logarithmic is obtained with any sufficiently smooth distribution
function. A different approach [77] also led to the conclusion that
the surface heterogeneity of platinum is of the microscopic type.
The possible occurrence of "induced heterogeneity" has been mentioned by Brummer and Cahill [78].

The question whether the usual thermodynamic meaning can be
attached to the observed dependence of θ on c, i.e., whether it can

be regarded as the result of establishment of equilibrium between adsorption and desorption processes (e.g., see [26, 49, 57]) is significant in the case of substances which undergo deep destructive conversion during adsorption. The opposite viewpoint is that the dependence of adsorption on concentration may be the consequence of a balance between the adsorption and hydrogenation rates (at low potentials) or between the rates of adsorption and of oxidation to carbon dioxide and water at high potentials (e.g., see the paper by Biegler and Koch [79]). In this case the adsorption isotherms should be regarded as stationary, reflecting the establishment of a steady state between these processes. Finally, it is possible that because of the slowness of adsorption on a platinum electrode (see below) and the dependence of the rate of this process on the concentration and potential, maximum coverages are not attained at low concentrations in the time intervals used in the determinations. In our opinion, the available experimental material does not yet provide a final answer to this question. For example, in the case of methanol, in addition to experimental data indicating that the adsorption process is irreversible,* the literature also contains references to the possibility of desorption of chemisorbed methanol [11, 12]† and exchange with particles in solution. For example, in experiments with methanol [83] containing H^3 and C^{14} it was found that the $H^3:C^{14}$ ratio of the dissolved methanol in $1 N$ H_2SO_4 (at 30 and 63°C) decreased during contact of the solution with platinum on open circuit. This result can be explained on the assumption that at least a part of the methanol is reversibly chemisorbed. However, additional investigations are needed for final elucidation of this problem.

---

*Biegler's investigation [80], in which it was found that the area and structure of the platinum surface changed during adsorption measurements in methanol solutions, is interesting in this connection. These effects may be the consequence of strong interaction between the adsorbed particles and the surface atoms, weakening the bonds between the latter and the metal.

†Beskorovainaya, Vasil'ev, and Bagotskii [81] carried out direct measurements of the rate of desorption of chemisorbed methanol from a smooth platinum electrode. However, as the authors point out, these experiments are not sufficient to justify the view that adsorption of methanol is a reversible process. Exchange between firmly adsorbed methanol labeled with C^{14} and methanol in solution was not observed in experiments with a Pt/Pt electrode [82].

3 .   A d s o r p t i o n   K i n e t i c s

o f   O r g a n i c   C o m p o u n d s

Adsorption of organic compounds on a platinum electrode, in distinction from a mercury electrode, proceeds at a relatively low rate which can be found by direct experiment.

The first attempt to investigate the kinetics of methanol adsorption on smooth platinum was made by Breiter [84]. However, his conclusions on the adsorption rate were drawn from indirect data obtained under nonstationary conditions: from the results of polarization and adsorption measurements with application of triangular pulses at the rate of 30 mV/sec to the electrode. Brummer and Makrides [32] used the cathodic galvanostatic pulse method for direct measurements of the degree of surface coverage as a function of time in formic acid solutions. In the φ_r range of 0.35-0.7 V the adsorption of HCOOH was slow and stationary coverages were attained about 2 min after the start of adsorption. The data were interpreted with the assumption of Langmuir kinetics of adsorption, although this resulted in appreciable deviations between the experimental results and the simple adsorption model used.

Some data on the adsorption kinetics of formic acid have also been reported by Breiter [68].

A systematic investigation of the kinetics of adsorption of various organic compounds on smooth platinum, rhodium, and iridium was carried out by Bagotskii, Vasil'ev, and co-workers [11-15, 34, 69, 85-87]. The measurements were made in $1 N$ H_2SO_4 at various potentials, temperatures, and volume concentrations of the organic compounds, by a combination of cathodic and anodic pulse methods. It was shown that in solutions with adsorbate concentrations higher than $10^{-3} M$ the rotation speed of the electrode does not influence the rate of adsorption of organic compounds, which is therefore considerably lower than the rate of diffusion to the electrode surface. Curves representing the dependence of coverage of the electrode surface with organic particles on the adsorption time in solutions with a constant concentration of the organic compound were obtained. Such relationships are known as kinetic adsorption isotherms. Typical kinetic isotherms of methanol adsorption are shown in Fig. 137. The coverage on platinum at

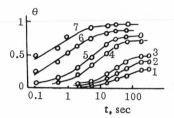

Fig. 137. Kinetic isotherms of methanol adsorption on smooth platinum at φ_r = 0.4 V in 1 N H_2SO_4 with additions of CH_3OH: 1) 10^{-3}; 2) $5 \cdot 10^{-3}$; 3) 10^{-2}; 4) $5 \cdot 10^{-2}$; 5) 10^{-1}; 6) $5 \cdot 10^{-1}$; 7) 1 M (from data in [11, 12]).

various methanol concentrations in the range of medium coverages between 0.1 and the stationary value of θ increases linearly with the logarithm of the adsorption time:

$$\theta = \text{const} + \frac{1}{\alpha f} \ln t, \qquad (X.7)$$

where $0 < \alpha < 1$.

The kinetic isotherms for different volume concentrations of methanol are parallel. The dependence of the logarithm of the adsorption rate at φ_r = 0.4 V for various constant surface coverages on the logarithm of the volume concentration, calculated from kinetic isotherms, is linear with a slope of 1. Therefore the adsorption rate at θ = const is directly proportional to the volume concentration of methanol. The logarithm of the adsorption rate decreases linearly with increase of surface coverage. Thus the rate V_{ads} of methanol adsorption on a smooth platinum electrode at constant potential is satisfactorily represented by the Roginskii – Zel'dovich equation [88-90]:

$$v_{ads} \equiv \frac{d\theta}{dt} = k_{ads} c \exp(-\alpha f \theta), \qquad (X.7a)$$

where k_{ads} is a constant.

The kinetic adsorption isotherms retain their logarithmic character and equal slope at different potentials (from 0.2 to 0.7 V). Isotherms determined at 0.4, 0.5, and 0.6 V virtually coincide. This means that the rate of methanol adsorption depends little on the potential in this range. The adsorption rate falls when the electrode potential is shifted into the hydrogen region. At constant coverage of the surface with the organic substance, this fall corresponds to the equation

$$v_{ads} = k \exp\left(\frac{\alpha F \varphi_r}{RT}\right). \tag{X.8}$$

Since, in the first approximation, in the hydrogen adsorption region

$$\varphi_r = \text{const} - \frac{RT}{F} f_H \theta_H, \tag{X.9}$$

therefore

$$v_{ads} = k' \exp\left(- \alpha_H f_H \theta_H\right), \tag{X.10}$$

where α_H, f_H, and θ_H refer to adsorption of hydrogen. Analogously, in the region $\varphi_r > 0.7$ V the rate of methanol adsorption also falls with increasing surface coverage with adsorbed oxygen:

$$v_{ads} = k'' \exp\left(- \alpha_0 f_0 \theta_0\right), \tag{X.11}$$

where α_0, f_0, and θ_0 refer to adsorption of oxygen.

Therefore the aggregate influence of various factors on the kinetics of methanol adsorption on platinum may be represented by the following equation:

$$v_{ads} = k_{ads} c \exp\left[- \alpha f \theta - \alpha_H f_H \theta_H - \alpha_0 f_0 \theta_0\right]. \tag{X.12}$$

Isosteric (corresponding to $\theta = \text{const}$) differential activation energies of adsorption were determined from kinetic isotherms of methanol adsorption for various temperatures (from 11 to 72 °C) by plotting the logarithm of the time required to attain a given surface coverage against $1/T$. The calculated activation energy E_a increases linearly with the surface coverage from 9.9 kcal/mole at $\theta = 0.1$ to 13 kcal/mole at $\theta = 0.7$ (Fig. 138). This dependence also indicates activation adsorption kinetics on a uniformly heterogeneous platinum surface:

$$E_a = E_a^0 + \alpha f R T \theta. \tag{X.13}$$

The value E_a^0, corresponding to $\theta = 0$, for adsorption of methanol is 9.5 kcal/mole.

Similar relationships were obtained in investigations of the adsorption kinetics of ethylene glycol ($E_a^0 = 16.4$ kcal/mole) and formic acid ($E_a^0 = 5.2$ kcal/mole) on a smooth platinum electrode, and for adsorption of glycolaldehyde ($E_a^0 = 12.8$ kcal/mole) [17, 18].

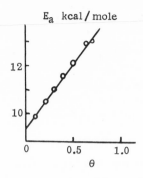

Fig. 138. Dependence of the activation energy of methanol adsorption on smooth platinum on the surface coverage (from data in [11, 12]).

Bagotskii et al. [91] determined isotherms of methanol adsorption at $\varphi_r = 0.4$ V at various temperatures and hence found the dependence of the heat of adsorption of methanol, q_a, on the surface coverage,

$$q_a = q_a^0 - fRT\theta , \qquad (X.14)$$

which was to be expected for a uniformly heterogeneous surface. Direct comparison of the activation energy and heat of adsorption of methanol on platinum showed that

$$\Delta E_a = -\alpha \Delta q_a , \qquad (X.15)$$

where the experimental value of $\alpha \sim 0.5$. Thus, the Brønsted— Polanyi—Temkin linearity relation [90] is fulfilled; according to this condition, the change of activation energy of adsorption in passing from one site on a heterogeneous surface to another is a certain constant fraction of the change of the heat of adsorption. The heat of adsorption determined by Bagotskii et al. [91] refers to the overall process

$$CH_3OH \rightarrow COH + 3H^+ + 3e . \qquad (X.16)$$

The relationships found for the adsorption kinetics of organic substances were used for quantitative interpretation of curves representing the shift of potential of a platinum electrode on open circuit after addition of methanol to the solution [92]. According to the concepts examined in Chapter VIII, the shift of potential is caused by hydrogenation of methanol, equilibrium being established between the hydrogen formed and hydrogen ions in solution. Thus, the shift of electrode potential is determined by the kinetics of methanol chemisorption. The theory put forward in

[92] presumes that adsorption equilibrium is established between the particles adsorbed on the surface and the substance in the bulk solution (however, see Section 2), and is based on certain simplifying assumptions regarding the nature of the interaction between the adsorbate and atomic hydrogen. In the range of potentials between 0.35 V and the φ_r' values established in presence of methanol, a linear relation exists, according to [92], between φ_r and log t:

$$\varphi_r = \text{const} - k \log t . \tag{X.17}$$

This is the consequence of the relationships between θ and t and between v_{ads} and θ, examined above. The linear dependence of the stationary potential on the logarithm of the volume concentration of the organic compound,

$$\varphi_r' = \text{const} - \beta \log c , \tag{X.18}$$

may be explained in the light of the logarithmic relationship between the amount of methanol adsorbed and its volume concentration. The theory explains such experimental facts as the independence of the stationary potential of the potential at which methanol is introduced, as long as this is in the double-layer region, and the linear dependence of the rate of the potential shift on the methanol concentration for addition at a constant potential.*

The kinetics of methanol adsorption on rhodium [13, 15] conforms to Roginskii − Zel'dovich equations with two different values of αf. This indicates that adsorption proceeds on a uniformly heterogeneous surface having two types of adsorption sites with different heterogeneity factors.

The dependence of the coverage θ of iridium by methanol on time [14, 15, 87] can be represented by the equation

$$\log \theta = \text{const} + (1/\alpha n)\log t \tag{X.19}$$

at moderate surface coverages ($1/\alpha n \simeq 0.36$; n = 6). This equation is analogous to the equation of Bangham and Burt [93] for adsorption on a surface having exponential heterogeneity. The kinetic adsorption isotherms for different volume concentrations of meth-

---

*Assumption of uniform heterogeneity of the platinum surface is not necessary for explanation of these results, because the relationships in question stem only from the fact that $\theta_H = 0$ at potentials of the double-layer region and $\theta = 0$ at t = 0.

anol are parallel, the shift of the isotherm with concentration being
represented by the equation

$$\log \theta = \text{const} + (1/\alpha n)\log c \,.\tag{X.20}$$

However, at $\theta = \text{const}$

$$v_{ads} = k_{ads}c \,.\tag{X.21}$$

Therefore

$$v_{ads} = k_{ads}c\theta(1 - \alpha n) \,,\tag{X.22}$$

which is analogous to Kwan's equation [94] for a surface with exponential distribution. The maximum rate of methanol adsorption
on iridium is observed at $\varphi_r = 0.3-0.4$ V. The rate of methanol
adsorption is lower by an order of magnitude on iridium than on
platinum. The activation energy of adsorption increases linearly
with the logarithm of the surface coverage:

$$E_a = E_a^0 + \alpha n \ln \theta \,.\tag{X.23}$$

All these results provide direct proof that the nature of the surface
heterogeneity has a determining influence on the adsorption process.

In an investigation of the adsorption of ethane, Gilman [95]
found that the adsorption rate is lower than the diffusion rate by
about two orders of magnitude. In his opinion, kinetic data for
low and moderate coverages are represented more satisfactorily
by the equations

$$\frac{d\theta}{dt} = kc\,(1 - \theta)^2\tag{X.24}$$

or

$$\frac{1}{1-\theta} = kct + 1,\tag{X.25}$$

where θ is the coverage of the surface by ethane. These equations mean that adsorption of ethane is a reaction of the second
order with respect to surface sites not occupied by ethane, and
conforms to Langmuir kinetics. On the basis of his results Gilman
suggests the following mechanism for the slow adsorption stage:

$$\begin{array}{c} C_2H_6 + 2\,Pt \rightarrow C_2H_5 + H \\ | | \\ Pt Pt \end{array}\tag{X.26}$$

Experiments on desorption of ethane at low potentials showed that this process depends on the electrode potential and the surface coverage, and conforms to the following kinetic law:

$$-\frac{d\theta}{dt} \equiv v_{ads} = H' \exp\left[\frac{\alpha n' F \varphi}{RT} + m\theta\right], \qquad (X.27)$$

where $\alpha \simeq 0.5$, n' $\simeq 2$, and H' and m are constants. Only one third of the adsorbate is removed from the surface during desorption, and the coverage decreases from 0.45 to 0.30.

The composition of the adsorbed ethane layer is the same in presence of Cl^- ions in the solution as in their absence. However, the specifically adsorbed anions lower the rate of ethane adsorption at the initial instant. The influence of chloride anions is represented by the equation

$$v_{ads} = kc \left(1 - \theta - \theta_{Cl^-}\right)^2, \qquad (X.28)$$

where θ_{Cl^-} is the coverage of the surface by Cl^- anions; i.e., this equation is consistent with the concept of simple physical blocking of adsorption sites [96]. The coverage θ_{Cl^-} decreases with increasing adsorption time and the adsorption rate tends to the value observed in absence of Cl^- anions.

Determinations of the kinetics of ethane adsorption on a porous platinum-black electrode [41] gave results in agreement with data for a smooth platinum electrode. At short adsorption times the amount of ethane adsorbed increases roughly linearly with time. For long adsorption times, in the opinion of Niedrach et al. [41], the adsorption rate corresponds better to the Roginskii−Zel'dovich equation. A similar conclusion was reached by Niedrach [42] in the case of methane.

According to Cairns et al. [97, 98], the rate of propane adsorption from hydrofluoric acid solutions may be represented by a third-order reaction equation, conforming to Langmuir kinetics:

$$v_{ads} = kc \left(1 - \theta\right)^3. \qquad (X.29)$$

The activation energy of adsorption at $\varphi_r \sim 0.3$ V is about 20 kcal/mole.

The rate of propane adsorption on a Pt/Pt electrode in $3 N$ H_2SO_4 at 90°C can be represented by the Roginskii−Zel'dovich equation [50].

Attempts have also been made to study the adsorption kinetics of other organic compounds: ethylene [54, 55], acetylene [55], propane [47], and methane [44] in phosphoric acid solutions, and of n-butane [45], n-octane [52], benzene [3, 9], carbon monoxide [28, 99, 36], hexene-1 [100], and n-hexane [51]. The adsorption rate of these compounds under the experimental conditions used (stationary electrodes) is controlled by diffusion during the first instant.

4. Relation between Adsorption and Kinetics of Anodic Oxidation of Organic Compounds on Platinum Electrodes

The participation of adsorbed organic substances in electrochemical oxidation processes had been presumed a long time ago by Müller and co-workers [101] and by Gerasimenko [102]. However, establishment of a relation between adsorption and anodic oxidation of organic compounds became possible only after direct experimental determinations of adsorption of organic compounds had been carried out and the results compared with kinetic data obtained under the same conditions. Such comparisons were performed for the first time by Gilman and Breiter [103] for methanol and by Gilman [28] for carbon monoxide; however, the surface coverages and polarization curves were determined under nonstationary conditions.

Direct proof of the part played by adsorption of organic compounds in steady-state anodic oxidation at a platinum electrode was obtained by Bockris et al. [104, 105] in the case of ethylene. A negative reaction order was found in an investigation of the dependence of the reaction rate on the ethylene partial pressure; this could be explained only if adsorption of ethylene on the electrode was taken into account.

Investigations of the mechanism of anodic oxidation processes have been reviewed in the literature [106-109]. In this section we confine ourselves to a brief examination of the main views on the role of adsorption of organic compounds in the overall process of anodic oxidation.

The most simple relation between adsorption and oxidation process is probably observed in the case of electrochemical oxidation of formic acid on palladium. No chemisorbed substance can be detected on the electrode during the stationary process; the reaction is probably controlled by the rate of adsorption of the acid, with subsequent rapid oxidation of the hydrogen formed during its catalytic decomposition [110].

The role of chemisorption of methanol on the electrode surface in anodic oxidation has been investigated in detail [11, 12, 23, 70, 79, 85, 111-125]. In some of the publications cited [11, 12, 70, 85, 117-120] concepts of adsorption and oxidation of organic compounds on a uniformly heterogeneous electrode surface [88-90] were used for interpretation of the anodic oxidation mechanism.

Detailed study of the electrochemical oxidation of methanol showed that a distinction must be made between the mechanism of the process on a free surface or on a surface covered only slightly with the chemisorbed substance, and the mechanism of electrochemical oxidation under stationary conditions.

It was shown in Chapter VIII that when methanol is brought into contact with a platinum electrode maintained at a constant potential a nonstationary current due to ionization of the atomic hydrogen formed during dehydrogenation of methanol is produced. Since the appearance of hydrogen occurs in the same elementary step as the appearance of chemisorbed organic species, the kinetics of which was examined in the preceding section, it can be concluded that the nonstationary current i_N is proportional to the rate of dissociative chemisorption of methanol:

$$i_N = k' \frac{d\theta}{dt} = kc \exp(-\alpha f\theta). \qquad (X.30)$$

Putting the dependence of θ on t represented by Eq. (X.7) into Eq. (X.30), we obtain

$$i_N = \frac{const}{t}. \qquad (X.31)$$

In fact it is found experimentally [85] that the dependence of the nonstationary current on time corresponds to Eq. (X.31) in the time interval in which the coverage increases from 0.1 to 0.7.

The dependence of the maximum nonstationary current observed immediately after contact between methanol and the electrode on the potential is shown in Fig. 139. At potentials of the double-layer region the electrode potential has little influence on the rate of methanol oxidation at a clean platinum surface; the slope of curve 1 in Fig. 139 is ~350-400 mV. At φ_r in the hydrogen region the oxidation rate depends strongly on the potential. In acid solution, in the φ_r range from 100 to 300 mV, the slope of the curve is ~130 mV, and the slope diminishes appreciably at lower φ_r. In alkaline solution the slope of the curve at the potentials of hydrogen adsorption is 40-50 mV. The change of slope in the transition from the double-layer to the hydrogen region may be qualitatively explained as follows. In the former case the reaction proceeds at a surface of low surface coverage with adsorbed hydrogen and oxygen, whereas in the latter the coverage by hydrogen varies, in the first approximation, linearly with the potential [see Eq. (X.9)]. The rate of methanol oxidation during the first instant varies linearly with the methanol concentration, and at constant φ_r it depends little on the solution pH. All these results

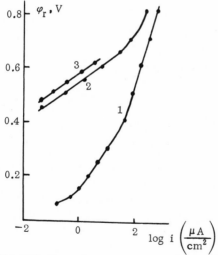

Fig. 139. Dependence of the rate of methanol oxidation on the potential of a Pt/Pt electrode free from organic adsorbate (1) and under stationary conditions (2) in 1 N H_2SO_4 + 0.1 M CH_3OH. 3) Polarization curve for oxidation of products of methanol chemisorption in 1 N H_2SO_4 (from data in [111, 113]).

support the view that the nonstationary current of methanol oxidation corresponds to the kinetics of methanol adsorption on the platinum surface. The high rates of methanol oxidation on the surface of platinum free from chemisorbed substance, resulting from dehydrogenation, make it possible to explain the peculiarities of the nonstationary curves obtained during linear variation of the electrode potential [121].*

The polarization curve for electrochemical oxidation of methanol under stationary conditions (Fig. 139, curve 2) is represented by the Tafel equation with a slope† of about 0.06 V in the φ_r range ~0.45-0.55 V. At low anodic potentials there is also a large quantitative difference between the rates of the nonstationary and stationary oxidation processes. For example, at φ_r = 0.4 V the rate of methanol dehydrogenation on a clean surface in acid solution is ~10^4 times the rate of the stationary electrochemical oxidation process. The difference between the two rates diminishes with increasing anodic φ_r.

A polarization curve for electrochemical oxidation of the products of methanol chemisorption (curve 3, Fig. 139) can be obtained with the aid of measurements of electrochemical oxidation in the adsorbed layer [111, 112, 128]. For this one can use either the potentials of the arrests corresponding to oxidation of chemisorbed methanol on the charging curves of a Pt/Pt(CH₃OH)ₐds electrode, determined at different current densities, or the initial regions of potentiostatic curves for electrochemical oxidation of the chemisorbed product, recorded with slow variation of the potential (Fig. 140).‡ It is seen in Fig. 139 that the slopes of curves 2 and 3 coincide. In acid solutions the overpotential of methanol oxidation (with methanol present in solution) and the overpotential of oxidation of the chemisorbed product depend little on the solution pH. The galvanostatic curves determined in alkaline solutions show two arrests corresponding to oxidation of the chemisorbed substance (Fig. 140). The first of these corresponds to the

*The results of this work were reproduced by Takahashi and Miyake [126].
†The slope depends on the "age" of the electrode [127].
‡These curves relate to oxidation of the products accumulated during anodic polarization of the electrode at 0.35-0.5 V. The amount of chemisorption products accumulated under these conditions is roughly double the amount accumulated on open circuit.

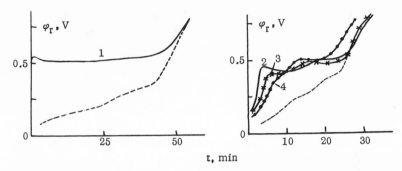

Fig. 140. Galvanostatic curves of electrochemical oxidation of substances chemisorbed during polarization of a Pt/Pt electrode at $i = 10^{-4}$ A/cm^2 in: 1) 0.1 N H$_2$SO$_4$ + 0.5 M CH$_3$OH; 2) 0.1 N KOH + 0.5 M CH$_3$OH; 3) 1 N KOH + 0.5 M CH$_3$OH; 4) 10.7 M KOH + 0.5 M CH$_3$OH. Dash lines represent charging curves in 0.1 N H$_2$SO$_4$ and 1 N KOH (from data in [113]).

prewave on the potentiodynamic curves. Appearance of a more easily oxidizable substance in alkaline solutions leads to decrease of the overpotential of stationary oxidation of methanol. With rise of pH in alkaline solutions the overpotential for the start of oxidation of the chemisorbed substance decreases; a roughly similar dependence of pH is also characteristic of the process in presence of methanol in the solution.

As has already been noted [111, 112], despite the parallelism noted above between the oxidation of products of stable chemisorption of methanol on the electrode and stationary oxidation processes, there is no justification for the view that the oxidation reaction proceeds entirely in accordance with the equation

$$\overset{I_1}{CH_3OH} \rightarrow \overset{I_2}{CHO} \rightarrow CO_2 \qquad\qquad (X.32)$$

In that case the total current I of methanol oxidation must be equal to $I_1 + I_2 = 2I_2$. In reality, comparison of the rate of oxidation of products of stable chemisorption of methanol with the rate of the stationary process with methanol present in the solution shows that the latter is greater than the former by about an order of magnitude [113, 127].

Direct comparison of the oxidation rates of chemisorbed products and of processes in presence of organic compounds in solution shows an even greater difference in the case of ethanol and

TABLE 26. Products of Electrochemical Oxidation
of Methanol at a Pt/Pt Electrode

Solution	Potential during electrolysis, φ_r (mV)	Current efficiency, %		
		HCHO	HCOOH	CO_2 (by difference)
1 N H_2SO_4 + 0.5 M CH_3OH	522-531	2.6	9.0	88.4
	519-527	3.2	10.8	86.0
1 N KOH + 0.5 M CH_3OH	453-465	1.0	68.0	31.0
	450-468	1.0	71.0	28.0

formic acid [129]. For example, the overpotential for electrochemical oxidation of the substance chemisorbed from both acid
and alkaline solutions containing HCOOH* is higher by about 200
mV than the overpotential of HCOOH oxidation in presence of formic acid in the solution. As the slope of the polarization curve of
formic acid oxidation is 60 mV, this indicates a difference between
the oxidation rates by a factor of ~10^3.

The role of the products of stable chemisorption of organic
substances in the stationary oxidation process can be elucidated
by analysis of the products formed by their oxidation on platinum
metals [23, 115, 123-125]. It has been shown [136-140] that formaldehyde[†] is formed as an intermediate reaction product during
oxidation of methanol. However, if it is assumed that only HCO
particles [Eq. (X.32)], which represent an intermediate oxidation
stage between formaldehyde and formic acid, undergo oxidation,
it is difficult to explain formation of formaldehyde during oxidation
of methanol.

The results of analysis of the products of methanol oxidation
on platinum, reported by Podlovchenko, Frumkin, and Stenin [23],

---

*The formate ion is oxidized on platinum at considerable rates at low anodic poten-
tials [129-133], although it is stated in some publications that oxidation of HCOO⁻
does not occur in alkaline solutions [134, 135].

†Formation of formaldehyde has also been detected during catalytic oxidation of
methanol [141]. However, Vereecken et al. [142] were unable to detect formation
of formaldehyde and formic acid during oxidation of methanol. Apparently, the
analytical method used was not sufficiently sensitive.

are given in Table 26. It is seen that carbon dioxide is not the only product (and in the case of alkaline solutions not even the main product) of electrochemical oxidation of methanol under stationary conditions.

On the basis of these results, the electrochemical oxidation of methanol under stationary conditions (as long as the degree of oxidation of the original substance is low) was represented by several parallel reactions:

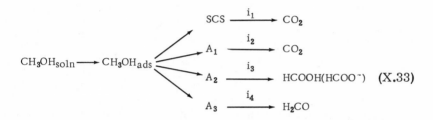

$$CH_3OH_{soln} \longrightarrow CH_3OH_{ads}$$

$$SCS \xrightarrow{i_1} CO_2$$
$$A_1 \xrightarrow{i_2} CO_2$$
$$A_2 \xrightarrow{i_3} HCOOH(HCOO^-) \quad (X.33)$$
$$A_3 \xrightarrow{i_4} H_2CO$$

where SCS is the stably chemisorbed substance detected on the electrode after washing; A_1, A_2, and A_3 are relatively weakly adsorbed species, present on the electrode in comparatively small amounts and removed by washing. The following relations between the rates of the individual reactions may be postulated: for acid solutions $i_2 > i_3$, i_4, and i_1; for alkaline solutions $i_3 > i_2 > i_1$ and i_4.

Thus, although the stable chemisorbed substance is one of the products of the current-determining reaction, it passivates electrochemical oxidation of methanol by covering the surface.

Breiter [125] emphasized that for elucidation of the role played in the stationary process by the products of stable chemisorption of methanol it is necessary to carry out simultaneous determination of the ratios

$$I_{CO_2}/I \text{ and } I_{CO_2}/I_2 ,$$

where I_{CO_2} is the rate of CO_2 formation (in electrical units), I is the stationary rate of methanol oxidation, and I_2 is the oxidation rate of the products of stable adsorption. According to Breiter, the overall process of electrochemical oxidation of methanol can be represented by the following scheme:

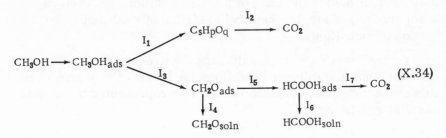

$$
\text{(X.34)}
$$

Experiments showed that the CO_2 yield is 90% at 0.01 M methanol concentration and $\sim 10\%$ in a 5 M methanol solution in 0.5 M H_2SO_4 at 0.5-0.7 V. Thus, the I_{CO_2}/I ratio is less than unity and depends on the methanol concentration; this indicates that the process occurs predominantly through the bottom branch of scheme (X.34).*

Éntina and Petrii [116] showed by analysis of the reaction products that oxidation of methanol at a platinum—ruthenium electrode proceeds by several parallel routes.

In a study of anodic oxidation of formic acid on platinum [49, 143], Brummer concluded that the stably adsorbed substance is an inactive by-product which blocks the electrode surface.

The results of analysis of the products of electrochemical oxidation of ethanol on platinum [144-146] were interpreted by Podlovchenko et al. [146] in terms of a number of parallel reactions involving different adsorbed species. It was found [146] that acetaldehyde is not formed during electrochemical oxidation of the products of stable chemisorption of ethanol on platinum in 1 N H_2SO_4, whereas in oxidation of ethanol under stationary conditions the acetaldehyde yield is between 25 and 50%. The second oxidation product is acetic acid, and the sum of the current efficiencies for CH_3CHO and CH_3COOH is close to 100%. The acetaldehyde and acetic acid current efficiencies are independent of time; this evidently indicates that these products are formed by parallel

---

*In practice this means that the process occurs mainly by consecutive formation of CH_2O and HCOOH, as has also been suggested in the literature [139, 140]. However, it is difficult to agree with the use of Langmuir isotherms for analysis of the mechanism of methanol oxidation on a heterogeneous surface of a platinum electrode [139, 140].

reactions:

$$
\begin{array}{c}
\phantom{C_2H_5OH_{soln} \rightarrow}
\overset{\displaystyle I_4}{CH_3CHO_{soln} \longrightarrow CH_3COOH_{soln}} \\[2pt]
\nearrow^{\displaystyle I_1} \\[-6pt]
C_2H_5OH_{soln} \longrightarrow C_2H_5OH_{ads} \xrightarrow{\;\;I_2\;\;} CH_3COOH_{soln} \qquad (X.35)\\[-4pt]
\searrow \\[-2pt]
SCS \xrightarrow{\;\;I_3\;\;} CO_2
\end{array}
$$

As long as the degree of oxidation of ethanol is not high, I_1 and $I_2 \gg I_3$ and I_4. Thus, the stably chemisorbed substance retards electrochemical oxidation of the alcohol by covering the surface.

The foregoing summarizes the existing views on the general reaction scheme for anodic oxidation of alcohols and formic acid on metals of the platinum group, and on the role of organic adsorption products. As regards the rate-determining step, the most probable, in view of the slope of the polarization curve for methanol oxidation on platinum and of the fact that the reaction rate is independent of pH at $\varphi_r = $ const,* is interaction of the adsorbed particles with OH_{ads}, e.g., by the following scheme [128]:†

$$
H_2O - \bar{e} \rightleftarrows OH_{ads} + H^+
$$
$$
OH_{ads} + RH_{ads.} \xrightarrow{\text{slow}} R_{ads} + H_2O \qquad (X.36)
$$

where RH_{ads} is the chemisorbed species involved in the rate-determining step. At constant surface coverage θ by organic substances the kinetic equations corresponding to this hypothesis are of the form

$$
i = k \exp\left(\frac{F\varphi_r}{RT}\right), \qquad (X.37)
$$

$$
\left(\frac{\partial \varphi_r}{\partial \log i}\right)_{pH,\,\theta} = 2.3\frac{RT}{F}, \qquad (X.38)
$$

$$
\left[\frac{\partial \ln i}{\partial\,(pH)}\right]_{\varphi_r,\,\theta} = 0. \qquad (X.39)
$$

---

*Aikazyan and Pleskov [147] first studied the influence of pH in order to demonstrate that OH_{ads} is involved in anodic oxidation of alcohols.

†In the case of rhodium electrodes [152] the scheme is apparently more complicated.

Evidence for the presence of adsorbed oxygen at the potentials of methanol oxidation in solutions of the supporting electrolyte has been examined by Marvet and Petrii [148]. However, as yet there are no direct data indicating the possible presence of adsorbed oxygen in the φ_r range under consideration when the adsorbed organic substance is present.

Bagotskii, Vasil'ev, and co-workers [11, 12, 117-120] examined kinetic equations corresponding to scheme (X.36) for cases where θ depends on the potential, in the light of the concept of uniform heterogeneity of the platinum surface.

The rate of electrochemical oxidation of methanol is diminished in presence of specifically adsorbed halide ions [125, 149]. This was attributed in the publications cited to the influence of the anions on the reaction of water discharge, which is slowed down in their presence [150].

At potentials $\varphi_r > 0.6$ V the rate of electrochemical oxidation of the products of methanol chemisorption increases so much that the rate of methanol adsorption on the electrode surface begins to influence the process; i.e., the slowness of the adsorption step becomes apparent. At $\varphi_r > 0.8$ V the adsorption rate falls as the result of stable adsorption of oxygen on the electrode. The rate of stationary oxidation of methanol falls correspondingly.

Participation of OH_{ads} in the electrochemical oxidation of methanol makes the reaction rate dependent on the relative amount of free surface. This probably accounts for the negative reaction order with respect to methanol at high methanol concentrations in solution [70] and for the peculiarities of charging curves in presence of chemisorbed methanol. The characteristic feature of such curves (Fig. 140) is independence of φ_r of the quantity of electricity passed at the initial region of the arrest corresponding to electrochemical oxidation of the adsorbate, or even appearance of a minimum on curves corresponding to fairly large surface coverage.

The form of manifestation of this effect greatly depends on the absolute magnitude of the free surface, the distribution of chemisorbed particles over the electrode surface, interaction between the particles, and other factors. This probably accounts for the observed complex dependence of the form of the $Pt(X_{ads})$ charging curves on the nature of the organic compound and on the experimental conditions (especially the solution pH, presence of specifically adsorbed anions in the electrolyte, and the nature of the electrode).

Bagotskii, Vasil'ev, et al. [118, 120] consider that this effect, described as the "high coverage effect,"* may also be attributed to a change of the slow step at high coverages of the electrode surface with the organic substance; under such conditions discharge of water may become the slow step.†

The effect described above appears to be similar in nature to the sharp increase of the rate of electrochemical oxidation of carbon monoxide with progressive oxidation of the adsorbate, reported by Gilman [28]. To explain this result, Gilman postulated a "reacting pair" mechanism, involving reaction of adsorbed CO with an H_2O molecule adsorbed on neighboring sites. The reaction rate therefore depends both on the surface coverage by the adsorbate and on the number of free sites on the surface. The "reacting pairs" are presumed to be located between "clusters" of adsorbed CO molecules and free surface sites. If it is assumed that oxidation occurs only at the boundaries of the clusters of adsorbed CO molecules, the following expression for the reaction rate at φ_r = const can be written, with certain simplifying assumptions:

$$i = k\,(\theta_{CO} - a)^p\,(\theta_F - b)^q, \qquad (X.40)$$

where θ_{CO} is the fraction of the surface covered with CO; θ_F is the fraction of the free surface; p and q are constants having values between 0.5 and 1; a and b are the fractions of "inactive" surface regions, where oxidation does not occur.

The relation between adsorption processes and electrochemical oxidation under stationary conditions has also been examined for methane and ethane [41-43] and for propane and hexane [48, 51]. According to the data in [42, 43], oxidation of methane under stationary conditions at potentials of 350-500 mV at 60° is determined

---

*This term is somewhat inaccurate, because under certain conditions an increase of the reaction rate on oxidation of a part of the adsorbate can also be observed at relatively low θ ($\sim$0.5-0.6).

†Retardation of electrochemical processes by adsorbed reacting substances has also been observed in reactions at mercury electrodes at high surface coverages. Various possible causes of the "high coverage effect" on mercury have been examined by Tedoradze and Ershler [151].

by the rate of dissociative adsorption of methane. This conclusion was based on studies of the dependence of the rate of methanol oxidation on the potential, solution pH, methane partial pressure, temperature, and presence of specifically adsorbed anions, and on a comparison of the rates of oxidation and adsorption of methane. In the case of ethane [42] the process is apparently determined by the rate of adsorption on a surface partially covered by adsorbed particles. However, at low φ_r the reaction rate is determined by the rate of oxidation of the products of ethane adsorption to CO_2 and H_2O.

Brummer and Turner [48, 51] used data on the dependence of the reaction rate on the partial pressure of the hydrocarbon for elucidation of the part played by adsorption in anodic oxidation of hydrocarbons. In the case of propane and hexane the current increases with increasing pressure, whereas the amount of O-type species (see Chap. VIII) remains almost unchanged. It was concluded from this that the role of the O-type species is mainly to block the electrode surface. The oxidation rates of CH-α and CH-β particles are too low to ensure a stationary process. The process is therefore controlled by the rate of adsorption of the hydrocarbon on a surface partially covered by oxidation by-products.

The mechanism of electrochemical oxidation of a large group of unsaturated hydrocarbons (ethylene, acetylene, propylene, butene-1, butene-2, allene, butadiene, cyclohexadiene, benzene), containing isolated and conjugated double bonds, has been studied in detail by Bockris and co-workers [104-106, 153-160]. Their experimental results can be summarized as follows.

1. At $\varphi_r = 0.4-0.75$ V in $1 N$ H_2SO_4 at 80° on smooth and platinized platinum electrodes $\left(\dfrac{\partial \varphi}{\partial \log i}\right)_{pH,p} \simeq 0.14-0.19$ V for unsaturated compounds containing double bonds, and ~ 0.07 V for acetylene (p is the partial pressure of the hydrocarbon).

2. In the Tafel region $\left(\dfrac{\partial \log i}{\partial p}\right)_{\varphi,pH}$ is between -0.11 and -0.2.

3. $\left(\dfrac{\partial \log i}{\partial pH}\right)_\varphi \simeq 0.8$ for acetylene and $\sim 0.39-0.47$ for other hydrocarbons.

4. The apparent activation energies for oxidation of different hydrocarbons are similar, ~20-24 kcal/mole.

5. The oxidation rates of the hydrocarbons increase in the following series: benzene < butadiene < butenes < propylene < ethylene < allene. However, the differences between the rate constants are within the range of one order of magnitude.

6. At φ_r > 0.85-1.05 V (dependent on the nature of the hydrocarbon) oxidation of hydrocarbons is retarded owing to adsorption of considerable amounts of oxygen (the retardation mechanism has been discussed in detail by Gilroy and Conway [161]).

7. The electrode can be activated for hydrocarbon oxidation by an anodic pulse, i.e., by increase of the electrode potential to 0.9 V. The gradual decrease of current after activation of the electrode depends on the nature and partial pressure of the hydrocarbon, and is caused by adsorption of the hydrocarbon on the surface.

8. Study of the rates of ethylene oxidation in light and heavy water (i^H and i^D) shows that on platinum $i^H/i^D \lessapprox 2.7$.

9. The main product of hydrocarbon oxidation on platinum is CO_2.

The scheme representing oxidation of hydrocarbons containing double bonds, proposed on the basis of these data, includes a slow stage of discharge of water molecules with formation of adsorbed OH particles, and rapid interaction of these particles with the hydrocarbon adsorbed on the electrode surface:

$$H_2O \xrightarrow{\text{slow}} OH_{ads} + H^+ + \bar{e} \qquad (X.41)$$

$$RH_{ads} + OH_{ads} \rightarrow \text{products} \qquad (X.42)$$

This scheme corresponds to the equation

$$i = k(1 - \theta_t) \exp\left(\frac{\beta F \varphi_r}{RT}\right), \qquad (X.43)$$

where θ_t is the total coverage of the electrode surface by OH_{ads} and RH_{ads}. This equation provides an explanation of the decrease of the oxidation rate with increasing partial pressure of the hydrocarbon in the system, because the fraction of the surface on which

reaction (X.41) occurs diminishes.*   However, according to Eq.
(X.43) $\left(\frac{\partial \log i}{\partial pH}\right)_\varphi = 0$, whereas it is found experimentally that
$\left(\frac{\partial \log i}{\partial pH}\right)_\varphi = 0.5$. To account for this result it is presumed
that the potential in Eq. (X.43) should be relative to the point of
zero charge of platinum and that φ_n of platinum varies with pH in
accordance with Eq. (IX.71). However, this suggestion is inconsistent [78, 108] with general concepts of the role of the point of
zero charge in electrochemical kinetics [162, 163, 163a].

In electrochemical oxidation of acetylene reaction (X.42) becomes the slow step, as apparently the heat of adsorption of
acetylene on platinum is greater than the heat of adsorption of
unsaturated hydrocarbons containing double bonds. Moreover,
the heat of dissociation of C−H bonds is also higher in acetylene
than in ethylene.

An electrochemical mechanism for oxidation of oxalic acid at
a platinum electrode is postulated by Bockris et al. [164], based
on the assumption that HC_2O_4 particles adsorbed on the electrode
are involved in the process, e.g.,

$$HC_2O_{4\,ads} \rightarrow 2CO_2 + H^+ + \bar{e}. \qquad (X.44)$$

The kinetic equations for this reaction were analyzed in the
light of different concepts of the properties of the electrode surface ("Langmuir" conditions and "Temkin" conditions).

The list of publications concerned with anodic oxidation of
organic substances in which adsorption is an essential stage could
be extended considerably (e.g., [166-168]) but this is not the purpose of the present review.

We will merely note in conclusion that adsorption of organic
substances plays an important part in the occurrence of periodic
phenomena during anodic oxidation reactions [152, 169, 170].

Thus, the existence of an interconnection of one kind or another between adsorption and the process kinetics is now assumed

---

*According to Burshtein, Pshenichnikov, and Tyurin [165], a possible explanation of
the negative reaction order is that at high surface coverages the conditions become
unfavorable for adsorption processes involving C_2H_4 (see the data in Chapter VIII
on the influence of $p_{C_2H_4}$ on the chemisorbed product).

in discussions of anodic oxidation of organic compounds. Therefore study of the relationships in adsorption of organic compounds
provides deeper insight into electrode reaction mechanisms.

5. Influence of the Preparation Conditions and Nature of the Electrode on Adsorption of Organic Compounds

Many catalytically active materials have now been tested in
reactions of electrochemical oxidation of various organic compounds (for a review, see, e.g., [108]), owing to the practical importance of this problem. However, direct data on adsorption
properties are at present accessible only for a limited range of
systems.

The greatest amount of data has been accumulated for smooth
and platinized platinum. It was noticed a relatively long time ago
that anodic−cathodic activation of smooth platinum is required for
reproducible results, whereas this is not essential in the case of
platinized platinum [150]. However, it has recently been shown
that the main purpose of the usual brief anodic−cathodic activation of smooth platinum is desorption of surface contaminations
[171-176]. Therefore the first real indication of a significant difference between the properties of smooth and platinized platinum
was probably Obrucheva's discovery [177] of the difference in
character between the oxide layers formed on these types of
platinum during far anodic polarization.

It is also reported in a number of publications [178-181] that
Adams' platinum or varieties of platinum made by reduction of
its oxides differ appreciably from other types of platinum catalysts. In particular, the isotherms of hydrogen adsorption on
Adams' platinum has certain peculiarities: two regions of adsorption sites are more pronounced, and heat treatment of the electrode
has relatively little influence on the shape of the isotherm [179,
180, 182].

Bagotskii et al. [183] and Stenin and Podlovchenko [184] compared the kinetics of methanol adsorption on smooth and platinized
platinum. It was noted [183] that the adsorption rate was lower
on platinized than on smooth platinum, although the chemisorption
kinetics and stationary surface coverages by methanol were the

same on both. Stenin and Podlovchenko [184] did not find any significant differences between the specific rates of methanol adsorption on smooth and platinized electrodes. Other investigations [43, 105, 183-187] also failed to reveal any specific effect
of platinizing (however, cf. [43] and [44]). Nevertheless, the strong
dependence of the adsorption properties of platinized platinum on
the "age" of the electrode (e.g., see [149]) and the influence of the
time of contact of smooth platinum with methanol on its properties
[80] stress the need for deeper investigation of the dependence of
adsorption on surface structure. The size of the crystallites in
the platinum electrode used for the investigation [188] and the conditions of preparation of platinized electrodes [186, 189-195] must
be taken into consideration.

Changes in adsorption properties are to be expected if the
platinum is deposited on various supports [181, 196-199], and if the
platinum surface is modified by sulfur or selenium [200, 201].

The results discussed above (Chapters VIII-X) indicate that
the adsorption behavior of electrodes is strongly dependent on the
nature of the metal. This is an important factor in comparison of
the catalytic activities of different metals in anodic oxidation reactions, because the surface coverages with the organic substance
at the same φ_r may be different [160].

Similar conclusions follow from examination of adsorption data
obtained for alloys [202]. The most interesting result for catalysts
of this type is the strong influence of heat treatment on the adsorption and catalytic properties of alloys. For example, heating of
platinum – ruthenium alloy at 800° in argon or hydrogen [202] results in loss of high catalytic activity, and a catalyst similar to
platinum in its behavior is obtained. This result can be explained
by diffusion of ruthenium atoms from the surface layer into the
volume. The same explanation of the influence of heat treatment
on the properties of alloys was given by Damjanovic and Brusić
[203], who investigated electrochemical reduction of hydrogen on
alloys of palladium and platinum with gold.

Changes in the composition of the surface layers of alloys
during use were presumed by Breiter [204] and by Pshenichnikov
et al. [205].

Factors determining the dependence of adsorption and catalytic
behavior on the nature of the electrode, and certain useful correla-

tions between the catalytic activity and properties of metals and alloys, have been discussed in detail by Bockris and co-workers [106, 153, 160]. To summarize their material, we note the following factors which must be taken into account in relation to adsorption phenomena on platinum metals.

1. Crystalline structure and microstructure of the surface (e.g., see [206]).

2. Type of heterogeneity of the electrode surface [13-16].

3. Presence of d-electron vacancies in the atoms of the metal or alloy [153, 207-210].

4. Adsorption of hydrogen, oxygen, and ions of the double layer on the electrode.

6. Adsorption of Organic Substances
at High Anodic Potentials

Adsorption phenomena at potentials on the positive side of the reversible oxygen potential are very complex; this is due to peculiarities of the surface state of the platinum electrode under these conditions (for a review, see [211]; nonstationary oxidation processes on a smooth platinum electrode have also been studied in various solutions in the range $\varphi_r \sim 0.8\text{-}2.8$ V [212, 213]).

There have been few direct adsorption measurements relating to this range of potentials. However, their results are of considerable interest for elucidation of the mechanism of the Kolbe electrosynthesis and of a number of other anodic processes involving organic compounds [107, 214]. The dependence of the Kolbe electrosynthesis on adsorption of organic compounds at high anodic potentials has been noted by Antropov [215].

Adsorption of inorganic anions at such potentials has certain distinctive features. For example, Kazarinov [216] demonstrated by the radioactive tracer method that halide ions suppress adsorption of sulfate ions on platinum in the range $\varphi_r = 1.5\text{-}2.0$ V; the order of their surface activity, the reverse of that observed at lower potentials, is $F^- > Cl^- > Br^- > I^-$. This effect is due to incorporation of F^- ions in the oxide film.

Girina et al. [217] investigated the surface state of the platinum electrode at potentials preceding the Kolbe electrosynthesis

by determination of differential capacity curves and with the aid of
radioactive tracers. Adsorption of acetate ions on platinum in the
potential range studied results in a sharp change in the form of the
differential capacity curve. The start of the capacity decrease
due to anion adsorption is shifted toward less anodic potentials in
presence of acetate ions. The radioactive tracer method provides
direct evidence of the higher adsorbability of CH_3COO^- anions in
comparison with SO_4^{2-} anions. Addition of only 0.02 g-eq of sodium
acetate per liter to $0.1\,N$ Na_2SO_4 solution reduces adsorption of SO_4^{2-}
anions roughly fivefold. Direct determination of the adsorption of
acetate ions on platinum [218] shows that in the range of potentials
preceding the Kolbe electrosynthesis adsorption of acetate anions
increases sharply (Fig. 141). The potential corresponding to the
sharp increase of adsorption corresponds to the start of the Kolbe
electrosynthesis. These data were used to substantiate the view
that adsorption of acetate ions retards oxygen evolution and thereby makes the Kolbe synthesis possible.

Fioshin et al. [219-231] investigated the effects of various
additives on anodic processes at high potentials. In addition to
determination of polarization curves, they determined the differential capacity of the platinum electrode from the potential decay
curves on open circuit; this enabled them to draw certain conclusions regarding adsorption of the additives on platinum. As an example, Fig. 142 shows differential capacity curves of a platinum
electrode in $1M$ CH_3COONa with additions of butadiene-1,3. In
presence of butadiene the electrode capacity is lowered considerably as the result of adsorption of the diene on oxidized platinum.
In the opinion of Fioshin et al., adsorption of the diene is due to the

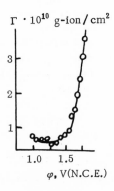

$\Gamma \cdot 10^{10}$ g-ion / cm^2

Fig. 141. Dependence of
adsorption of acetate ions
on the potential of a
smooth platinum electrode
in $0.1\,N$ CH_3COONa +
$0.1\,N$ CH_3COOH (from
data in [218]).

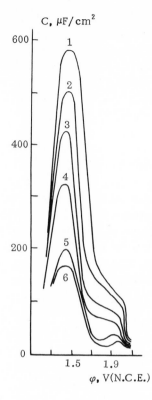

Fig. 142. Dependence of the capacity of a platinum electrode on the potential in 1 M CH_3COONa with additions of butadiene-1,3: 1) 0; 2) 8.35 · 10^{-5}; 3) 4.175 · 10^{-4}; 4) 8.35 · 10^{-4}; 5) 4.175 · 10^{-3}; 6) 8.35 · 10^{-3} M (from data in [224]).

presence of an easily polarized system of conjugated double bonds, the π-electrons of which interact with positive charges of the electrode surface, in the molecule. According to the authors, this is confirmed by the fact that adsorption effects can also be observed in presence of cyclohexene-3 and benzene, whereas additions of hexane or cyclohexane have no influence on the anodic processes at high anodic potentials. In many other cases compounds containing conjugated double bonds are generally used in preparative electrosynthesis at these potentials. Adsorption of the above-mentioned additives eliminates the influence of extraneous inorganic anions on the anodic condensation reactions.

Isotherms of butadiene adsorption, based on the results of capacity measurements, are of the logarithmic type.

Methanol and other alcohols also have a substantial influence on the electrode capacity at high anodic potentials; this indicates that they are adsorbed on the surface of strongly oxidized platinum. Adsorbability of alcohols increases with their chain length.

Fioshin et al. refer to the possibility of formation of a peculiar type of chemisorbed compound consisting of surface oxides with occluded radicals formed in the course of the electrochemical processes, e.g.,

$$RCOO^- - e \rightarrow RCOO \cdot \qquad (X.45)$$

A similar suggestion was made by Dickinson and Wynne-Jones [232].

For elucidation of the mechanism of electrochemical decarboxylation and of the Kolbe reaction, and of the part played by adsorption in these processes, Conway et al. [233-244]* investigated a number of relatively simple model reactions, not complicated by various side processes. The reactions chosen were electrochemical decarboxylation of formate in anhydrous formic acid, acetonitrile, and propylene carbonate, and the Kolbe reaction in solutions of trifluoroacetate in trifluoroacetic acid. Addition of water to these systems, in which evolution of oxygen and formation of surface oxides are excluded, makes it possible to investigate the influence of oxide formation on the anodic processes. It was concluded from the result of measurements by the galvanostatic and potential decay method that the transition to decarboxylation processes is preceded by formation of a monolayer of adsorbed intermediate reaction products on the electrode surface: $HCOO_{ads}$ in the case of formate, and CF_3COO_{ads} in the case of trifluoroacetate. These products are formed by an electrochemical mechanism:

$$HCOO^- \rightarrow HCOO_{ads} + \bar{e}$$
$$CF_3COO^- \rightarrow CF_3COO_{ads} + \bar{e}. \qquad (X.46)$$

Subsequent processes in nonaqueous formate solutions are represented by the equations

$$HCOO_{ads} \rightarrow CO_2 + H_{ads}$$
$$H_{ads} \rightarrow H^+ + \bar{e} \qquad (X.47)$$

---

*Some critical comments on [240-243] have been made by Eberson [245].

or

$$HCOO_{ads} \rightarrow CO_2 + H^+ + \bar{e}. \qquad (X.48)$$

Conway et al. [233-239] have examined methods for determination of adsorption pseudocapacity resulting from changes of the surface coverage by adsorbed reaction intermediates with variation of the electrode potential, with the aid of galvanostatic and potentiostatic measurements, and measurements of potential decay.

References

1. A. N. Frumkin, Dokl. Akad. Nauk SSSR, 154:1432 (1964).
2. J. O'M. Bockris, M. Green, and D. A. J. Swinkels, J. Electrochem. Soc., 111:742 (1964).
3. W. Heiland, E. Gileadi, and J. O'M. Bockris, J. Phys. Chem., 70:1207 (1966).
4. E. Gileadi, Comptes Rendus, Deuxieme Symposium Européen sur les Inhibiteurs de Corrosion, Ferrara, 1965, Vol. 2 (1966), p. 543.
5. E. Gileadi, J. Electroanal. Chem., 11:137 (1966).
6. G. A. Bogdanovskii, L. G. Feoktistov, and A. I. Shlygin, Nauchn. Dokl. Vysshei Shkoly, Khim. i Khim. Tekhnol., 3:443 (1958).
7. G. M. Beloslyudova and D. V. Sokol'skii, Élektrokhimiya, 2:704 (1966).
8. D. V. Sokol'skii, Hydrogenation in Solutions [in Russian], Izd. Akad. Nauk Kaz.SSR, Alma-Ata (1962).
9. E. Gileadi, L. Duic, and J. O'M. Bockris, Electrochim. Acta, 13:1915 (1968).
10. O. A. Khazova, Yu. B. Vasil'ev, and V. S. Bagotskii, Élektrokhimiya, 1:82 (1965).
11. V. S. Bagotskii and Yu. B. Vasil'ev, Electrochim. Acta, 11:1439 (1966).
12. V. S. Bagotskii and Yu. B. Vasil'ev, Advances in the Electrochemistry of Organic Compounds [in Russian], Izd. "Nauka," Moscow (1966), p. 38.
13. O. A. Khazova, Yu. B. Vasil'ev, and V. S. Bagotskii, Recent Progress in the Electrochemistry of Organic Compounds [in Russian], Izd. "Nauka" (1967), p. 19.
14. S. S. Sedova, Yu. B. Vasil'ev, and V. S. Bagotskii, Recent Progress in the Electrochemistry of Organic Compounds [in Russian], Izd. "Nauka" (1967), p. 76.
15. O. A. Khazova, S. S. Sedova, Yu. B. Vasil'ev, and V. S. Bagotskii, The Double Layer and Adsorption on Solid Electrodes [in Russian], Izd. Tartusk. Gos. Univ., Tartu (1968), p. 171.
16. S. S. Sedova, Yu. B. Vasil'ev, and V. S. Bagotskii, Élektrokhimiya, 4:1113 (1968).
17. L. Formaro and S. Trassatti, Chim. Ind. (Milan), 48:706 (1966).
18. S. Trassatti and L. Formaro, J. Electroanal. Chem., 17:343 (1968).
19. R. E. Smith, H. B. Urbach, J. H. Harrison, and N. L. Hatfield, J. Phys. Chem., 71:1250 (1967).
20. B. I. Podlovchenko, O. A. Petrii, and E. P. Gorgonova, Élektrokhimiya, 1:182 (1965).

21.   B. I. Podlovchenko and E. P. Gorgonova, Dokl. Akad. Nauk SSSR, 156:673 (1964).

22.   B. I. Podlovchenko and V. F. Stenin, Élektrokhimiya, 3:649 (1967).

23.   B. I. Podlovchenko, A. N. Frumkin, and V. F. Stenin, Élektrokhimiya, 4:339 (1968).

24.   M. W. Breiter and S. Gilman, J. Electrochem. Soc., 109:622 (1962).

25.   Hira Lal, O. A. Petrii, and B. I. Podlovchenko, Élektrokhimiya, 1:316 (1965).

26.   V. S. Éntina and O. A. Petrii, Élektrokhimiya, 3:1237 (1967).

27.   M. W. Breiter, J. Electrochem. Soc., 109:42 (1962).

28.   S. Gilman, J. Phys. Chem., 66:2657 (1962); 67:78, 1898 (1963).

29.   M. W. Breiter, Electrochim. Acta, 8:447, 457 (1963).

30.   V. E. Kazarinov, Élektrokhimiya, 2:1170 (1966).

31.   C. W. Fleischmann, G. K. Johnson, and A. T. Kuhn, J. Electrochem. Soc., 111:602 (1964).

32.   S. B. Brummer and A. C. Makrides, J. Phys. Chem., 68:1448 (1964).

33.   S. B. Brummer, J. Phys. Chem., 69:562 (1965).

34.   N. Minakshisundaram, Yu. B. Vasil'ev, and V. S. Bagotskii, Élektrokhimiya, 3:193, 283 (1967).

35.   M. W. Breiter, J. Phys. Chem., 72:1305 (1968).

36.   S. Gilman, J. Phys. Chem., 71:4330, 4339 (1967).

37.   B. I. Podlovchenko, V. F. Stenin, and A. A. Ekibaeva, Advances in the Electrochemistry of Organic Compounds [in Russian], Izd. "Nauka" (1967), p. 3.

38.   B. I. Podlovchenko, V. F. Stenin, and A. A. Ekibaeva, Élektrokhimiya, 4:1004 (1968).

39.   S. B. Brummer and M.J. Turner, J. Phys. Chem., 71:3902 (1967).

40.   L. W. Niedrach, Hydrocarbon Fuel Cell Technology (S. Baker, editor), Academic Press, New York—London (1965), p. 377.

41.   L. W. Niedrach, S. Gilman, and J. Weinstock, J. Electrochem. Soc., 112:1161 (1965).

42.   L. W. Niedrach, J. Electrochem. Soc., 113:645 (1966).

43.   R. V. Marvet and O. A. Petrii, Élektrokhimiya, 3:153 (1967).

44.   A. H. Taylor and S. B. Brummer, J. Phys. Chem., 72:2856 (1968).

45.   R. J. Flannery and D. C. Walker, Hydrocarbon Fuel Cell Technology (S. Baker, editor), Academic Press, New York—London (1965), p. 335.

46.   S. B. Brummer, J. I. Ford, and M. J. Turner, J. Phys. Chem., 69:3424 (1965).

47.   S. B. Brummer and M. J. Turner, Hydrocarbon Fuel Cell Technology (S. Baker, editor), Academic Press, New York—London (1965), p. 409.

48.   S. B. Brummer and M. J. Turner, J. Phys. Chem., 71:2825 (1967).

49.   S. B. Brummer, J. Electrochem. Soc., 113:1041 (1966).

50.   A. G. Pshenichnikov, A. M. Bograchev, and R. Kh. Burshtein, Élektrokhimiya (in press).

51.   M. J. Turner and S. B. Brummer, J. Phys. Chem., 71:3494 (1967).

52.   M. L. Savitz, K. J. Januszeski, and G. R. Frysinger, Hydrocarbon Fuel Cell Technology (S. Baker, editor), Academic Press, New York—London (1965), p. 443.

53.   H. Dahms, M. Green, and J. Weber, Nature, 196:1310 (1962).

54. E. Gileadi, B. T. Rubin, and J. O'M. Bockris, J. Phys. Chem., 69:3335 (1965).
55. S. Gilman, Trans. Faraday Soc., 62:466, 481 (1965).
56. S. Gilman, Hydrocarbon Fuel Cell Technology (S. Baker, editor), Academic Press, New York—London (1965), p. 349.
57. S. Gilman, J. Electrochem. Soc., 113:1036 (1966).
58. V. S. Tyurin, A. G. Pshenichnikov, and R. Kh. Burshtein, Élektrokhimiya (in press).
59. A. A. Michri, A. G. Pshenichnikov, and R. Kh. Burshtein, Élektrokhimiya, 4:508 (1968).
60. L. D. Burke, F. A. Lewis, and C. Kemball, Trans. Faraday Soc., 60:913, 919 (1964).
61. Y. Matsuda and H. Tamura, J. Chem. Soc. Japan, 70:2121 (1967).
61a. T. Takamura and F. Mochimaru, Denki Kagaku, 36:233 (1968).
62. K. Penttinen and J. J. Lindberg, Suomen Kem., 13:41, 128 (1968).
63. M. Green and H. Dahms, J. Electrochem. Soc., 110:466 (1963).
64. H. Dahms and M. Green, J. Electrochem. Soc., 110:1075 (1963).
65. M. Petit, Nguyen Van Huong, and J. Clavilier, Compt. Rend., 266C:300 (1968).
66. A. Hamelin and G. Valette, Compt. Rend., 267C:127 (1968).
67. A. Hamelin and G. Valette, Compt. Rend., 267C:211 (1968).
68. M. W. Breiter, Electrochim. Acta, 10:503 (1965).
69. Ya. Veber, Yu. B. Vasil'ev, and V. S. Bagotskii, Élektrokhimiya, 2:515, 522 (1966).
70. V. S. Bagotskii and Yu. B. Vasil'ev, Electrochim. Acta, 9:869 (1964).
71. N. A. Balashova and V. E. Kazarinov, Dokl. Akad. Nauk SSSR, 157:1174 (1964); Usp. Khim., 34:1721 (1965).
72. O. A. Petrii, A. N. Frumkin, and Yu. G. Kotlov, Élektrokhimiya (in press).
73. B. J. Piersma, in: Electrosorption (E. Gileadi, editor), Plenum Press, New York (1967), p. 19.
74. Yu. A. Chizmadzhev and V. S. Markin, Élektrokhimiya, 3:127 (1967).
75. Yu. A. Chizmadzhev and V. S. Markin, Élektrokhimiya, 4:3 (1968).
76. V. S. Markin and Yu. A. Chizmadzhev, Élektrokhimiya, 4:123 (1968).
77. B. I. Podlovchenko, Hira Lal, and O. A. Petrii, Élektrokhimiya, 1:744 (1965).
78. S. B. Brummer and K. Cahill, Discussion Faraday Soc., No. 45 (1968).
79. T. Biegler and D. F. A. Koch, J. Electrochem. Soc., 114:904 (1967).
80. T. Biegler, J. Electrochem. Soc., 114:1261 (1967).
81. S. S. Beskorovainaya, Yu. B. Vasil'ev, and V. S. Bagotskii, Élektrokhimiya, 1:691 (1965).
82. B. I. Podlovchenko, V. F. Stenin, and V. E. Kazarinov, The Double Layer and Adsorption on Solid Electrodes [in Russian], Izd. Tartusk. Gos. Univ., Tartu (1968), p. 121.
83. R. E. Smith, H. B. Urbach, and N. L. Hatfield, J. Phys. Chem., 71:4121 (1967).
84. M. W. Breiter, J. Electrochem. Soc., 110:449 (1963).
85. S. S. Beskorovainaya, Yu. B. Vasil'ev, and V. S. Bagotskii, Élektrokhimiya, 2:167 (1965).
86. S. S. Beskorovainaya, Yu. B. Vasil'ev, and V. S. Bagotskii, Élektrokhimiya, 1:1020 (1965).

87.   S. S. Sedova, Yu. B. Vasil'ev, and V. S. Bagotskii, Élektrokhimiya, 4:1221 (1968).

88.   M. I. Temkin, Zh. Fiz. Khim., 15:296 (1941).

89.   S. Z. Roginskii, Adsorption and Catalysis on Heterogeneous Surfaces [in Russian], Izd. Akad. Nauk SSSR (1948).

90.   S. L. Kiperman, Introduction to the Kinetics of Heterogeneous Catalytic Reactions [in Russian], Izd. Akad. Nauk SSSR (1964).

91.   I. Gonz, S. S. Beskorovainaya, Yu. B. Vasil'ev, and V. S. Bagotskii, Élektrokhimiya, 4:315 (1968).

92.   S. S. Beskorovainaya, Yu. B. Vasil'ev, and V. S. Bagotskii, Élektrokhimiya, 2:44 (1966).

93.   D. H. Bangham and F. F. Burt, Proc. Roy. Soc., A105:481 (1924).

94.   T. Kwan, J. Phys. Chem., 60:1033 (1956).

95.   S. Gilman, Trans. Faraday Soc., 61:2546, 2561 (1965).

96.   S. Gilman, J. Phys.Chem., 71:2424 (1967).

97.   E. J. Cairns and A. M. Breitenstein, J. Electrochem. Soc., 114:764 (1967).

98.   E. J. Cairns, A. M. Breitenstein, and A. J. Scarpellino, J. Electrochem. Soc., 115:569 (1968).

99.   S. Gilman, J. Phys. Chem., 70:2880 (1966).

100. M. L. Savitz, R. L. Carreras, and G. R. Frysinger, Extended Abstracts, Battery Division, Electrochemical Society Meeting, Philadelphia, Vol. 11 (1966), p. 7.

101. E. Müller, Z. Elektrochem., 29:264 (1923); E. Müller and S. Takegami, ibid., 34:704 (1928); S. Tanaka, ibid., 35:38 (1929).

102. N. Gerasimenko, Ukr. Khim. Zh., 4:439 (1929).

103. S. Gilman and M. W. Breiter, J. Electrochem. Soc., 109:1099 (1962).

104. J. O'M. Bockris, Proceedings of the First Australian Conference on Electrochemistry, 1963, Pergamon Press (1965), p. 691.

105. H. Wroblowa, B. J. Piersma, and J. O'M. Bockris, J. Electroanal. Chem., 6:401 (1963).

106. E. Gileadi and B. Piersma, in: Modern Aspects of Electrochemistry, Vol. 4 (J. O'M. Bockris, editor), Plenum Press, New York (1966), p. 47.

107. A. P. Tomilov, S. G. Mairanovskii, M. Ya. Fioshin, and V. A. Smirnov, Electrochemistry of Organic Compounds [in Russian], Izd. "Khimiya," Leningrad (1968).

108. O. A. Petrii, Progress in the Electrochemistry of Organic Compounds [in Russian], Izd. "Nauka," Moscow (1969).

109. V. S. Bagotskii and Yu. B. Vasil'ev, Advances in Electrochemistry and Electrochemical Engineering (P. Delahay, editor) (in press).

110. A. G. Polyak, Yu. B. Vasil'ev, and V.S.Bagotskii,Élektrokhimiya,4:535(1968).

111. Hira Lal, O. A. Petrii, and B. I. Podlovchenko, Élektrokhimiya, 1:316 (1965).

112. B. I. Podlovchenko, O. A. Petrii, and E. P. Gorgonova, Élektrokhimiya, 1:182 (1965).

113. O. A. Petrii (Petry), B. I. Podlovchenko, A. N. Frumkin, and Hira Lal, J. Electroanal. Chem., 10:253 (1965).

114. O. A. Petrii and B. I. Podlovchenko, Fuel Cells [in Russian], No. 2, Izd. "Nauka," Moscow (1968).

115. V. F. Stenin and B. I. Podlovchenko, Advances in the Electrochemistry of Organic Compounds [in Russian], Izd. "Nauka," Moscow (1967), p. 4.

116. V. S. Éntina and O. A. Petrii, Élektrokhimiya, 4:678 (1968).

117. O. A. Khazova, Yu. B. Vasil'ev, and V. S. Bagotskii, Izv. Akad. Nauk SSSR, Ser. Khim., 1778, 1931 (1965).

118. O. A. Khazova, Yu. B. Vasil'ev, and V. S. Bagotskii, Élektrokhimiya, 2:267 (1966).

119. S. S. Beskorovainaya, O. A. Khazova, Yu. B. Vasil'ev, and V. S. Bagotskii, Élektrokhimiya, 2:932 (1966).

120. V. S. Bagotskii and Yu. B. Vasil'ev, Electrochim. Acta, 12:1323 (1967).

121. Hira Lal, O. A. Petrii, and B. I. Podlovchenko, Dokl. Akad. Nauk SSSR, 158:1416 (1964).

122. M. R. Andrew, Electrochim. Acta, 11:1425 (1966).

123. M. W. Breiter, International Symposium on Fuel Cells, Dresden (1967).

124. M. W. Breiter, Z. Phys. Chem. (in press).

125. M. W. Breiter, Discussions Faraday Soc., No. 45 (1968).

126. S. Takahashi and Y. Miyake, Denki Kagaku, 36:223 (1968).

127. V. F. Stenin and B. I. Podlovchenko, Élektrokhimiya, 3:481 (1967).

128. A. N. Frumkin and B. I. Podlovchenko, Dokl. Akad. Nauk SSSR, 150:349 (1963).

129. B. I. Podlovchenko, O. A. Petrii (Petry), A. N. Frumkin, and Hira Lal, J. Electroanal. Chem., 11:12 (1966).

130. K. Schwabe, Z. Elektrochem., 61:743 (1957).

131. Yu. B. Vasil'ev and V. S. Bagotskii, Dokl. Akad. Nauk SSSR, 148:132 (1963).

132. R. A. Munson, J. Electroanal. Chem., 5:292 (1963).

133. J. Eckert, Electrochim. Acta, 12:307 (1967).

134. R. Buck and L. Griffith, J. Electrochem. Soc., 109:1005 (1962).

135. C. Liang and T. C. Franklin, Electrochim. Acta, 9:517 (1964).

136. T. O. Pavela, Ann. Acad. Sci. Fennicae, Ser. A, II, No. 59 (1954).

137. M. J. Schlatter, Fuel Cells (G. J. Young, editor), Reinhold Publishing Co., New York (1963), p. 190; 145th National Meeting, American Chemical Society, Division of Fuel Chemistry, Vol. 7, No. 4, New York (1963).

138. C. E. Heath and W. J. Sweeney, Fuel Cells (W. Mitchell, editor), Academic Press, New York—London (1963), p. 65.

139. M. Prigent, O. Bloch, and J.-C. Balaceanu, Bull. Soc. Chim., France, 368 (1963).

140. S. Yoshizawa, Z. Takehara, and A. Tokuoka, Extended Abstracts of the 19th CITCE Meeting, Detroit, Mich. (1968), p. 33.

141. T. C. Franklin and M. Kawamata, J. Phys. Chem., 71:4213 (1967).

142. J. Vereecken, C. Capel-Boute, and C. Decroly, Electrochim. Acta, 13:645 (1968).

143. S. B. Brummer, Élektrokhimiya, 4:243 (1968).

144. R. A. Rightmire, R. L. Rowland, D. L. Boos, and D. L. Beals, J. Electrochem. Soc., 111:242 (1964).

145. M. Höllnagel and U. Lohge, Z. Phys. Chem., 232:237 (1966).

146. B. I. Podlovchenko, V. F. Steinin, and M. V. Avramenko, Élektrokhimiya (in press).

147. É. A. Aikazyan and Yu. V. Pleskov, Zh. Fiz. Khim., 31:205 (1957).

148. R. V. Marvet and O. A. Petrii, Élektrokhimiya, 3:901 (1967).

149. V. N. Kamath and Hira Lal, J. Electroanal. Chem., 19:249 (1968).

150. A. N. Frumkin, Advances in Electrochemistry and Electrochemical Engineering (P. Delahay, editor), Vol. 3, Interscience (1963), p. 287.

151. G. A. Tedoradze and A. B. Érshler, Usp. Khim., 34:1865 (1965).

152. O. A. Petrii and N. Lokhanyai, Élektrokhimiya, 4:656 (1968).

153. J. O'M. Bockris and H. Wroblowa, J. Electroanal. Chem., 7:428 (1964).

154. M. Green, J. Weber, and V. Dražić, J. Electrochem. Soc., 111:721 (1964).

155. H. Dahms and J. O'M. Bockris, J. Electrochem. Soc., 111:728 (1964).

156. J. O'M. Bockris, B. J. Piersma, E. Gileadi, and B. D. Cahan, J. Electroanal. Chem., 7:487 (1964).

157. E. Gileadi and S. Srinivasan, J. Electroanal. Chem., 7:452 (1964).

158. J. O'M. Bockris, H. Wroblowa, E. Gileadi, and B. J. Piersma, Trans. Faraday Soc., 61:2531 (1965).

159. J. O'M. Bockris and S. Srinivasan, J. Electroanal. Chem., 11:350 (1966).

160. A. T. Kuhn, H. Wroblowa, and J. O'M. Bockris, Trans. Faraday Soc., 63:1458 (1967).

161. D. Gilroy and B. E. Conway, J. Phys. Chem., 69:1259 (1965).

162. A. N. Frumkin, Élektrokhimiya, 1:394 (1965).

163. R. Parsons, Surface Sci., 2:418 (1964).

163a. S. D. Argade and E. Gileadi, Electrosorption (E. Gileadi, editor), Plenum Press, New York (1967), p. 87.

164. J. W. Johnson, H. Wroblowa, and J. O'M. Bockris, Electrochim. Acta, 9:639 (1964).

165. V. S. Tyurin, A. G. Pshenichnikov, and R. Kh. Burshtein, Élektrokhimiya (in press).

166. J. O'M. Bockris, B. J. Piersma, and E. Gileadi, Electrochim. Acta, 9:1329 (1964).

167. J. W. Johnson and L. D. Gilmartin, J. Electroanal. Chem., 15:231 (1967).

168. J. W. Johnson, J. L. Reed, and W. J. James, J. Electrochem. Soc., 114:572 (1967).

169. H. F. Hunger, J. Electrochem. Soc., 115:492 (1968).

170. J. Wojtowicz, N. Marincic, and B. E. Conway, J. Chem. Phys., 48:4333 (1968).

171. S. Gilman, J. Phys. Chem., 67:78 (1963); Electroanalytical Chemistry (A. J. Bard, editor), Vol. 2, Marcel Dekker, Inc., New York (1967), p. 111.

172. M. W. Breiter, Electrochim. Acta, 11:905 (1966).

173. G. C. Barker, J. Electrochem. Soc., 113:1024 (1966).

174. M. Z. Kronenberg, J. Electroanal. Chem., 12:122, 168 (1966).

175. S. D. James, J. Electrochem. Soc., 114:1113 (1967).

176. O. A. Petrii and I. G. Shchigorev, Élektrokhimiya, 4:370 (1968).

177. A. D. Obrucheva, Zh. Fiz. Khim., 26:1448 (1952).

178. M. Fukuda, C. L. Rulfs, and P. J. Elving, Electrochim. Acta, 9:1563 (1964).

179. R. V. Marvet and O. A. Petrii, Élektrokhimiya, 3:518 (1967).

180. R. V. Marvet and O. A. Petrii, Élektrokhimiya, 3:591 (1967).

181. E. J. Cairns and E. J. McInerney, J. Electrochem. Soc., 114:980 (1967).

182. O. A. Petrii, R. V. Marvet, and Zh. N. Malysheva, Élektrokhimiya, 3:962, 1141 (1967).
183. O. A. Khazova, Yu. B. Vasil'ev, and V. S. Bagotskii, Élektrokhimiya, 3:1020 (1967).
184. V. F. Stenin and B. I. Podlovchenko, Vestn. Mosk. Gos. Univ. Ser. Khim., No. 4, 21 (1967).
185. M. H. Gottlieb, J. Electrochem. Soc., 111:465 (1964).
186. R. Thacker, Hydrocarbon Fuel Cell Technology (S. Baker, editor), Academic Press, New York–London (1965), p. 525.
187. R. Woods, Electrochim. Acta, 13:1967 (1968).
188. O. M. Poltorak and V. S. Boronin, Zh. Fiz. Khim., 40:2990 (1966).
189. M. Prigent, Rev. Gen. Elec., 74:69 (1965).
190. D. F. A. Koch, Extended Abstracts, Electrochemical Society Meeting, Dallas, Vol. 5 (1967), p. 67.
191. G. Bianchi, Ann. Chim. (Rome), 40:222 (1950).
192. H. T. Beans and L. P. Hammett, J. Am. Chem. Soc., 47:1215 (1925).
193. P. Popov, A. H. Kunz, and R. D. Snow, J. Phys. Chem., 32:1056 (1928).
194. A. I. Shlygin, Uch. Zap. Kishinevsk. Gos. Univ., 7:3 (1958).
195. R. Thacker, Nature, 212:182 (1966).
196. J. D. Voorhies, J. S. Mayell, and H. P. Landi, Hydrocarbon Fuel Cell Technology (S. Baker, editor), Academic Press, New York–London (1965), p. 455.
197. W. Vielstich, Hydrocarbon Fuel Cell Technology (S. Baker, editor), Academic Press, New York–London (1965), p. 79.
198. H. Schmidt and W. Vielstich, Z. Anal. Chem., 224:84 (1967).
199. W. T. Grubb and D. W. McKee, Nature, 210:192 (1966).
200. H. Binder, A. Köhling, and G. Sandstede, Advanced Energy Conversion, 7:77 (1967).
201. H. Binder, A. Köhling, and G. Sandstede, Nature, 214:268 (1967).
202. V. S. Éntina and O. A. Petrii, Élektrokhimiya, 4:111 (1968).
203. A. Damjanovic and V. Brusić, J. Electroanal. Chem., 15:29 (1967).
204. M. W. Breiter, Trans. Farady Soc., 62:503 (1966); J. Phys. Chem., 69:3377 (1965).
205. A. A. Michri, A. G. Pshenichnikov, R. Kh. Burshtein, and V. I. Bernard, Élektrokhimiya (in press).
206. T. B. Warner, J. Electrochem. Soc., 115:615 (1968).
207. M. L. B. Rao, A. Damjanovic, J. O'M. Bockris, J. Phys. Chem., 67:2508 (1963).
208. A. Damjanovic, V. Brusić, and J. O'M. Bockris, J. Phys. Chem., 71:2741 (1967).
209. G. L. Padyukova, A. B. Fasman, and V. Kh. Khizhnyak, Élektrokhimiya, 4:191 (1968).
210. Z. N. Novikova and A. B. Fasman, Élektrokhimiya, 4:307 (1968).
211. A. N. Frumkin, Electrochim. Acta, 5:265 (1961).
212. M. Fleischmann, J. R. Mansfield, and W. F. K. Wynne-Jones, J. Electroanal. Chem., 10:511, 522 (1965).
213. Yu. M. Tyurin, G. N. Afon'shin, and V. K. Goncharuk, The Double Layer and Adsorption on Solid Electrodes [in Russian], Izd. Tartusk. Gos. Univ., Tartu (1968), p. 162.

214.   G. Atherton, M. Fleischmann, and F. Goodridge, Trans. Faraday Soc., 63:1468 (1967).

215.   L. I. Antropov, Sb. Tr. Erevansk. Politekhn. Inst., No. 2, 109 (1946).

216.   V. E. Kazarinov, Élektrokhimiya, 2:1389 (1966).

217.   G. P. Girina, M. Ya. Fioshin, and V. E. Kazarinov, Élektrokhimiya, 1:478 (1965).

218.   V. E. Kazarinov and G. P. Girina, Élektrokhimiya, 3:167 (1967).

219.   M. Ya. Fioshin and Yu. B. Vasil'ev, Dokl. Akad. Nauk SSSR, 134:879 (1960); Izv. Akad. Nauk SSSR, Ser. Khim., 437 (1963).

220.   A. P. Tomilov and M. Ya. Fioshin, Usp. Khim., 32:60 (1963).

221.   M. Ya. Fioshin and A. I. Kamneva, Khim. Prom., No. 5, 359 (1960).

222.   M. Ya. Fioshin, G. P. Girina, Yu. B. Vasil'ev, M. V. Khrulev, M. K. Polievktov, and A. G. Artem'ev, Dokl. Akad. Nauk SSSR, 140:1388 (1961).

223.   G. P. Girina and M. Ya. Fioshin, Izv. Akad. Nauk SSSR, Ser. Khim., 1387 (1964).

224.   L. A. Mirkind and M. Ya. Fioshin, Dokl. Akad. Nauk SSSR, 154:1163 (1964).

225.   L. A. Mirkind, M. Ya. Fioshin, and V. I. Romanov, Zh. Fiz. Khim., 38:2223, 2840 (1964).

226.   L. A. Mirkind, M. Ya. Fioshin, and D. Constantinescu, Élektrokhimiya, 2:193, 1095 (1966).

227.   I. A. Avrutskaya and M.Ya. Fioshin, Electrokhimiya, 2:920 (1966); 3:1288 (1967).

228.   A. P. Tomilov and M. Ya. Fioshin, Advances in the Electrochemistry of Organic Compounds [in Russian], Izd. "Nauka" (1966), p. 65.

229.   A. G. Kornienko, L. A. Mirkind, and M. Ya. Fioshin, Élektrokhimiya, 3:1370 (1967); 4:49 (1968).

230.   M. Ya. Fioshin, Progress in Science, Chemistry Series, Electrochemistry, 1965 [in Russian], Izd. Vsesoyuzn. Inst. Nauchn. Tekhn. Inform., Moscow (1967), pp. 152, 169.

231.   M. Ya.Fioshin and L. A. Mirkind, Progress in Science, Chemistry Series, Electrochemistry, 1966 [in Russian], Izd. Vsesoyuzn. Inst. Nauchn. Tekhn. Inform., Moscow (1968), p. 114.

232.   T. Dickinson and W. Wynne-Jones, Trans. Faraday Soc., 58:382, 388, 400 (1962).

233.   B. E. Conway and M. Dzieciuch, Nature, 189:914 (1961).

234.   B. E. Conway and M. Dzieciuch, Can. J. Chem., 41:21, 38, 55 (1963).

235.   B. E. Conway, E. Gileadi, and M. Dzieciuch, Electrochim. Acta, 8:143 (1963).

236.   B. E. Conway and E. Gileadi, Modern Aspects of Electrochemistry, Vol. 3 (J. O'M. Bockris and B. E. Conway, editors), Plenum Press, New York (1964), p. 347.

237.   B. E. Conway, E. Gileadi, and J. Angerstein-Kozlowska, J. Electrochem. Soc., 112:341 (1965).

238.   B. E. Conway, Theory and Principles of Electrode Processes, Ronald Press, New York (1965).

239.   B. E. Conway, M. Marincic, D. Gilroy, and E. Rudd, J. Electrochem. Soc., 113:1144 (1966).

240. B. E. Conway and A. K. Vijh, J. Org. Chem., 31:4283 (1966).

241. B. E. Conway and A. K. Vijh, Electrochim. Acta, 12:102 (1967).

242. B. E. Conway and A. K. Vijh, Z. Anal. Chem., 224:149, 160 (1967); 230:81 (1967).

243. B. E. Conway and A. K. Vijh, J. Phys. Chem., 71:3637, 3655 (1967).

244. A. K. Vijh and B. E. Conway, Chem. Rev., 67:623 (1967).

245. L. Eberson, Electrochim. Acta, 12:1473 (1967).

Index

*F, T, or * found after a page number indicates that the subject referred to is to be
found in a figure, table, or footnote, respectively.

488

INDEX